T0241868

Undergraduate Lecture Notes in Physics

Undergraduate Lecture Notes in Physics (ULNP) publishes authoritative texts covering topics throughout pure and applied physics. Each title in the series is suitable as a basis for undergraduate instruction, typically containing practice problems, worked examples, chapter summaries, and suggestions for further reading.

ULNP titles must provide at least one of the following:

- An exceptionally clear and concise treatment of a standard undergraduate subject.
- A solid undergraduate-level introduction to a graduate, advanced, or non-standard subject.
- A novel perspective or an unusual approach to teaching a subject.

ULNP especially encourages new, original, and idiosyncratic approaches to physics teaching at the undergraduate level.

The purpose of ULNP is to provide intriguing, absorbing books that will continue to be the reader's preferred reference throughout their academic career.

Series editors

More information about this series at http://www.springer.com/series/8917

Alessandro Bettini

A Course in Classical Physics 1—Mechanics

Alessandro Bettini
Dipartimento di Fisica e Astronomia
Università di Padova
Padova
Italy

ISSN 2192-4791 ISSN 2192-4805 (electronic)
Undergraduate Lecture Notes in Physics
ISBN 978-3-319-29256-4 ISBN 978-3-319-29257-1 (eBook)
https://doi.org/10.1007/978-3-319-29257-1

Library of Congress Control Number: 2016934941

Printed on acid-free paper

This Springer imprint is published by Springer Nature
The registered company is Springer International Publishing AG Switzerland

Preface

This is the first in a series of four volumes, all written at an elementary calculus level. The complete course covers the most important areas of classical physics such as mechanics, thermodynamics, statistical mechanics, electromagnetism, waves, and optics. The volumes result from a translation, an in depth revision and update of the Italian version published by Decibel-Zanichelli. This first volume deals with classical mechanics, including an introduction to relativity.

The laws of Physics, and more in general of Nature, are written in the language of mathematics. The reader is assumed to know already the basic concepts of calculus: functions, limits, and the differentiation and integration operations. We shall however, without mathematical rigor, give the necessary information on vectors and matrices.

Physics is an experimental science, meaning that it is based on the experimental method, which was developed by Galileo Galilei in the seventeenth century. He taught us, in particular, that to try to understand a phenomenon one must simplify as much as possible the relevant working conditions, understanding which of the aspects are secondary and eliminating them as far as possible. The understanding process is not immediate, but rather it proceeds by trial and error, in a series of experiments, which might lead, with a bit of fortune and a lot of thinking, to discover the governing laws. Induction of the physics laws process goes back from the observed effects to their causes, and, as such, cannot be purely logic. Once a physical law is found, it is necessary to consider all its possible consequences. This is now a deductive process, which is logical and similar to the mathematical one. Each of the consequences, the predictions, of the law must then be experimentally verified. If only one prediction is found to be false by the experiment, even if thousands of them had been found true, it is enough to prove that the law is false. This implies that we can never be completely sure that a law is true; indeed the number of its possible predictions does not have limits, and in any historical moment not all of them have been controlled. However, this is the price we must pay in choosing the experimental method, which has allowed humankind to advance in the past four centuries much more than in all the preceding millennia.

Classical Mechanics is one of the big intellectual constructions of Physics. Its laws are well established as well as their limits of validity. Consequently, it can be exposed in an axiomatic way, as a chapter of mathematics. We can start from a set of propositions whose axioms are assumed to be true by definition, and deduce from them a number of theorems using only logics, as from the Euclid postulates the Euclidean geometry theorems are deduced.

We shall not follow this path. The reason is that, while it allows a shorter and quicker treatment and is also logically more satisfactory for somebody, it also hides the inductive trial-and-error historical process through which the postulates and the general laws have been discovered. These are arrival, rather than starting points. This path has been complex, laborious, and highly nonlinear. Errors have been made, hypotheses have been advanced that turned out to be false, but finally the laws were discovered. The knowledge of at least a few of the most important aspects of this process is indispensable to develop the mental capabilities that are necessary to anybody contributing to the progress of natural sciences, whether they pursue applications or teach them. This is one of the reasons for which we shall read and discuss several pages of the two scientists that built the foundations of physics, Galileo Galilei and Isaac Newton. A second reason is that reading the geniuses is always an enlightening experience.

The Galilei and Newton mechanics that we shall discuss in this book is a coherent set of laws able to describe a great number of physical phenomena. These laws, however, have a limited validity. One type of limitations does not have a fundamental nature. Some of the laws, as for example the laws of friction or the elastic force are, consciously we can say, approximate. In other words, they provide a description that we know to be valid only in a first approximation and provided that the values of certain quantities are within some definite intervals (for example, for the elastic force, for not too large strains). We shall always clearly state those limits.

The limits of the second type are of a fundamental nature. A first limit of the Galilei-Newton laws is met when the velocities are very high, high enough to get close to the speed of light.

The latter is so high, 300,000 km/s, that the speeds of all objects of a common experience, planets included, are extremely small in comparison. However, we can reach velocities close to that of light in experiments with microscopic particles, like atomic nuclei and electrons. In the universe there are double stars and double black-holes, which are extremely dense and rotate about each other at very high speeds, close to the speed of light. We observe that in these conditions the predictions of Newtonian mechanics are in contradiction with experience. Newtonian mechanics is an approximation valid at velocities substantially smaller than the speed of light. The theory that generalizes Newtonian mechanics, including high-speed phenomena, is relativistic mechanics, which was developed between the end of the nineteenth and the beginning of the twentieth centuries by, principally, Hendrik Lorentz, Henry Poincaré, and Albert Einstein. We discuss the basic elements of relativistic mechanics in Chap. 6 . They are not necessary for understanding of the following ones.

Newton or relativistic mechanics, depending on the velocities of the problem, is called classical mechanics. However, not even this is true in every circumstance; the laws of classical mechanics do not describe correctly the very small-scale phenomena, like vibrations and rotations of molecules, those of the electrons inside atoms, the nuclear, and subnuclear phenomena. As a matter of fact, the bodies at these microscopic scales behave in a completely different way than those of everyday experience. The theory able to describe all the known phenomena both at small and large scale is quantum physics. Its limit for large scales is classical mechanics. The study of quantum physics not only requires mathematical instruments more advanced than classical physics, but, even more importantly, cannot be profitably studied without an in-depth knowledge of classical physics. Consequently, this course is limited to classical physics. We shall however warn the reader of the limit of validity, whenever necessary.

In this book we deal with the mechanics of a material point and of extended bodies, in particular of the rigid ones. The mechanics of fluids will be one of the objects of study in the second volume, together with their thermal properties. Mechanical oscillations are treated here only in their most elementary aspects. A deeper discussion will be given, together with electric oscillations in the fourth volume.

We start the first chapter with introductory elements: the measurement of physical quantities, the measurement units and their internationally adopted system, the International System, reference frames, and basic concepts on vectors and matrices. The second part of the first chapter deals with kinematics, which is the mathematical description of motion, without reference to its causes. The second chapter contains the fundamental laws of the material point (the simplest body) and the basic concepts of mass (both the inertial and the gravitational masses), of force, of momentum, of moment of a force, and of angular momentum. We introduce also the concepts of work of a force, of energy, of power, and the energy conservation principle. We work on these arguments mainly considering the two most usual examples of force, weight and friction. At this point we have acquired the basic laws of mechanics. Historically, these are the result of the work of G. Galilei and I. Newton. It is important to have some knowledge of how these great authors came to establish the laws of mechanics. For this purpose a few of their fundamental pages, describing experiments and mathematical arguments, are reproduced and discussed. The reader will see also how both authors expose the concepts in a scientific superb language.

The third chapter describes the different forces, gives their mathematical expressions, and discusses their limits of validity. We discuss important examples of motion, in particular the circular and the oscillatory ones. We know now that the different forces that we see in nature and that look at first sight very different can be reduced to a very limited number of fundamental forces. The forces present at macroscopic level, the level of classical mechanics, are different manifestations of two basic ones: the gravitational and the electromagnetic forces. The latter will be studied in the third volume of this course, the former in the fourth chapter of this first book. As a matter of fact the study of gravitation has enormous historic and cultural importance. It underlines our comprehension of the universe in which we live. For this reason we recall the most important steps in the historical development of the universal gravitation theory.

The description of any motion depends on the frame to which it is referred. In particular it is different in two frames moving one relative to the other. The study of this issue is the object of the fifth chapter, in the limit of velocities much smaller than that of light. We shall meet with the extremely important principle of relativity, a universally valid principle established already by Galilei. The relations between reference frames at speed comparable to that of light and the critical analysis of the concepts of time and space intervals leading to the relativistic mechanics are dealt with in Chap. 6 .

In the last two chapters we study the mechanics of extended bodies. We start Chap. 7 with systems made of only two different material points. We show that in any case in which a force acts on a body, this is due to another body, which in turn is acted upon by a force due to the first one. In other words the forces are always due to the interaction between bodies. Having studied the issue on two-body systems, we proceed in the second part of the chapter with the study of material systems in full generality, finding the fundamental laws of their motion. In the last chapter we study the principal aspects of the motion of particular, and importantly, material systems, namely rigid bodies. Their motion is described by well-defined differential equations. Their solution is an important mathematical problem, which is however outside the scope of this course.

Each chapter of the book starts with a brief introduction to a scope that will give to the reader a preliminary idea of the arguments he/she will find. There is no need to fully understand these introductions, at the first reading, as all the arguments are fully developed in the following pages.

At the end of each chapter the reader will find a number of queries on which to check his/her level of understanding of the arguments of the chapter. The difficulty of the queries is variable; some of them are very simple, some more complex, a few are true numerical exercises. On the other hand, the book does not contain a sequence of full exercises, considering the existence of very good textbooks dedicated specifically to that.

The answers to a large majority of the queries are included. However, the solution of numerical exercises (without looking at the answers) is mental gymnastics that is absolutely necessary for understanding the subject. Only the effort to apply what has been learned to specific cases allows us to master them completely. The reader should be conscious of the fact that the solution of numerical exercises requires mental mechanisms different from those engaged in understanding a text. The latter, indeed, has been already organized by the author; solving a problem requires much more active initiative from the student. This is just the type of initiative, a creative activity that is needed, for advancing scientific knowledge and its technical applications as well. Consequently, the student should work on exercises alone, without looking at the solutions in the book. Even failed attempts to autonomously reach the solution, provided they are undertaken with sufficient persistence, give important returns, because they develop processing skills. If after several failed attempts the solution has not yet been reached, it is a better practice to momentarily abandon the exercise, rather than looking at the solution, going to another one, and coming back later.

The following working scheme is methodologically advisable:

1. Examine at depth the conditions posed by the problem. If it is possible, make a drawing containing the essential elements.
2. Solve the problem using letters, not numbers, in the formulas, then develop them until the requested quantities are expressed in terms of the known ones. Only then should you put numbers in the formulas.
3. Confirm the correctness of the physical dimensions (see Sect. 1.3).
4. When necessary transform all the data into the same system of units (preferably SI, see Sect.1.2). Use scientific notation, for example 2.5×10^3 rather than 2500, 2.5×10^{-3}rather than 0.0025. In general two or three significant figures are enough.
5. Once you have the final result, always verify if it is reasonable. For example the mass of a molecule cannot turn out to be 30 mg, the speed of a bullet cannot be 10^6 m/s, the distance between two towns cannot be 25 mm, etc.

Acknowledgments

The pages from Isaac Newton's, *Phylosophyae Naturalis Principia* are from the English translation from Latin by Andrew Motte (1729) modernized by the author.

The pages from G. Galilei's *Dialogue concerning two chief world systems* are a translation into English by the author from the Edizione Nazionale delle Opere, edited by Antonio Favaro; Florence, tip. Barbèra, 1890–1909.

The pages from G. Galilei's *Dialogues and mathematical demonstrations concerning two new sciences* are adapted from the English translation from Italian and Latin by Henry Crew and Alfonso de Salvio; McMillan 1914.

Figure 4.18 is from the National Aeronautics and Space Administration at http://www.compadre.org/Informal/images/features/Jupitmoons12-20-072.jpg

Figure 4.21 is from the European Space Agency at http://www.esa.int/var/esa/storage/images/esa_multimedia/images/2007/05/globular_cluster_ngc_28082/9535369-4-eng-GB/Globular_Cluster_NGC_2808.jpg

Figure 4.22 is from the National Aeronautics and Space Administration at http://hubblesite.org/newscenter/archive/releases/2007/41/image/a//

Contents

Symbols and Units

Table 1 Symbols for the principal quantities

Acceleration	$\mathbf{a}$, a_s
Angular acceleration	$\boldsymbol{\alpha}$, α
Angular frequency	ω
Angular momentum	$\mathbf{l}$, $\mathbf{L}$
Density (mass)	ρ
Dynamic friction coefficient	μ_d
Force	$\mathbf{F}$
Frequency	ν
Gravitational field	$\mathbf{G}$
Gravitational mass	m_g
Gravity acceleration	$\mathbf{g}$
Impulse	$\mathbf{i}$
Inertia radius	ρ
Inertial mass	m_i
Kinetic energy	U_K
Mass	m, M
Moment of a force	$\boldsymbol{\tau}$
Moment of inertia about a -axis	I_a
Momentum	$\mathbf{p}$
Newton constant	G_N
Normal constraint reaction	$\mathbf{N}$
Period	T
Plane angle	θ
Polar angle	θ, ϕ
Polar coordinates (space)	ρ, θ, ϕ
Position vector	$\mathbf{r}$
Potential	ϕ
Potential energy	U_p

(continued)

Table 1 (continued)

Power	w
Pressure	p
Reduced mass	μ
Spring constant	k
Static friction coefficient	μ_s
Time	t
Tension	$\mathbf{T}$
Total angular momentum	$\mathbf{L}_{\mathrm{tot}}$
Total (mechanical) energy	U_{tot}
Total moment	$\mathbf{M}$
Total momentum	$\mathbf{P}$
Young module	E
Weight	$\mathbf{F}_w$
Work	W
Mean value, of x	$<x>$
Angular velocity	$\boldsymbol{\omega}$, $\boldsymbol{\Omega}$
Velocity of light (in vacuum)	c
Velocity	$\mathbf{v}$,υ
Velocity divided by light velocity	$\boldsymbol{\beta}$
Unit vector of $\mathbf{v}$	$\mathbf{u}_\upsilon$
Unit vectors of the axes	$\mathbf{i}$, $\mathbf{j}$, $\mathbf{k}$
Volume	V

Table 2 Base units in the SI

Quantity	Unit	Symbol
Length	metre/meter	m
Mass	kilogram	kg
Time	second	s
Current intensity	ampere	A
Thermodynamic temperature	kelvin	K
Amount of substance	mole	mol
Luminous intensity	candela	cd

Table 3 Decimal multiples and submultiples of the units

Factor	Prefix	Symbol	Factor	Prefix	Symbol
10^{24}	yotta	Y	10^{-1}	deci	d
10^{21}	zetta	Z	10^{-2}	centi	c
10^{18}	exa	E	10^{-3}	milli	m
10^{15}	peta	P	10^{-6}	micro	μ

(continued)

(continued)

Factor	Prefix	Symbol	Factor	Prefix	Symbol
10^{12}	tera	T	10^{-9}	nano	n
10^{9}	giga	G	10^{-12}	pico	p
10^{6}	mega	M	10^{-15}	femto	f
10^{3}	kilo	k	10^{-18}	atto	a
10^{2}	hecto	h	10^{-21}	zepto	z
10	deka	da	10^{-24}	yocto	y

Table 4. Fundamental constants

Quantity	Symb.	Value	Uncertainty
Speed of light in vacuum	c	299 792 458 m s^{-1}	Definition
Newton constant	G_N	$6.67384(80) \times 10^{-11}$ m^3kg^{-1} s^{-2}	120 ppm
Astronomical unit	$a.u.$	149 597 870 700	Definition
Avogadro number	N_A	$6.022\ 1415(10) \times 10^{23}$mole^{-1}	170 ppb

Tropical year = time interval between two consecutive passages of the sun at the spring equinox

Table 5. Solar planets orbits

Planet	Mean distance from sun (a.u.)	Sidereal period (tropical year)	Angle with ecliptic	Eccentricity
Mercury	0.387099	0.24085	7°00'14"	0.2056
Venus	0.723332	0.61521	3°23'39"	0.0068
Earth	1	1.00004	0	0.0167
Mars	1.523691	1.88089	1°50'59"	0.0934
Jupiter	5.202803	11.86223	1°18'19"	0.0484
Uranus	19.181945	84.01308	0°46'23"	0.0472
Neptune	30.057767	164.79405	1°46'26"	0.0086
Pluto	39.51774	248.4302	17°08'38"	0.2486

Table 6. Data on some bodies of the solar system

Body	Mean radius (Mm)	Radius (Earth radiuses)	Mass (Earth masses)	Mean density (kg/m^3)
Mercury	2.44	0.38	0.055	5430
Venus	6.05	0.95	0.815	5250
Earth	6.37	1	1	5520
Moon	1.74	0.27	0.012	3360
Mars	3.38	0.53	0.108	3930
Jupiter	71.49	11.19	317.9	1330

(continued)

(continued)

Body	Mean radius (Mm)	Radius (Earth radiuses)	Mass (Earth masses)	Mean density (kg/m^3)
Saturn	60.27	9.46	95.18	710
Uranus	25.56	3.98	14.54	1240
Neptune	24.76	3.81	17.13	1670
Pluto	1.12	0.176	0.0026	1990
Sun	696	109.3	330,000	1400

Table 7. Greek alphabet

alpha	α	A	iota	ι	I	rho	ρ	P
beta	β	B	kappa	κ	K	sigma	σ, ς	Σ
gamma	γ	Γ	lambda	λ	Λ	tau	τ	T
delta	δ	Δ	mu	μ	M	upsilon	υ	Y, ϒ
epsilon	ε	E	nu	ν	N	phi	ϕ, φ	Φ
zeta	ζ	Z	xi	ξ	Ξ	chi	χ	X
eta	η	H	omicron	ο	O	psi	ψ	Ψ
theta	θ, ϑ	Θ	pi	π	Π	omega	ω	Ω

Chapter 1
Space, Time and Motion

Physics is an experimental science that gives a quantitative, mathematical description of natural phenomena. This means that physical laws are mathematical relations amongst physical quantities (such as position, velocity, force, energy, etc.). These relations are to be considered true only if they correspond to experience. Physical laws must always be experimentally verified. Experiment is the sole judge of scientific truth. Consequently, any physical quantity must be measurable, namely the set of operations that must be performed to measure it must be defined. First, a system of units of measurement must be defined. We shall see in the first three sections how this is done. The choice of units is a priori arbitrary; the physical laws depend on Nature, not on our choices, In practice, however, having standardized choices is extremely important to make the results understandable to everybody. International agreements have defined the system of units to be named, in French, Système International (International System).

Some of the physical quantities, like mass and temperature, are represented by a single number and are called scalar. Other, like velocity and force, are more complex; specifying how big they are is not sufficient, also their direction must be given. Mathematically, an ordered set of real numbers represents them; they are vector quantities. We shall study in the sections from Sects. 1.5 to 1.8 the elementary mathematical properties of vectors and of the operations (sum, difference, products) amongst them. In Sect. 1.9 we shall introduce some elements that will be useful in the following on another mathematical object, matrices.

In the second part of the chapter we shall move to physics, dealing with the kinematics of the point-like particle, namely the study of its motion, independently of its causes. We shall introduce the concepts of velocity, angular velocity and acceleration. These are vector quantities, in general depending on time. Section 1.13 is again of mathematical type, presenting a formula that will be very useful in the following, the time derivative of a vector.

A few types of motion are particularly important: circular motion, studied in Sects. 1.11 and 1.12, motion on a plane in Sect. 1.14 and free fall of weights in Sect. 1.16.

A. Bettini, *A Course in Classical Physics 1—Mechanics*,
Undergraduate Lecture Notes in Physics, DOI 10.1007/978-3-319-29257-1_1

1.1 Measurement of Physical Quantities

Physics gives a quantitative description of natural phenomena (or, better, of the known part of them). Measurement of the relevant physical quantities leads to discovery of the physical laws, which are mathematical relations amongst those quantities (for example, the law of the free fall, the Kepler laws, etc.).

All natural phenomena take place in space and have a temporal duration; some of them happen before, others afterwards. Consequently, space and time are fundamental concepts. Physical objects are characterized by quantities like length, area, volume, color, hardness, mass, temperature, etc. All these concepts result from a common experience and are present in common language. However, Physics must give to each quantity a rigorous definition, in order to be able to give it numerical values. In this definition process, the concept may become rather different from a common language.

Consider for example the length of an object or the distance between two places. If we want to designate a number we must first define a unit of length. Indeed we say: "That bar is 5 m long", or, if we are in England: "That city is 20 miles away". The measure of the length of an object is the ratio between its length and the length of another object we have chosen as unit. "A bar is 5 m long," means that its length is equal to that of 5 one-meter long rules in a line. "The mass of a body is 8 kg," means that it is equal to that of eight bodies of 1 kg together.

The measurement of physical quantity is the ratio between that quantity and its measurement unit.

The measurement operation allows associating to each physical quantity a number. The symbols that appear in the physical laws representing the various physical quantities are just these numbers. For example, when we write $F = ma$ we mean that the ratio between the force we are considering and the force taken as unit, is equal to the ratio of the mass of the object and the mass of the object taken as unit, times the ratio between the designated acceleration and the unit acceleration.

Every physical quantity must be measurable and its definition must be precise and rigorous. The operational definition is the most effective way to define a physical quantity. This is defined *as the set of operations needed to measure that quantity.*

This procedure has two important implications. The first implication is that quantities that are not, even in principle, measurable are not physical quantities. This does not imply that such quantities cannot be used. Indeed they are often useful in the mathematical developments of a theory. Any theory, however, if it has to be a physical theory, must lead to predictions that are experimentally testable. The experimentally testable predictions are mathematical relations amongst physical quantities, meaning measurable ones. The auxiliary, non-measurable, quantities, should not appear in the final theoretical expression.

The second implication is that scale matters: quantities may be small or large. Consider for example the length. If we want to measure distances, say, from millimeters to kilometers we can use graduated bars or rules (like yardsticks, measuring tapes, calipers, gauges, etc.). If we need to measure distances of tens or hundreds of kilometers, as for example between two mountaintops or the height of Mount Everest, the procedure is very different and we must perform triangulations. If the distances are very much larger, as those of the galaxies, the procedures change completely again. And different procedures are required to measure small distances such as the diameter of an atom or of an atomic nucleus. In every range of orders of magnitude, the set of procedures to measure a length is different. To be rigorous we would need to talk of many different lengths. This would lead to a terrible confusion. Fortunately, we experimentally verify that, in the large intervals in which two or more methods work contemporarily, the results are equal, and we can define a single length concept. However, the above arguments tell us to be careful. Suppose that a physical law is well experimentally verified for objects of sizes between meters and kilometers. We tend to think the same law to be valid also for objects much smaller and much larger than that. But we have no guarantee that the extrapolation is true. On that, as always, only experiment can judge. For example the Newton laws valid at the speeds of ordinary experience are no longer valid at speeds comparable to the speed of light. The laws of classical mechanics are simply not valid for atoms and smaller objects.

Let us go back to the measuring operation. For each quantity we need a measurement unit. The choice is in principle arbitrary but is far from being so in practice. If every Country, for example, would choose a different unit for lengths or areas, the exchanges, not only the scientific ones, but also the commercial ones, would be extremely complex. The units must be standardized. The issue is so important that both units and procedures are made compulsory by law in the majority of Countries.

1.2 The International System (SI)

The modern international standardization of units started with the French Revolution. In 1791 the Decimal Metric System was officially announced, but it took almost a century for its substantial diffusion and acceptance (and, most important, Napoleon to impose it; in Britain, where his reach was insufficient, the Imperial System is still used, as it is also in the USA). In May 1875, at the "Metre convention", the representatives of 17 Nations signed an international treaty in Paris. National and international laboratories were created with the mission to develop measurement standards and procedures. This is a very important sector of physics, know as *metrology*.

International Organizations were created to foster international standardization of weights and measures in the world. The International Conference of Weights and Measures, CGPM for brief using the initials in French, which meets every several years, is the main decision-making body. It decides on the evolution of the internationally adopted system of units, named in French *Sistème International* or SI for brief. In 1971 the European Community issued a directive to the member states for the legal adoption of the SI.

There are two classes of units: *base units* and *derived units*. The base units are given by definition. Each derived unit is obtained using a physics law, namely a mathematical expression that links it to quantities of the basic units. The choice of the basic units, and even their number, is, from a logic point of view, arbitrary. The choices are based on convenience, taking quantities for which measurements can be as much as possible precise and reproducible.

Let us consider an example. Take the "physical laws": (1) the area S of a rectangle of sides of lengths a and b is proportional to the product of the lengths, (2) the area A of a circle is proportional to the square of the length R of its radius, (3) the space s covered by a body moving in absence of any force is proportional to the time t employed and to its velocity υ. The mathematical expression for these laws would be

$$S = kab \quad A = k'R^2 \quad s = k''\upsilon t. \tag{1.1}$$

where k, k' and k'' are purely numerical constants. They depend on the choice of measurement units. We might take both length and area as base quantities and as units the meter and square foot respectively. The k and k' would then have definite values. Our measuring system would be simpler taking length as the base unit, say 1 m, and the area as derived. Still, however, some arbitrariness remains. For example, we can choose the units in such a way to have $k = 1$ or, differently, to have $k' = 1$. In the first option the unitary area is the square of 1 m side, in the second one it is the circle of 1 m radius. The second choice gives $k = 1/\pi$, and appears funny. The choice $k = 1$ appears to be the obvious one, and is the universally used one, but, in principle, it is not necessary.

Similarly, in the third equation we make $k'' = 1$ by choosing as measuring unit of velocity the velocity of a body that covers the unit length in the unit of time.

As already mentioned, the internationally accepted system of units is the SI. It is the easiest to use and the most rational one. In the SI the base units are seven: length, mass, time, electric current intensity, thermodynamic temperature, amount of substance, luminous intensity. For each of them, the name of the unit (e.g. "meter") and its symbol (e.g. "m") are fixed, as in Table 1.1. Notice that the initial of the name of a unit is always lower case, including when it is the name of a scientist (e.g. "ampere"). Most important, the SI gives a precise and clear definition for each unit. Notice that these may change with time, as a consequence of the progress of metrology, after formal approval by the CGPM. We shall give here the definitions of the first three units, which are the only ones needed in this textbook. The other ones will be defined in the other volumes of the series, when needed.

The *metre (meter* is the distance travelled by light in vacuum in a time interval of 1/229 792 458 of a second.

The *kilogram* is the mass of the international prototype kilogram (located in the Pavillon de Breteuil at Sèvres).

Table 1.1 The base quantities, their units and symbols

Quantity	Unit	Symbol
Length	metre/meter	m
Mass	kilogram	kg
Time	second	s
Current intensity	ampere	A
Thermodynamic temperature	kelvin	K
Amount of substance	mole	mol
Luminous intensity	candela	cd

The *second* is the duration of 9192631770 periods of the radiation corresponding to the transition between the two hyperfine levels of the ground state of the Cesium 133 atom.

The SI defines the names and symbols of all the derived units. We shall introduce them when we meet them for the first time. The SI further defines names and symbols of multiples and submultiples of the units. This is done in steps in general of three orders of magnitudes, of one order for the first three, as in Table 1.2. With the exception of da, h, and k, all multiple prefix symbols are upper case; all submultiple prefix symbols are lower case letters.

The derived measurement units are defined, as mentioned, using a physical law in order to have a definition as simple as possible. Hence, the unit for areas is the square of 1 m side, the unit of volume is the cube of 1 m side, the unit of velocity is the velocity of a body travelling 1 m in one second, etc.

The (mean) acceleration is the change of velocity Δv divided by the time interval Δt in which that change happens, namely $a = \Delta v / \Delta t$. The acceleration unit is the acceleration of a body, the velocity of which varies by a unit (1 m/s or 1 m s^{-1}) in the unit of time (1 s). It is consequently the meter per second per second (m/s^2 or m s^{-2}).

Let us now observe, as an example, that all the plane figures, triangles, rectangles, circles etc. are expressed as a numerical factor times the product of two lengths. Namely, all areas have a physical dimension of length squared. If we change the unit of length, for example from meter to centimeter, the measures of all

Table 1.2 Decimal multiples and submultiples

Factor	Prefix	Symbol	Factor	Prefix	Symbol
10^{24}	yotta	Y	10^{-1}	deci	d
10^{21}	zetta	Z	10^{-2}	centi	c
10^{18}	exa	E	10^{-3}	milli	m
10^{15}	peta	P	10^{-6}	micro	μ
10^{12}	tera	T	10^{-9}	nano	n
10^{9}	giga	G	10^{-12}	pico	p
10^{6}	mega	M	10^{-15}	femto	f
10^{3}	kilo	k	10^{-18}	atto	a
10^{2}	hecto	h	10^{-21}	zepto	z
10	deka	da	10^{-24}	yocto	y

the areas change by the same factor: 100^2 in the example. The physical dimensions of velocity are length divided by time, of acceleration of length divided by time squared, etc. The corresponding mathematical expressions are called *dimensional equations* which are of the type

$$[A] = [L^2], \quad [\upsilon] = [LT^{-1}], \quad [a] = [LT^{-2}]. \tag{1.2}$$

Dimensional equations are very useful in practice. Consider any relationship amongst physical quantities, for example $F = ma$ or $A + B = C$. All the terms must have the same physical dimensions. Otherwise, a change of units will cause the different terms to change in different ways; the validity of the relation would depend on the choice of units, which is arbitrary. This is the so-called *homogeneity principle*. It is very useful to check analytical expressions obtained with more or less complex calculations. If we find that some of the terms have different dimensions, we must conclude that we have made some mistake.

Notice that there are also physical quantities having nil dimensions, namely $[L^0T^0M^0]$, they are pure numbers. An important example is the angle. In radians (rd) it is the ratio between the arc of a circumference and its radius. If we change the unit of length, the ratio between two of them does not change.

Finally notice that a physical law may contain mathematical functions, for example $x = \sin\alpha, y = \exp(-\beta)$ or $z = \ln\gamma$. These expressions make sense only if both the functions themselves (x, y, z) and their arguments (α, β, γ) have no physical dimensions. All of them must be pure numbers.

1.3 Space and Time

Our study begins with the study of the motion of bodies. Motion of a body means that its position in space varies in time. The notion of motion is relative: a passenger in a plane sitting in his chair has a fixed position relative to the plane, but moves at, say 800 km/h relative to a person standing on earth. The latter moves at 800 km/h relative to the passenger, in the opposite direction. To describe the motion we then need a reference frame.

We normally live standing on earth and such are the laboratories in which we do our experiments. Let us then start by choosing a reference frame fixed on the earth. The possible choices are still infinite.

The position of a body is defined when we know were it is. The simplest case is when we deal with a particle, a body that is so small that it can be considered point-like. It is called a *material point*. Let us see how we can define the position of a material point. For an extended body the positions of all its points should be similarly defined.

To know the position of a point in space we need three numbers, one for each of its dimensions. To define its position on a given surface, two numbers are needed

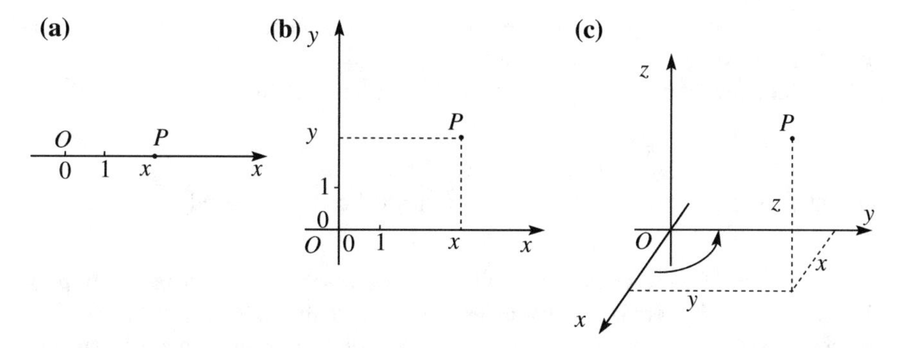

Fig. 1.1 Orthogonal co-ordinate frames. **a** One dimension, **b** two dimensions, **c** three dimensions

(as for example longitude and latitude on the earth surface). To know the position on a given curve, one number is needed.

Let us start by considering a point P that can move only on a straight line (see Fig. 1.1a). To define its position: (1) we choose one of the two directions and call it positive, (2) we choose a point on the line and call it the origin of the co-ordinates (O in Fig. 1.1a), (3) we choose a unit length. The oriented line, with an origin and a measuring unit is called a *co-ordinate axis*. The position of the generic point P is given by a real number, called the co-ordinate of the point (x in the figure), which is the distance of P from O, taken as positive if P is on the right of O, negative if it is on the left.

Let us now assume that point P can move on a plane (Fig. 1.1b). We now need two co-ordinate axes, which should not be parallel. It is usually convenient to take them perpendicular, the origin at the point in which they cross and the same unit length for both (none of these choices is compulsory, they are just generally the most convenient). The position of P is given by its two co-ordinates, which is an ordered pair of real numbers (x, y).

Consider now a point in space. The reference frame shown in Fig. 1.1c is called a *Cartesian rectangular* right-handed frame, after René Descartes (1596–1650). It is made of three co-ordinate axes, called x, y and z. They cross in a single point, the origin of the frame. All the angles between the (three) pairs of axes are right. The length units on the three axes are equal. Finally we must choose positive orientations of the axes. There are two basic possibilities. Let us assume that we have already defined the positive directions of x and y. We have two possible choices for the positive direction of z. Figure 1.1c shows one of them; an observer standing with his feet on the xy plane lying along the z axis and looking down, willing to move the x axis on the y axis by a 90° rotation, sees this rotation happening anticlockwise. The second possibility is the opposite sign of z. The two frames are called right-handed and left-handed respectively.

Now consider the inversion of the axes. If we start from a right-handed frame and invert one axis, that is a mirror reflection and we get a left-ended frame. The same happens if we invert all three axes. The inversion of two axes gives, on the contrary, the same result as a rotation of 180° around the third axis: the initial and final frame have the same "handness".

To define the reference frame we have made a series of choices, which we recall:

(1) choice of the origin
(2) choice of the directions of the axes
(3) choice of the positive directions (left-handed or right-handed)
(4) choice of the units.

While each of these choices is arbitrary, we can ask whether there is any privileged choice, or if there is one that is better posed, are the physics laws independent of these choices? The answers cannot come from logics or mathematics, but only from an experiment. Let us consider each of them.

(1) Are the physics laws independent of the origin of the axes? To check the point, let us build two identical apparatuses. Let each of them contain inclined planes with balls rolling on them, pendulums, flywheels, gears, etc., all identical. We position the two apparatuses in two different locations. We prepare them to be in exactly the same initial state: the pendulums are out of equilibrium at the same distance, the spheres are at the same heights on the inclined planes, the gears and the flywheels are in the same positions. We let them go contemporarily and observe their evolutions. Do the two systems evolve in the same way? Do they assume the same configurations at the same times? As a matter of fact the answer is not always yes. However, every time some difference is noticed, it is possible to identify the reason for that in some physical condition that is different in the two locations. For example, the gravitational acceleration might be a bit different in the two sites and consequently the periods of the pendulums are a bit different too. In any case, experiments show that, once all the local effects are eliminated, or accounted for, the apparatuses evolve in the same manner, i.e. going through the same configurations at the same instants. The very important conclusion is: *The physical laws are independent on position*. In other words all positions are equivalent, or space is homogeneous. Let us repeat that this is an experimental conclusion. No experiment up to now has found it wrong. One can state that the *physical laws are invariant*, meaning that they do not vary, *under space translations*.
(2) Are the physics laws independent of directions of the axes? We now take our two identical apparatuses and rotate one to the other. For example, in one case the z-axis is vertical, in the other is at 45° with the vertical. Do the two systems evolve through the same states? Certainly not! Indeed, for example, pendulums oscillate around a vertical axis in one case, around an inclined one in the other. In this case a privileged direction exists, the direction of weight. But, think a moment. If we were far from earth, or in absence of weight, the privileged direction would not exist. That direction is not a property of the space, but is the "local" effect of a body, the earth. In other words, if we want to compare the two experiments in the same conditions, we should also rotate the earth in the second case. If all the external conditions are properly taken

into account, all the experiments show that *the physical laws are independent on the directions of the axes*. In other words, no privileged direction exists, or, space is isotropic. Still in other words, *physics laws are invariant under rotations*.

(3) Are the physics laws, independent of the orientation, left-handed or right-handed? Experiments have shown that all physics laws at the macroscopic level are independent of the choice. But this is no truer at a microscopic level. A class of radioactive phenomena, like beta decays, is due to a fundamental force called weak interaction. Its laws distinguish between the left and right cases. Namely, not all the physics laws are invariant under inversion of the axes.

(4) Are the physics laws independent of the scale of length? This time we build two apparatuses that are identical but for having all their dimensions different, scaled by the same factor. Do the two evolve in the same manner? The answer was discovered by Galileo Galilei (Italy, 1564–1642) and is NO.

Consider for example a beam made of a certain homogeneous material. The beam has a certain length, and its cross section, which we assume to be rectangular, has a certain width and a certain height. We lay it on two supports near to its extremes on a horizontal plane. Suppose the beam to be in equilibrium. We now take a beam geometrically similar to the first one but ten times longer, ten times wider, ten times higher. Again we lay it on two supports near to its extremes. We observe that the beam breaks down in its middle point. The reason is the following. The weight of the beam is a force applied in its middle point directed downward. The weight tends to break the beam, the cohesion forces between molecules tend to keep it together. The weight, which is proportional to the volume, is for the second beam one thousand times larger than for the first one. The resistance to fracture is proportional to the area of the cross section and for the second beam is one hundred times larger than for the first one. Consequently, above a certain dimension the beam breaks down under its own weight. For the same reason the animals cannot be too big. The bones of the legs of a hypothetical horse ten times bigger than the real ones would break under their own weights. We know now that the fundamental reason for that is that substances are made of molecules and atoms, which have a definite size. Certainly we cannot build one of the above-considered apparatuses so small to be made of a few molecules.

As another example, consider the heavenly bodies. Stars and the largest planet emit light, the smaller planet, like earth, do not. Only if the size, hence the mass of the body, is large enough, the pressure and temperature in its core, which are due to the action of the gravitational forces between its parts, are large enough to fire the thermonuclear fusion reactions that produce light.

In conclusion: *The physics laws are NOT invariant under changes of scale*.

We now come back to mathematics of reference frames. In the following we shall need to use another type of, equivalent, co-ordinates, the *spherical polar coordinates*. Figure 1.2a shows such co-ordinates on the plane, Fig. 1.2b in space, together with the orthogonal coordinates in both cases. On the plane, the two polar

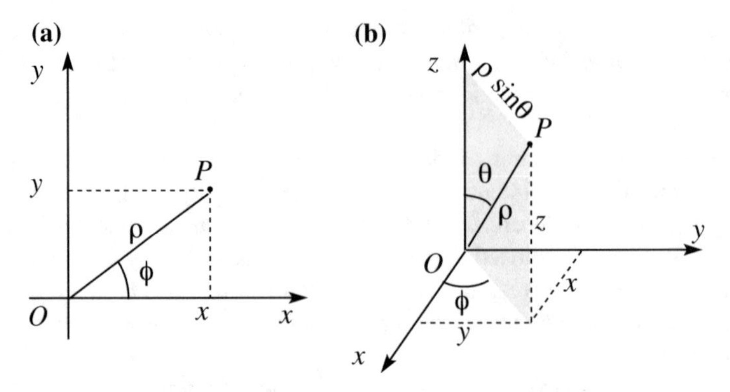

Fig. 1.2 Polar co-ordinates. **a** Two dimensions, **b** three dimensions

co-ordinates of the generic point P are its distance from the origin ρ, which is a non-negative number, called a *radius*, and the angle ϕ, between the x-axis and the segment OP, called an *azimuth*. It is measured in anticlockwise direction and varies between 0 and 2π, namely

$$\rho \geq 0, \quad 0 \leq \phi < 2\pi. \tag{1.3}$$

We can easily see from the figure that the relations between polar and rectangular co-ordinates are

$$x = \rho \cos \phi, \quad y = \rho \sin \phi, \tag{1.4}$$

and the inverse ones

$$\rho = \sqrt{x^2 + y^2}, \quad \phi = \arctan \frac{y}{x}. \tag{1.5}$$

Figure 1.2b shows polar co-ordinates in three dimensions. The first co-ordinate of the generic point P is again its distance r from the origin (*radius*), the second co-ordinate is the angle ϕ between the plane through the z and P and the plane xz (*azimuth*), the third co-ordinate is the angle θ between the segment OP and the z axis (*zenith angle*). Again r is a non-negative number. The angle θ varies from 0 to π, covering in such a way the semi-plane shown in the figure. This semi-plane rotates around z when ϕ varies between 0 and 2π. Hence

$$\rho \geq 0, \quad 0 \leq \theta \leq \pi, \quad 0 \leq \phi < 2\pi. \tag{1.6}$$

The relations with the orthogonal co-ordinates are

$$x = \rho \sin \theta \cos \phi, \quad y = \rho \sin \theta \sin \phi, \quad y = \rho \cos \theta, \tag{1.7}$$

and the inverse ones

$$\rho = \sqrt{x^2+y^2+z^2}, \quad \phi = \arctan\frac{y}{x}, \quad \theta = \arctan\frac{\sqrt{x^2+y^2}}{z}. \tag{1.8}$$

If the point P is on the xy plane, namely if $\theta = 0$, Eq. (1.8) become

$$x = \rho\cos\phi, \quad y = \rho\sin\phi,$$

which are equal to Eq. (1.4).

To know the motion of a body we need to know its position in different time instants. Consequently we most measure the time. More precisely, we measure intervals of time, rather than an absolute time. In practice we choose a certain instant and define it as the *origin of times*, for which $t = 0$. We next choose a time interval and define it as the unit of time. In the SI it is the second. In principle we should also choose one of the two directions as positive, but the choice is obvious. It is, we can say, imposed by Nature: the positive direction of time is from past to future. Consequently, the time of an event is negative if it happened before $t = 0$, positive if after that.

We now ask: is the origin of times arbitrary? As always we must apply to the experiment. Let us go back to one of our experimental apparatuses and let us repeat the experiment starting from the same initial state, for example in the morning, then in the afternoon, and again in the night, etc. For each trial we take the origin of time as the initial instant. We observe that, once all the spurious elements are taken care of (e.g. light in the day, dark in the night) all the experiments evolve in the same way. The origin of times is arbitrary, time is homogeneous. The physics laws are invariant under translations in time. In addition, similarly to space, no fundamental time interval exists.

We have said that the choice of the positive direction of time is imposed by Nature. Several books have been written on this issue, the "arrow of time". We shall not enter in this discussion. We only state here that in the purely mechanical phenomena no arrow of time exists. Suppose we hit a billiard ball and shoot a movie of its motion hitting other balls, the walls, etc. If we now play the movie backwards we observe a perfectively legitimate evolution. We cannot know if it is backwards of forwards. But, wait; this is not true forever. Indeed, if the movie is long enough we see that, when plaid forwards, the speeds of the balls gradually slow down and finally they stop. If it is plaid backwards, the balls are initially steady and start moving alone. The natural arrow of time is the one in which the kinetic energy diminishes. When we study thermodynamics in the second volume of this course we shall see how it explains the arrow of time.

1.4 Vectors

Many physical quantities, such as temperature and atmospheric pressure, are represented by a single number. This is not the case of other ones, such as velocity, acceleration, force, etc. For example to know the velocity of a car is not sufficient to know how fast it moves (that is the number we read on the speedometer), but also in which direction (towards South, North or other). Another example is a displacement in space. To know it we need to know how long it is and in which direction it happens. These physical quantities are represented by vectors.

A vector is a mathematical entity. To define it, let us start considering line segments. A segment is called *oriented*, if one of its two senses is chosen as positive. Two oriented segments are said to be equipollent if they have the same length, the same direction and the same sense. A *vector is the class of all the oriented segments equipollent to a given one.* It is graphically represented with an arrow. It is characterized by the length, called *magnitude*, the *direction* and the *sense*. Differently from the oriented segment, it is not characterized by its position. The velocities of two cars moving at 100 km/h heading West, one near Paris, one near London are the same.

Once a reference frame is chosen, we can represent a vector by an *ordered triple of real numbers*, which are its *components* in that reference. However, an ordered triple of real numbers is not necessarily a vector. To be so the following important property must be satisfied. Indeed, if we change the reference frame, for example rotating the axes, the components of the vector change, but the vector does not. Vector is a definite object; its components are the way to see it in one or another frame. To satisfy these properties, the vector components, namely the ordered triples in the two frames, must be connected by well-defined relations, which we shall now find.

Figure 1.3 shows a reference frame and a point P of co-ordinates x, y, z. Consider the oriented segment from the origin O to P and the corresponding vector $\mathbf{r}$ (namely the class of equipollent oriented segments). It is called a *position vector*, and is, we can say, the prototype of all vectors. Its components in the (Cartesian) reference frame we are considering are simply the coordinates of P, i.e. the ordered triple (x, y, z). Let us now take another reference with the same origin and axes rotated by an angle θ. The point P does not move and $\mathbf{r}$ does not change. But its components

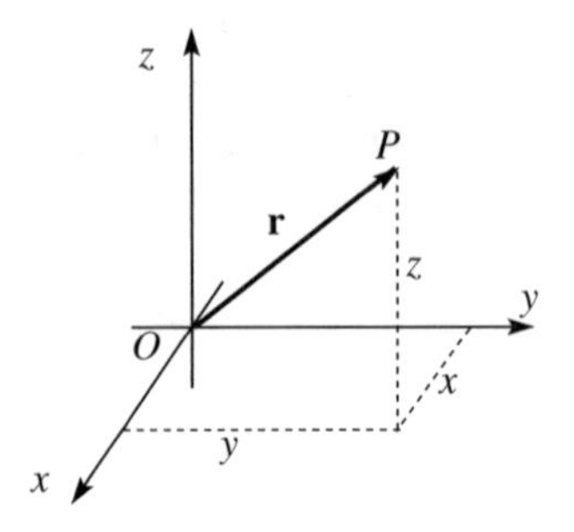

Fig. 1.3 The orthogonal co-ordinates and the position vector

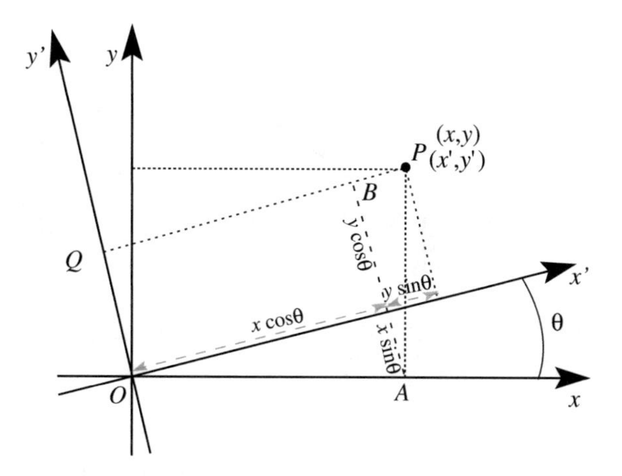

Fig. 1.4 A rotation of a Cartesian reference frame around the common z axis

(x', y', z') are different. The general relation between the two triplets is rather complex. For simplicity we shall consider two frames with the same origin and the same z-axis, as shown in Fig. 1.4.

Let us consider the point P in the figure of co-ordinates (x, y) in one frame, (x', y') in the other. We must express x' and y' as functions of x, y and θ. One relation is obvious, $z' = z$. In practice, we are reduced to two dimensions.

We now draw perpendiculars from P to all the axes. We also draw the segment AB perpendicular to PQ. The figure shows that x' is the sum of two lengths along the x' axis and y' the difference of two lengths along AB. We obtain

$$\begin{aligned} x' &= x\cos\theta + y\sin\theta \\ y' &= -x\sin\theta + y\cos\theta \\ z' &= z, \end{aligned} \tag{1.9}$$

where, to be complete, we included also the third co-ordinate. Notice that these relations are both the relations between the components of the position vector in the two frames and the relations between the co-ordinates in the two frames. As a matter if fact they analytically define the rotation of the axes.

We now state that a vector *is an ordered triple of real numbers that under rotations of the reference frame transforms* (changes) *in the same way as the triple representing the position vector, namely as co-ordinates.*

Figure 1.5a represents, in a plane for simplicity, a generic vector **A**, which we can think as of drawn starting from the origin, because all the equipollent segments are the same vector, and its components in the two frames.

By definition, the relations amongst its components are equal to Eq. (1.9), namely

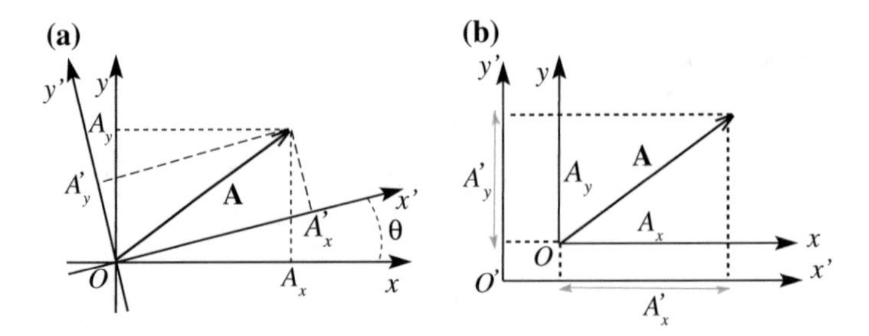

Fig. 1.5 Components of the vector **A** in two frames different for **a** a rotation **b** a translation

$$\begin{aligned} A'_x &= A_x \cos\theta + A_y \sin\theta \\ A'_y &= -A_x \sin\theta + A_y \cos\theta \\ A'_z &= A_z. \end{aligned} \tag{1.10}$$

The inverse relations, namely the expressions of (A'_x, A'_y, A'_z) as functions of (A_x, A_y, A_z) and θ can be obtained in two ways: inverting the system (1.10) or, which is simpler, thinking that the first reference is obtained from the second by a rotation of an angle $-\theta$. Consequently we have

$$\begin{aligned} A_x &= A'_x \cos\theta - A'_y \sin\theta \\ A_y &= A'_x \sin\theta + A'_y \cos\theta \\ A_z &= A'_z. \end{aligned} \tag{1.11}$$

We have considered two frames differing for a rotation of the axes, with a common origin. Consider now two frames differing for a translation, namely with parallel axes and different origins, as shown in Fig. 1.5b, again for simplicity in a plane. We see that the components of the vector **A** in the two frames are equal.

1.5 Operations with Vectors

A quantity represented by a number, like temperature or pressure, is called a *scalar*. Scalars are invariant under rotations of the axes. In two reference frames rotated one to the other a scalar has the same value. Notice that not every quantity is scalar. For example the x component of a vector is not, because it changes under rotations.

We shall represent a vector with its components in a given frame with $\mathbf{A} = (A_x, A_y, A_z)$.

Given the vector **A** and the scalar k their product is the vector $k\mathbf{A} = (kA_x, kA_y, kA_z)$. Namely the components of $k\mathbf{A}$ are k times those of **A**. To be sure, we must verify that the just given definition agrees with the definition of vector. Indeed, it is immediate to check that the oriented triple (kA_x, kA_y, kA_z) transforms like a vector.

Geometrically, $k\mathbf{A}$ is the vector with the same direction as **A**, the magnitude $|k|$ times the one of **A** and the sense of **A** or opposite depending on k being positive or negative respectively.

The product of **A** times the reciprocal of its magnitude is a vector with the direction of **A** and unitary magnitude. A vector of unitary magnitude is called a *unit vector* or *versor*. We shall use the symbol $\mathbf{u}_A$ for the unit vector of **A**.

The product of the vector **A** and –1 is called the *opposite* of **A**. It has the same magnitude and direction of **A** and opposite sense.

Consider now two vectors **A** and **B**, which in a given reference frame have the components (A_x, A_y, A_z) and (B_x, B_y, B_z) respectively. Consider the triple of numbers that are the sums of the homologous components of **A** and **B**, namely (A_x + B_x, A_y + B_y, A_z + B_z). Is it a vector? Let us check. Knowing that **A** are **B** vectors we know that

$$\begin{array}{ll} A'_x = A_x \cos\theta + A_y \sin\theta & B'_x = B_x \cos\theta + B_y \sin\theta \\ A'_y = -A_x \sin\theta + A_y \cos\theta & B'_y = -B_x \sin\theta + B_y \cos\theta \\ A'_z = A_z & B'_z = B_z. \end{array}$$

By summing member to member we have

$$\begin{aligned} A'_x + B'_x &= (A_x + B_x)\cos\theta + \left(A_y + B_y\right)\sin\theta \\ A'_y + B'_y &= -(A_x + B_x)\sin\theta + \left(A_y + B_y\right)\cos\theta \\ A'_z + B'_z &= A_z + B_z. \end{aligned}$$

We see that the answer is positive. We can then define as the vector sum of two vectors the vector with components equal to the sums of their homologous components. Notice that the just found properties are immediate consequences of the component transformations being linear operations.

It is immediate to verify that the sum of vectors has the usual properties of the sum, namely commutative

$$\mathbf{A} + \mathbf{B} = \mathbf{B} + \mathbf{A}. \tag{1.12}$$

and associative

$$(\mathbf{A} + \mathbf{B}) + \mathbf{C} = \mathbf{A} + (\mathbf{B} + \mathbf{C}). \tag{1.13}$$

Figure 1.6 shows the geometric meaning of the vector sum. In Fig. 1.6a the sum is made putting the tail of **B** on the head of **A**; the sum is the vector from the tail of **A** to the head of **B**, as one immediately understands thinking to the components. For the commutative property, we might have done vice versa, namely start from **B** and putting the tail of **A** on the head of **B**. We should have reached the same point.

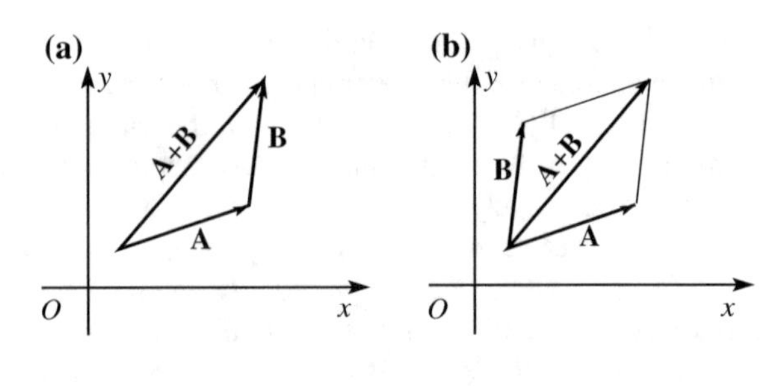

Fig. 1.6 The sum of two vectors

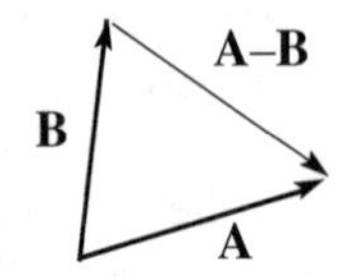

Fig. 1.7 The difference between two vectors

Figure 1.6b shows an equivalent way to sum, the parallelogram rule. We put both vectors with the tails in the same point and we draw the parallelogram having them as sides.

The vector difference between the two vectors **A** and **B** is the vector of components equal to the differences between the homologous components or, equivalently, the sum of **A** and –**B**. The geometrical meaning is shown in Fig. 1.7.

The properties of vector sums, or composition, which we have just discussed looks to be obvious, but they are not. Indeed they are valid if the space is flat, not if it has any curvature. To make things simpler, consider two dimensions. A plane surface is flat, but not a spherical one or a saddle shaped one. As a matter of fact the surface on which we live, the surface of the earth, is flat only if we consider distances substantially smaller than the earth radius, which has a mean value $R = 6\ 371$ km, and only in a first approximation.

Let us consider the following example of vector addition. Consider a vector with the tail in A at 45° in latitude and 0° in longitude and the head B on the same meridian at 46° latitude. The length of one degree along a meridian is everywhere 10 000 km/90 = 111 km. The second vector has the tail in B and the head on the same parallel 100 km towards West, say in C. Now we commute the operations. We start with a vector 100 km long from A to, say, D on its parallel at 100 km to West. Then we add a 111 m long vector to the North with tail in D and head, say, in C'. Will C' be equal to C? The answer is NO. This is because the distance between two meridians along a parallel is different at different latitudes. Indeed the radius of the parallel at the latitude λ is $r(\lambda) = R \cos \lambda$, that is 7071 km at 45° and 6947 km at 46°, which is 1.8 % shorter. Consequently, C' is 1.8 km West of C.

Question Q 1.2. Repeat the calculation with the same vector lengths starting at 65° latitude.

Question Q 1.1. Repeat the calculation with vector lengths of 1 km starting at 45° latitude.

The question whether or not space has a curvature should be answered experimentally, and experiments show this being the case. In particular, the measurement of the mean curvature of the Universe over cosmological distances is one of the important objects of contemporary cosmology. All measurements are compatible with zero mean curvature, within their uncertainties. However, we should mention that space curvature exists in another context. General relativity describes local gravitational effects in terms of a modification of geometry in the space surrounding a massive object. Its predictions are confirmed by observations. We shall not deal with this topic in this book.

1.6 Scalar Product of Two Vectors

There two ways to take the product of two vectors, called *dot product* and *cross product* respectively. We start here with the former.

Consider the two vectors **A** and **B**. Their dot product is indicated with a dot between them, namely $\mathbf{A} \cdot \mathbf{B}$. In a given reference frame the dot product is, by definition, the sum of the products of the homologous components

$$\mathbf{A} \cdot \mathbf{B} = A_x B_x + A_y B_y + A_z B_z. \tag{1.14}$$

The dot product has the important property to be scalar, namely invariant under rotations of the axes. It is consequently also called a *scalar product*. Let us show the property, namely that

$$A'_x B'_x + A'_y B'_y + A'_z B'_z = A_x B_x + A_y B_y + A_z B_z.$$

For simplicity, let us consider only a rotation around the *z-axis*. The components of **A** in the rotated frame as functions of its components in the starting one are given by Eq. (1.10) and similarly for **B**. We can write

$$\begin{aligned}
A'_x B'_x + A'_y B'_y + A'_z B'_z &= \left(A_x \cos\theta + A_y \sin\theta\right)\left(B_x \cos\theta + B_y \sin\theta\right) \\
&\quad + \left(-A_x \sin\theta + A_y \cos\theta\right)\left(-B_x \sin\theta + B_y \cos\theta\right) + A_z B_z \\
&= A_x B_x \cos^2\theta + A_x B_y \cos\theta \sin\theta + A_y B_x \sin\theta \cos\theta \\
&\quad + A_y B_y \sin^2\theta + A_x B_x \sin^2\theta - A_y B_y \sin\theta \cos\theta \\
&\quad - A_y B_x \cos\theta \sin\theta + A_y B_y \cos^2\theta + A_z B_z. \\
&= A_x B_x + A_y B_y + A_z B_z.
\end{aligned}$$

We see that the product is invariant.

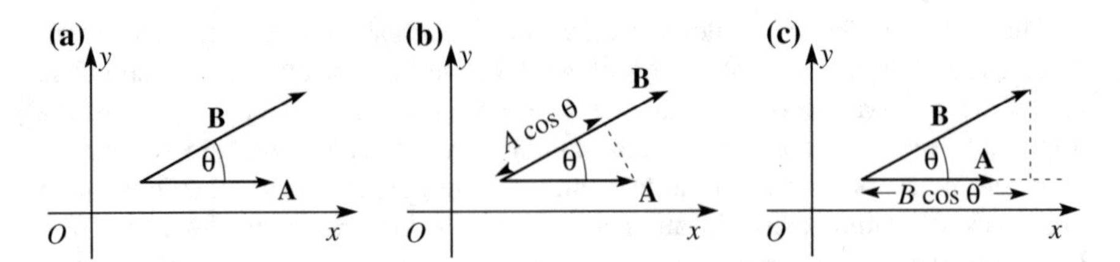

Fig. 1.8 Geometric meanings of scalar product

It is easy to show that both the commutative and distributive properties are valid for the dot product.

$$\mathbf{A}\cdot(\mathbf{B}+\mathbf{C}) = \mathbf{A}\cdot\mathbf{B}+\mathbf{A}\cdot\mathbf{C}. \tag{1.15}$$

We shall see now the geometric meaning of the scalar product. We can profit from it being invariant to choose convenient axes. We take x in the direction of **A** and y in the plane defined by **A** and **B** (Fig. 1.18a). If θ is the angle between the vectors, the components are $\mathbf{A} = (A, 0, 0)$ and $\mathbf{B} = (B\cos\theta, B\sin\theta, 0)$. Their dot product is then

$$\mathbf{A}\cdot\mathbf{B} = AB\cos\theta. \tag{1.16}$$

In words, the scalar product of two vectors is the product of their magnitudes times the cosine of the angle between them. There are also two other interpretations that may be useful. The scalar product is the product of the magnitude of the first vector times the projection of the second vector on the first one (Fig. 1.8b), or, the same with inverted roles (Fig. 1.8c).

The dot product is zero if the vectors are perpendicular, positive if the angle is acute, and negative if obtuse.

A particular and interesting case is the product of a vector by itself

$$\mathbf{A}\cdot\mathbf{A} = A_x^2+A_y^2+A_z^2 = A^2. \tag{1.17}$$

By definition the square of a vector is the dot product of the vector times itself and is equal to the square of its magnitude and also to the sum of the squares of its components. The latter property is an immediate consequence of the Pythagorean theorem. It is also called the *norm* of the vector. The norm is obviously the same in any reference.

Figure 1.9 shows a Cartesian reference frame in which three important vectors are drawn, the unit vectors of the coordinate axes, **i**, **j** and **k**. They have unit magnitude and are mutually normal. Consequently

$$\begin{aligned} \mathbf{i}\cdot\mathbf{i}=1, \quad \mathbf{j}\cdot\mathbf{j}=1, \quad \mathbf{k}\cdot\mathbf{k}=1 \\ \mathbf{i}\cdot\mathbf{j}=0, \quad \mathbf{j}\cdot\mathbf{k}=0, \quad \mathbf{k}\cdot\mathbf{i}=0. \end{aligned} \tag{1.18}$$

The components of any vector can be written in terms of the three unit vectors. Indeed, the x component of the vector **A** is its dot product with **i**, because the

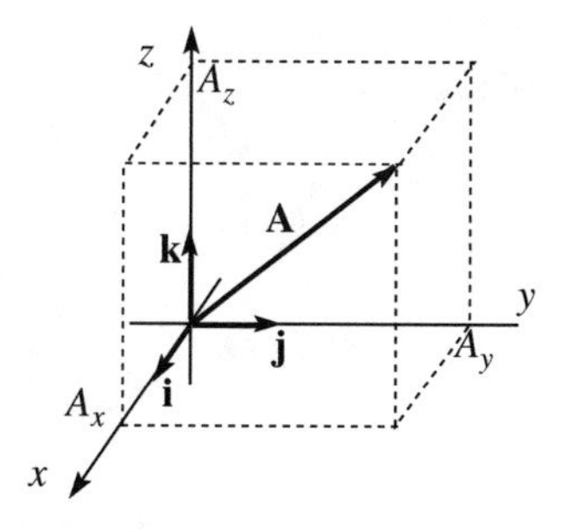

Fig. 1.9 The unit vectors of the Cartesian axes

magnitude of the latter is 1, and similarly for the other components. We then can write the vector as

$$\mathbf{A} = A_x\mathbf{i} + A_y\mathbf{j} + A_z\mathbf{k}, \tag{1.19}$$

namely as the sum of three vectors having the directions of the axes. These are called the vector components.

In particular the position vector can be written as

$$\mathbf{r} = x\mathbf{i} + y\mathbf{j} + z\mathbf{k}. \tag{1.20}$$

1.7 Vector Product of Two Vectors

Given the two vectors $\mathbf{A} = (A_x, A_y, A_z)$ and $\mathbf{B} = (B_x, B_y, B_z)$, their cross product is defined as the ordered triple of real numbers

$$\begin{aligned} \mathbf{C} &= C_x\mathbf{i} + C_y\mathbf{j} + C_z\mathbf{k} = \mathbf{A} \times \mathbf{B} \\ &= \left(A_yB_z - A_zB_y\right)\mathbf{i} + (A_zB_x - A_xB_z)\mathbf{j} + \left(A_xB_y - A_yB_x\right)\mathbf{k}. \end{aligned} \tag{1.21}$$

We now show that the cross product transforms as a vector under rotations of the axes and is also called the *vector product*. We show that for the x' component, the demonstration for the other two are exactly the same.

$$\begin{aligned} C'_x &= (\mathbf{A} \times \mathbf{B})'_x = A'_yB'_z - A'_zB'_y = \left(-A_x \sin\theta + A_y \cos\theta\right)B_z - A_z\left(-B_x \sin\theta + B_y \cos\theta\right) \\ &= (A_zB_x - A_xB_z)\sin\theta + \left(A_yB_z - A_zB_y\right)\cos\theta = C_y \sin\theta + C_x \cos\theta. \end{aligned}$$

The vector product is not commutative and the order of the factors matters. We have immediately from the definition that

$$\mathbf{B} \times \mathbf{A} = -\mathbf{A} \times \mathbf{B}. \tag{1.22}$$

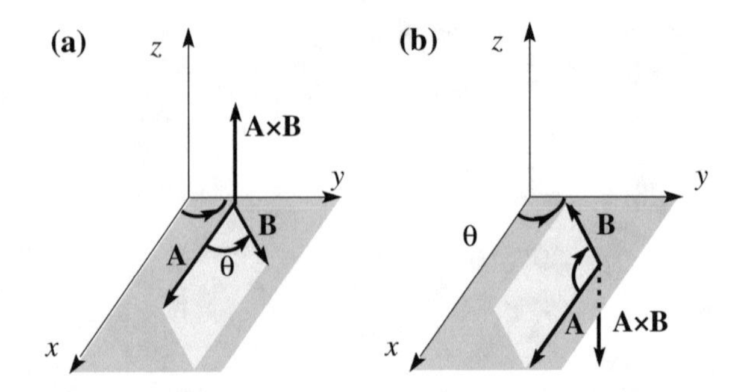

Fig. 1.10 The vector product of two vectors

Inverting the order of the factors the product changes sign. The property is called *anticommutative*.

It is easy to see that the vector product is distributive to the sum

$$\mathbf{A} \times (\mathbf{B} + \mathbf{C}) = \mathbf{A} \times \mathbf{B} + \mathbf{A} \times \mathbf{C}. \tag{1.23}$$

We now see the geometric meaning of the cross product using the same frame as in the previous section. We draw the two vectors as starting from the same point and take the x axis in the direction and sense of **A**, the y axis in the plane of the two vectors and the z axis to complete the right-handed reference (Fig. 1.10). The components are $\mathbf{A} = (A, 0, 0)$ and $\mathbf{B} = (B\cos\theta, B\sin\theta, 0)$. The cross product has only the z component different from zero

$$\mathbf{A} \times \mathbf{B} = \mathbf{k}AB\sin\theta. \tag{1.24}$$

Hence, the cross product is in the positive direction of the z-axis if θ is in one of the first two quadrants (Fig. 1.10a), in the negative one if in the third and fourth ones (Fig. 1.10b).

In conclusion, the geometric meaning of the vector product, independently of the reference frame, is the following. Its magnitude is equal to the area of the parallelogram having the two vectors as sides. Alternatively, we can also say that its magnitude is the magnitude of the first (A) times the projection of the second on the normal to the first (B sin θ) or vice versa. The direction of the product is perpendicular to the plane of the two vectors. Its sense is the one seeing the first factor going to the second through the smaller angle in anticlockwise direction.

Notice that we have followed here the same convention we used to define the positive direction of the z-axis. In a left-handed frame, the sense of the vector product would have changed too.

The cross product is zero if one of the vectors is zero or if the two are parallel. In particular the product of a vector times itself is zero.

Each of the unit vectors of the axes is the cross product of the other two

$$\mathbf{i} \times \mathbf{j} = \mathbf{k}, \mathbf{j} \times \mathbf{k} = \mathbf{i}, \mathbf{k} \times \mathbf{i} = \mathbf{j}. \tag{1.25}$$

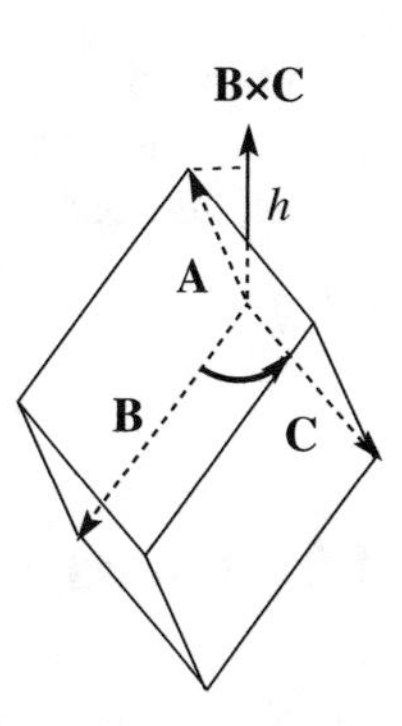

Fig. 1.11 The scalar triple product

The expressions of this type can be easier remembered thinking that each of them is obtained from the previous one by cyclic permutation. We now define the *scalar triple product* of three vectors, in the order **A**, **B** and **C**. It is the dot product of the first vectors times the cross product of the second times the third:

$$\mathbf{A}\cdot(\mathbf{B}\times\mathbf{C}) = A_x\left(B_yC_z - B_zC_y\right) + A_y(B_zC_x - B_xC_z) + A_z\left(B_xC_y - B_yC_x\right). \quad (1.26)$$

To see the geometrical meaning, we take the three vectors starting from the same point as in Fig. 1.11.

We can consider them as the sides of a parallelepiped. As we know the magnitude of $\mathbf{B}\times\mathbf{C}$ is equal to the area of the parallelogram having the two vectors as sides, which is a face of the parallelepiped. Its direction is the normal to that plane and the positive sense is the one that sees **B** going to **C**, rotating through the smaller angle, in anticlockwise direction. Let assume that **A** lies on the same side of the plane made by **B** and **C** as $\mathbf{B}\times\mathbf{C}$. The dot product of **A** times $\mathbf{B}\times\mathbf{C}$ is the product of the projection of **A** on the direction of $\mathbf{B}\times\mathbf{C}$ hence on the direction perpendicular to the plane of **B** and **C** times the magnitude of $\mathbf{B}\times\mathbf{C}$. But this projection is just the height h of the parallelepiped. In conclusion the triple product is equal to the volume of the parallelepiped having the three vectors as sides. In this case we are considering that this is true in absolute value and sign. It is the opposite of this volume if case **A** lies on the opposite side of the plane made by **B** and **C** that $\mathbf{B}\times\mathbf{C}$.

The following properties are immediately demonstrated: the triple scalar product is zero if the three vectors are coplanar, hence, in particular, if two or three are parallel. The triple product does not vary if the factors are circularly permuted

$$\mathbf{A}\cdot(\mathbf{B}\times\mathbf{C}) = (\mathbf{A}\times\mathbf{B})\cdot\mathbf{C}. \quad (1.27)$$

Obviously also

$$\mathbf{A}\cdot(\mathbf{A}\times\mathbf{C}) = 0. \quad (1.28)$$

A second triple product is the *triple vector product*, which is the cross product of the first vector times the cross product of the second and third ones. By direct verification one shows that

$$\mathbf{A} \times (\mathbf{B} \times \mathbf{C}) = \mathbf{B}(\mathbf{A} \cdot \mathbf{C}) - \mathbf{C}(\mathbf{A} \cdot \mathbf{B}). \qquad (1.29)$$

1.8 Bound Vectors, Moment, Couple

The forces are vectors. However, to completely characterize a force we need also to know its *application point*. If we push an object with our finger, we not only exert on it an action of a certain intensity and in a certain direction, but also we do that in a certain point. If we change that point, the effect of the force would change. A vector with an associated application point is called a *bound vector*. The line with the direction of the force through the application point is called the *line of action*.

Figure 1.12 shows the vector **A** and its point of application P. It may be a force for example. We arbitrarily choose a point Ω, which we call the *pole*. The *moment* of **A** about Ω is defined as the vector product of the vector leading from the pole to the application point of **A**, namely

$$\boldsymbol{\tau}_{\Omega} = \mathbf{\Omega P} \times \mathbf{A}. \qquad (1.30)$$

Let us see its geometrical meaning. The direction of the moment of the vector **A** is perpendicular to the plane defined by the segment **ΩP** and **A**. To see its positive direction we imagine **A** to be a force and **ΩP** a rigid bar. If we see the force turning the bar in an anticlockwise direction, we are on the positive side of the moment. The magnitude of the moment is given by the product of magnitude of the distance (h in the figure) of the pole Ω from the action line of **A**. In particular, if Ω lies on the action line the moment is zero.

The importance of the moments will be clear when we study the mechanics of the extended bodies in Chap. 7. We now consider a simple and particularly important case, the *couple* of vectors. A couple is a pair of bound vectors equal in

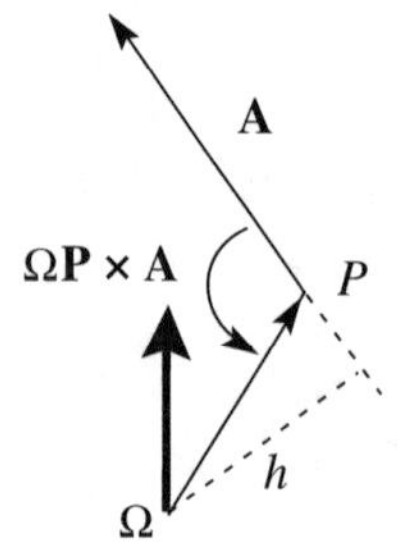

Fig. 1.12 The moment of vector **A** about the pole **Ω**

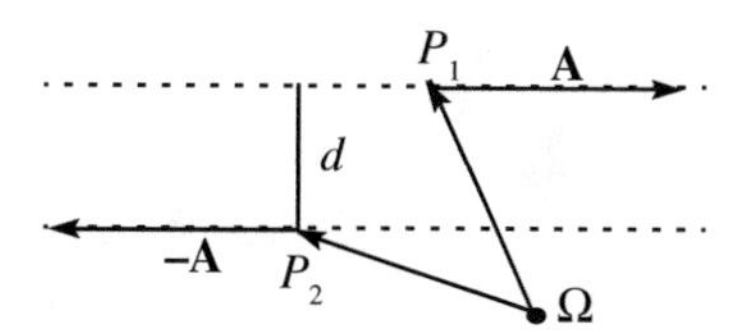

Fig. 1.13 A couple of bound vectors

magnitude in equal and opposite direction. The distance between the two action lines is called the *arm* of the couple.

A very important property of the couple is that their moment is independent of the pole. This may be called the *moment of the couple* or a *couple torque*. The two terms are synonymous.

Consider for simplicity the pole Ω lying in the plane of the couple, as in Fig. 1.13 (but the argument is valid in general). The two vectors are **A** and –**A**. P_1 and P_2 the application points respectively. The total moment, i.e. the sum of the two moments about Ω is

$$\boldsymbol{\tau}_\Omega = \boldsymbol{\Omega}\mathbf{P}_1 \times \mathbf{A} - \boldsymbol{\Omega}\mathbf{P}_2 \times \mathbf{A} = (\boldsymbol{\Omega}\mathbf{P}_1 - \boldsymbol{\Omega}\mathbf{P}_2) \times \mathbf{A} = \mathbf{P}_2\mathbf{P}_1 \times \mathbf{A},$$

which is independent of the pole. We can also see that the magnitude of the couple moment (or torque) is the product of the magnitude A of the vectors times the arm d of the couple, namely

$$\tau = d \cdot A. \tag{1.31}$$

Its direction is perpendicular to the plane of the couple, positive on the side seeing the couple rotate in an anticlockwise direction.

1.9 Matrices

Matrices are properly studied in mathematics courses. In this textbook only a few simple concepts and definitions will be needed and are recalled here.

A matrix A is an array of numbers ordered in rows and columns, say M lines and N columns

$$A = \begin{pmatrix} a_{11} & a_{12} & \dots & a_{1N} \\ a_{21} & a_{22} & \dots & a_{2N} \\ \dots & \dots & \dots & \dots \\ a_{M1} & a_{M2} & \dots & a_{MN} \end{pmatrix}. \tag{1.32}$$

The matrix is said to be *square* if the numbers of rows and column are equal; this number is called the *order* of the matrix. The generic element of the matrix is a_{ij}

where the first index i (i = 1, …, M) is the row index, the second j (j = 1, …, N) the column index.

Matrices with the same numbers of rows and column can be added. The sum $S = A + B$ of such matrices A and B is the matrix having as elements the sums of the corresponding elements of A and B, namely $s_{ij} = a_{ij} + b_{ij}$.

If the number of columns of the matrix A is equal to the number of rows of matrix B the product $P = A\ B$ is defined as follows. Be M the number of rows and N the number of columns of A, N the number of rows and L the number of columns of B. The product matrix P has M rows and L columns and its generic element is

$$p_{ij} = \sum_{k=1}^{N} a_{ik} b_{kj}. \tag{1.33}$$

We can use the concept of matrix product to re-write Eq. (1.10) for the transformation of a vector between two reference frames in compact form:

$$\begin{pmatrix} A'_x \\ A'_y \\ A'_z \end{pmatrix} = \begin{pmatrix} \cos\theta & \sin\theta & 0 \\ -\sin\theta & \cos\theta & 0 \\ 0 & 0 & 1 \end{pmatrix} \begin{pmatrix} A_x \\ A_y \\ A_z \end{pmatrix}. \tag{1.34}$$

We see that vectors are represented by a matrix with one column and three rows, while the rotation is represented by a three-by-three matrix.

Continuing with the definitions, the *minor* A_{ij} of the generic element a_{ij} is defined as the matrix one obtains from A suppressing row i and column j (i.e. the row and the column to which the element we are considering belongs).

For square matrices, say A, the *determinant* can be defined. It is a number, indicated with $\|A\|$ or with det A. The definition is recurrent. If the order of the matrix is one, its determinant is its only element. If the order is two,

$$A = \begin{pmatrix} a_{11} & a_{12} \\ a_{21} & a_{22} \end{pmatrix}, \quad \|A\| = a_{11}a_{22} - a_{12}a_{21}. \tag{1.35}$$

If the matrix order is three or larger, one starts choosing a row (or a column). It can be shown that the choice is arbitrary. We then choose the first row. Then we multiply each element of the row times the determinant of its minor, keeping it as it is, if the sum of the indices is even (11, 13, 15,…), changing its sign, if it is odd (12, 14, 16,…). Finally we sum all these numbers. The determinant of the 3×3 matrix

$$A = \begin{pmatrix} a_{11} & a_{12} & a_{13} \\ a_{21} & a_{22} & a_{23} \\ a_{31} & a_{32} & a_{33} \end{pmatrix}$$

is

$$\|A\| = \sum_{k=1}^{3} (-1)^{1+k} a_{1k} \|A_{1k}\| = a_{11}\|A_{11}\| - a_{12}\|A_{12}\| + a_{13}\|A_{13}\|$$
$$= a_{11}(a_{22}a_{33} - a_{23}a_{32}) - a_{12}(a_{21}a_{33} - a_{23}a_{31}) + a_{13}(a_{21}a_{32} - a_{22}a_{31}). \tag{1.36}$$

It is easy to show that if two (or more) rows or two columns are equal, or simply proportional, the determinant is null. It is also shown that the determinants of two matrices differing only for the exchange of two contiguous rows or two contiguous columns are equal and opposite.

The scalar triple product of three vectors, say **A**, **B** and **C**, can be usefully expressed as the determinant of a 3 × 3 matrix of their components

$$\mathbf{A} \times (\mathbf{B} \times \mathbf{C}) = \det \begin{pmatrix} A_x & A_y & A_z \\ B_x & B_y & B_z \\ C_x & C_y & C_z \end{pmatrix}$$
$$= A_x(B_yC_z - B_zC_y) - A_y(B_zC_x - B_xC_z) + A_z(B_xC_y - B_yC_x), \tag{1.37}$$

that is Eq. (1.26). The just mentioned properties of the determinant correspond to the known properties of the triple product: it is null if two factors are equal or parallel, i.e. with proportional components; inverting two factors the triple product changes sign.

Finally, also the vector product of two vectors can be written formally as the determinant of the matrix having in the first row the unit vectors of the axes, and second and third rows the components of the two vectors in the order. Indeed

$$\mathbf{A} \times \mathbf{B} = \det \begin{pmatrix} \mathbf{i} & \mathbf{j} & \mathbf{k} \\ A_x & A_y & A_z \\ B_x & B_y & B_z \end{pmatrix}$$
$$= \mathbf{i}(A_yB_z - A_zB_y) - \mathbf{j}(A_zB_x - A_xB_z) + \mathbf{k}(A_xB_y - A_yB_x), \tag{1.38}$$

that is Eq. (1.21).

1.10 Velocity

We shall now study the motion of the simplest body, the *material point* or *particle*. This is the case when its dimensions are small compared to the distances from other objects. This is clearly an idealization but it works often in practice. For example the planets are certainly not point-like, however in the mathematical description of

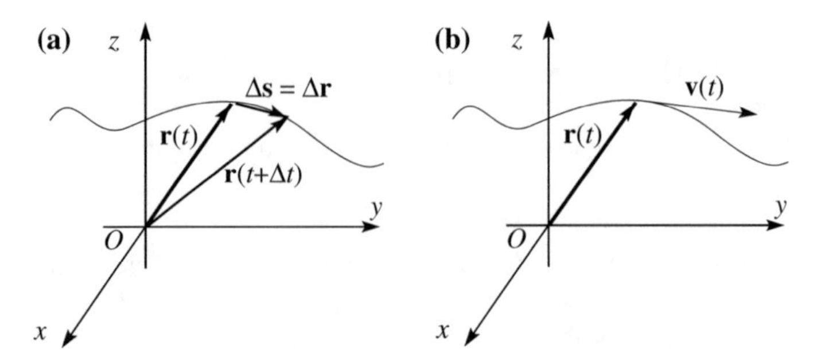

Fig. 1.14 **a** The trajectory of a particle, **b** the velocity

their motions around the sun they can be considered as such in a good approximation, as long as we do not consider the rotations about their axes, or the variations of the directions of those axes, or the tides on their surfaces. A ship can be considered a point when she is far from shore, but when she enters a harbor her dimension must be precisely known.

As we have already stated, the motion has to be studied in a given reference frame. The particle describes in its motion a curve, which is called the *trajectory*, as shown in Fig. 1.14a. The position vector is a function of time $\mathbf{r}(t)$ or, in other words, the co-ordinates are three functions of time $x(t)$, $y(t)$, $z(t)$. If we know these functions we completely know the motion of the particle. We say that the system has three degrees of freedom.

Let us consider the position vector at the instant of time t, $\mathbf{r}(t)$ as represented in Fig. 1.14a and an immediately following instant $t + \Delta t$, $\mathbf{r}(t + \Delta t)$, where Δt is a short time interval. In this time interval the particle has moved by $\Delta \mathbf{s}$, which is a step in the space having a magnitude and a direction, namely it is a vector. Looking at the figure one immediately sees that $\Delta \mathbf{s}$ is equal to the difference between the two vectors $\mathbf{r}(t + \Delta t)$ and $\mathbf{r}(t)$. This is the variation of the vector $\mathbf{r}$ in the time interval Δt. Hence

$$\Delta \mathbf{s} = \Delta \mathbf{r} = \mathbf{r}(t + \Delta t) - \mathbf{r}(t). \tag{1.39}$$

The *average velocity* in the time interval Δt is the vector obtained by dividing the displacement by the time interval in which it happens:

$$\langle \mathbf{v} \rangle = \frac{\Delta \mathbf{s}}{\Delta t} = \frac{\Delta \mathbf{r}}{\Delta t} = \frac{\mathbf{r}(t + \Delta t) - \mathbf{r}(t)}{\Delta t}. \tag{1.40}$$

or, for the components

$$\langle \upsilon_x \rangle = \frac{x(t + \Delta t) - x(t)}{\Delta t}, \ \langle \upsilon_y \rangle = \frac{y(t + \Delta t) - y(t)}{\Delta t}, \ \langle \upsilon_z \rangle = \frac{z(t + \Delta t) - z(t)}{\Delta t}. \tag{1.41}$$

Velocity is the limit for $\Delta t \to 0$ of the average velocity, namely

$$\mathbf{v} = \frac{d\mathbf{r}}{dt}. \tag{1.42}$$

In words, the velocity is the time derivative of the position vector. Its components are the derivatives of the coordinates

$$\upsilon_x = \frac{dx}{dt}, \; \upsilon_y = \frac{dy}{dt}, \; \upsilon_z = \frac{dz}{dt}. \tag{1.43}$$

In the limit $\Delta t \to 0$ the direction of $\Delta\mathbf{s}$ becomes tangent to the trajectory, in every point of the trajectory the direction of velocity is that of the tangent in that point (Fig. 1.14b).

The physical dimensions of velocity are those of a length divided by a time; the unit is consequently the meter per second (m/s or ms^{-1}).

The motion is said to be *uniform*, if the magnitude of velocity does not vary in time. In a uniform motion however, the velocity is not necessarily constant, because its direction may vary. The direction of velocity does not vary if the motion is rectilinear. Hence a motion with constant velocity is *rectilinear uniform.*

Example E 1.1 The motion of a particle is known when its three co-ordinates as functions of time are known. Consider the motion given by the equations

$$x(t) = bt + c, \quad y(t) = 0, \quad z(t) = 0,$$

where a and b are constants.

The co-ordinates y and z are always zero. Consequently the motion is along the *x-axis*, hence rectilinear. In the initial instant ($t = 0$) the particle is in the position $x(0) = c$. It is called the *initial position.* As time varies the position varies in proportion, b being the proportionality constant. The particle moves in the positive x direction (increasing x) if $b > 0$, in the negative one (decreasing x) if $b < 0$. The velocity has only one component different from zero, $\upsilon_x = \frac{dx}{dt} = b$. Hence, the motion is also uniform.

Example E 1.2 Consider the motion given by the equations

$$x(t) = b_1 t + c_1, \quad y(t) = b_2 t + c_2, \quad z(t) = 0.$$

Now the motion takes place in the xy plane, because the z co-ordinate is always zero. The initial position is

$$x(0) = c_1, \quad y(0) = c_2, \quad z(0) = 0.$$

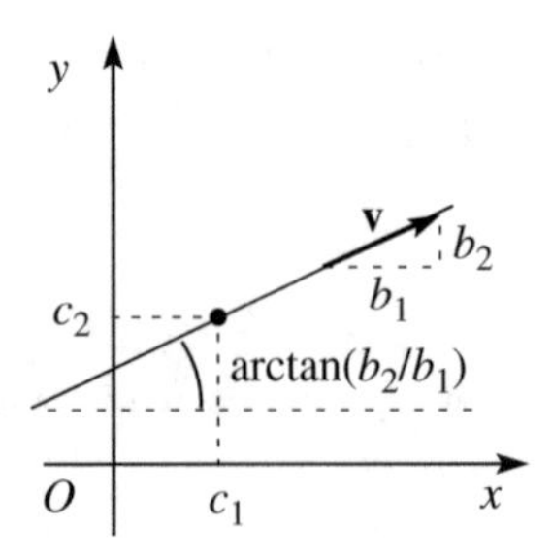

Fig. 1.15 Geometry of the motion of E 1.2

In order to find the equation of the trajectory we may take the ratio of the distances travelled in the same time along y and x. We find

$$\frac{y(t) - y(0)}{x(t) - x(0)} = \frac{b_2}{b_1},$$

which is a constant. This means that the trajectory is the straight line through the point ($c1$, $c2$) and making with *x-axis* the angle arctan (b_2/b_1).

Hence, the motion is rectilinear as shown in Fig. 1.15.

The components of velocity in the plane of the motion are $\upsilon_x = \frac{dx}{dt} = b_1$ and $\upsilon_y = \frac{dy}{dt} = b_2$. The velocity vector is then $\mathbf{v} = (b_1, b_2, 0)$, having the same direction as the (rectilinear) trajectory. It is consequently rectilinear uniform.

Example E 1.3 Consider the motion

$$x(t) = bt^2 + ct + d, \quad y(t) = 0, \quad z(t) = 0 \tag{1.44}$$

and let us calculate the velocity. There is only one non-zero component, namely $\upsilon_x = \frac{dx}{dt} = bt + c$. The velocity is not constant but increases (decreases) linearly with time if $b > 0$ ($b < 0$). The motion is rectilinear but not uniform.

As we have already seen, the motion of the bodies is always relative to the assumed reference frame. Consequently also the velocity is relative to the frame. In Chap. 5 we shall study in detail the relations between the kinematic quantities (position, velocity, acceleration, etc.) in different frames in relative motion. We anticipate here a simple concept, the *relative velocity*.

The velocity of a body relative to another one is the vector difference between their two velocities. Indeed, let $\mathbf{r}_1$ be the position vector of the first body and $\mathbf{r}_2$ that of the second. The position of the second body relative to the first is the vector $\mathbf{r}_{12} = \mathbf{r}_2 - \mathbf{r}_1$. The time derivative of this vector is the velocity of 2 relative to 1, which is the velocity of 2 seen by an observer travelling with 1. Calling it $\mathbf{v}_{12}$ we have

$$\mathbf{v}_{12} = \mathbf{v}_2 - \mathbf{v}_1. \tag{1.45}$$

The velocity of a passenger walking on the deck of a ship relative to the vessel is the difference between the velocity vectors of the passenger and of the ship relative to the sea.

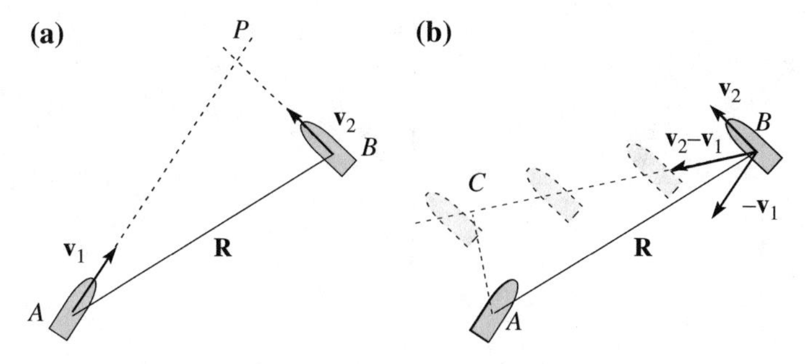

Fig. 1.16 Motion relative to the sea and of one sheep relative to the other

Notice that the position of 1 relative to 2 is the opposite of the position of 2 relative to 1. The same is true for the velocities.

Example E 1.4 Consider two ships, *A* and *B*, which at a certain instant are in the position shown in Fig. 1.16. Their velocities are $\mathbf{v}_1$ and $\mathbf{v}_2$ respectively. The two courses intercept in the point o *P*. Will the ships collide in *P* if they move with constant velocities?

The answer is immediate in a frame fixed with one of the two vessels, for example with *A* as in Fig. 1.16b. In this frame, all the relevant velocities, including that of the sea, are obtained from those relative to the sea by subtracting $\mathbf{v}_1$. Hence *A* does not move (by definition) and *B* moves with velocity $\mathbf{v}_2 - \mathbf{v}_1$. The vector **R** leading from *A* to *B* is the same in the two frames (they differ by a translation). Ship *B*, as seen by *A*, moves on the course shown in the figure. Hence the minimum distance she will pass from *A* is *AC*, namely the distance of *A* from the straight line *B* travels. In conclusion, they will pass close but will not collide.

Notice on purpose that a passenger *A* sees *B* moving sideway, not in the direction of bow. Indeed, we have a strange impression when we cross closely another ship, particularly offshore, when any reference to ground is missing. She looks to be travelling in a not “natural” direction.

1.11 Angular Velocity

An important motion is the circular one, in which the trajectory is a circle. Let *R* be its radius. It is always convenient to choose the reference frame taking profit from the symmetry of the problem, if any is present. We take the origin in the center of the circle and the *z*-axis perpendicular to its plane. The motion is then in the *xy* plane as shown in Fig. 1.17a.

We further choose the origin of time in the moment in which the point crosses the positive *x*-axis. Let $\phi(t)$ be the angle between the position vector and the *x* axis at time *t*, taken as positive in anticlockwise direction and let $s(t)$ be the length of the arc subtended by $\phi(t)$, taken with the same sign as ϕ, namely $s(t) = R\,\phi(t)$. Let $d\mathbf{s}$ be the infinitesimal movement in dt (Fig. 1.17b). The infinitesimal changes of *s* and ϕ

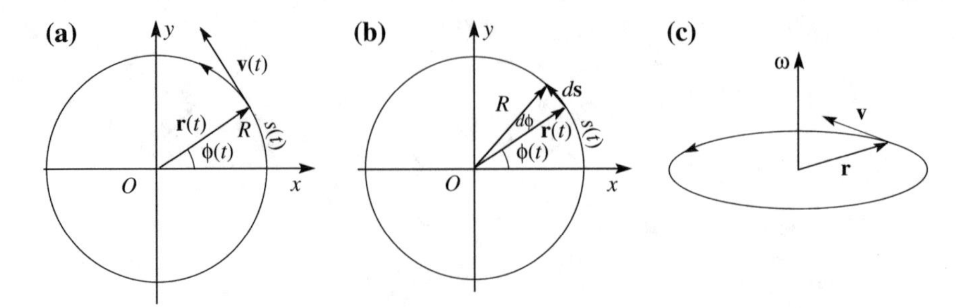

Fig. 1.17 **a** The circular motion, **b** an infinitesimal movement, **c** angular velocity **ω**

are linked by the relation $ds = R\,d\phi$, where in our notation ds is the magnitude of $d\mathbf{s}$ if the motion is anticlockwise (as in Fig. 1.17), and is opposite if clockwise. The angular velocity measures the rate of change of the angle. We then consider the time derivative

$$\omega_z = \frac{d\phi}{dt}. \tag{1.46}$$

This quantity has magnitude and a sign, depending on the sense of rotation. In fact, it is the z component of the angular velocity, which is a vector. Its magnitude is the absolute value of Eq. (1.46), its direction is perpendicular to the plane of the motion, taken positive on the side seeing the motion is anticlockwise. This is the z-axis in Fig. 1.17c.

The physical dimensions of the angular velocity are the inverse of time; its unit is radians per second (rad/s)

In a circular motion, the magnitudes of velocity $\upsilon = |ds|/dt$ and the magnitude of the angular velocity ω are related by

$$\upsilon = \omega R. \tag{1.47}$$

The relation between the corresponding vectors, as immediately seen from Fig. 1.17c is

$$\mathbf{v} = \boldsymbol{\omega} \times \mathbf{r}. \tag{1.48}$$

Let us consider the case in which the magnitude υ of the velocity is constant. The motion is circular and uniform, the arcs and the corresponding angles are proportional to the times taken to travel them, namely $\phi(t) = s(t)/R = \pm\upsilon t/R$ (where, as usual the sign is positive if the direction is anticlockwise and vice versa). Hence we have the equations of motion in polar co-ordinates:

$$r(t) = R, \quad \theta(t) = \frac{\pi}{2}, \quad \phi(t) = \pm\frac{\upsilon t}{R} = \omega_z t. \tag{1.49}$$

The equations of motion in Cartesian co-ordinates are

$$x(t) = R\cos\omega_z t, \quad y(t) = R\sin\omega_z t, \quad z(t) = 0. \tag{1.50}$$

As an exercise we can check that the trajectory is indeed a circle. Taking the squares of the members and summing we have $x^2(t) + y^2(t) + z^2(t) = R^2$ which is the equation of a circumference. Notice that the two Cartesian co-ordinates x and y are not independent but if we know one we know also the other. In fact the particle is bound to travel onto a prefixed trajectory. The system has one degree of freedom. This is evident in polar co-ordinates, Eq. (1.49). Two of them are constant.

We now express the Cartesian components of velocity

$$\upsilon_x(t) = \frac{dx}{dt} = -\omega_z R\sin\omega_z t, \quad \upsilon_y(t) = \frac{dy}{dt} = \omega_z R\cos\omega_z t, \quad \upsilon_z(t) = \frac{dz}{dt} = 0 \tag{1.51}$$

The components of the velocity vector change in time: when the particle moves on the circle its direction continuously varies even if its magnitude is constant. Indeed, the magnitude is

$$\upsilon = \sqrt{\upsilon_x^2 + \upsilon_y^2 + \upsilon_z^2} = \omega R\sqrt{\cos^2\omega_z t + \sin^2\omega_z t} = \omega R, \tag{1.52}$$

which is a constant.

As a further exercise, let us check that the velocity is always tangent to the trajectory, i.e., perpendicular to the position vector everywhere. To see that we take their scalar product and get

$$\mathbf{r}(t)\cdot\mathbf{v}(t) = x(t)\upsilon_x(t) + y(t)\upsilon_y(t) = -\omega_z R^2\cos\omega_z t\sin\omega_z t + \omega_z R^2\sin\omega_z t\cos\omega_z t = 0$$

We now make the following observation that will be useful in the following. In the case we have noted that we have two vectors: the position vector and the velocity. The x and y components of the first vector are proportional to the cosine (Eq. 1.50) and the sine of the angular co-ordinate respectively, those of the second to the opposit of its sine and to its cosine respectively (Eq. 1.51). When this happens the two vectors are perpendicular.

Both the co-ordinates and the components of velocity are proportional to the circular functions $\cos\omega t$ or $\sin\omega t$, which are periodic. In fact the motion is periodic, meaning that if position and velocity have some values in the instant t they have again the same values at the instants $t + T$, $t + 2T$, etc., for every t. The time T is called the *period* of the motion. It is inversely proportional to the angular velocity

$$T = \frac{2\pi}{\omega}. \tag{1.53}$$

1.12 Acceleration

The motion of a body in which the velocity varies with time in magnitude or direction is called *accelerated.* If the change of velocity in the time interval Δt is $\Delta \mathbf{v}$, the average acceleration in that time interval is the ratio

$$\langle \mathbf{a} \rangle = \frac{\Delta \mathbf{v}}{\Delta t}. \tag{1.54}$$

The instantaneous acceleration at time t is its limit for $\Delta t \to 0$, namely the time derivative of the velocity

$$\mathbf{a} = \frac{d\mathbf{v}}{dt} \quad \left[a_x = \frac{dv_x}{dt}, a_y = \frac{dv_y}{dt}, a_z = \frac{dv_z}{dt} \right]. \tag{1.55}$$

In the particular case of the rectilinear motion, when the direction of the velocity is constant, the acceleration direction is also on the line and its magnitude and sign are

$$a = \frac{dv}{dt}. \tag{1.56}$$

Example E 1.5 Consider again the motion of Example E 1.3, namely

$$x(t) = bt^2 + ct + d, \quad y(t) = 0, \quad z(t) = 0.$$

The motion is along the x-axis with velocity $v_x = bt + c$. The x component of the acceleration, the only different from zero, is then $a_x = \frac{dv_x}{dt} = \frac{d^2x}{dt^2} = b$. The acceleration is constant in magnitude and direction. Such motions are called *uniformly accelerated.*

We now consider a *uniform circular* motion in which the velocity vector has a constant magnitude and varies in direction with constant angular velocity. In order to find the acceleration, consider the auxiliary diagram of Fig. 1.18a (we assume an anticlockwise rotation direction). The axes of the figure are the x and y components of the velocity vector that we think of as having its tail in the origin. It is analogous to the position vector in the xy plane. The analogy is complete because both vectors rotate with constant angular velocity ω. In other words, the head of the velocity vector A describes a circularly uniform motion in the velocity plane, having a radius equal to its magnitude v.

Clearly the "velocity" of point A is just the acceleration of particle P because the displacement of A in the time interval dt is $d\mathbf{v}$ and consequently its "velocity" is $\mathbf{a} = \frac{d\mathbf{v}}{dt}$. This vector is tangent to the circle and consequently perpendicular to the velocity (Fig. 1.18a). More precisely, the direction of acceleration is obtained from that of the velocity by a rotation of 90° in an anticlockwise direction. Going back to the representation of the motion in the xy plane in Fig. 1.18b, the acceleration,

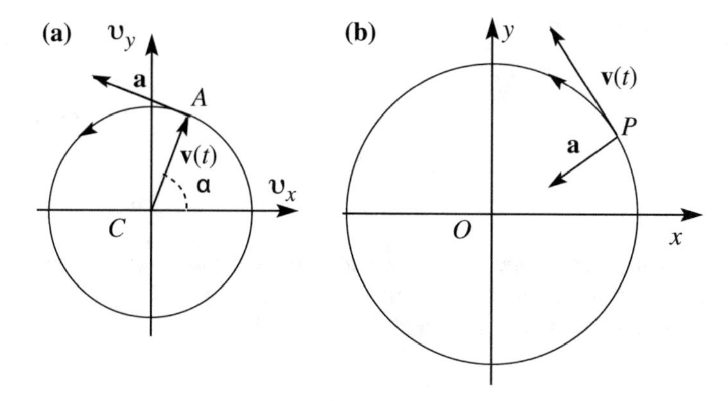

Fig. 1.18 Uniform circular motion

which we just saw to be at 90° from the velocity anticlockwise, is radial directed towards the center. It is then called *centripetal acceleration*.

We immediately find the magnitude of the acceleration. We denote by α the angle between the vector **v** and the abscissa axis in Fig. 1.18a, and $d\alpha$ its variation in the time dt. Considering that the vector rotates with constant angular velocity ω, we have $d\alpha = \omega\, dt$. On the other hand the change in velocity is $|d\mathbf{v}| = \upsilon d\alpha$ and we get the important relation

$$a = \left|\frac{d\mathbf{v}}{dt}\right| = \upsilon\frac{d\alpha}{dt} = \omega\upsilon = \frac{\upsilon^2}{R} = \omega^2 R. \tag{1.57}$$

Summing up, if the velocity varies only in magnitude, the acceleration is parallel to velocity, if the velocity varies only in direction, the acceleration is perpendicular to the velocity, directed towards the center of the trajectory. We shall see in Sect. 1.14 that in the general case in which both magnitude and direction of velocity vary, acceleration has two components one parallel and one perpendicular to velocity.

1.13 Time Derivative of a Vector

In the study of uniform circular motion we have dealt with the position vector **r** and the velocity **v**. Both are constant in magnitude and vary in direction with time, rotating at the angular velocity ω. We have seen that the magnitudes of their time derivatives are respectively ωr and $\omega\upsilon$, namely, in both cases the magnitude of the vector times ω. In both cases the direction of the derivative vector is at 90° forward to the original vector. The result is valid also if the angular velocity is not constant. Indeed, we did not use this assumption.

We now generalize the argument as follows, with reference to Fig. 1.19. Consider the vector function of time $\mathbf{A}(t)$, constant in magnitude, varying only in

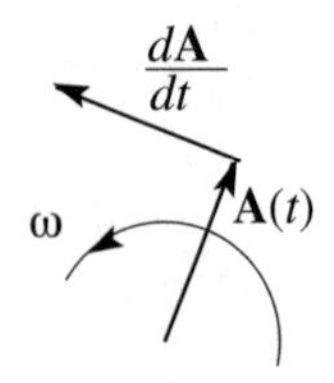

Fig. 1.19 A rotating vector and its time derivative

direction. At the generic time instant the vector rotates with angular velocity $\boldsymbol{\omega}$, not necessarily constant. Let $\mathbf{u}_p$ be the unit vector rotated by $\pi/2$ relative to $\mathbf{A}$ in the direction of the rotation. The time derivative of $\mathbf{A}$ is

$$\frac{d\mathbf{A}}{dt} = \omega A \mathbf{u}_p \tag{1.58}$$

or better

$$\frac{d\mathbf{A}}{dt} = \boldsymbol{\omega} \times \mathbf{A}. \tag{1.59}$$

This important formula that we shall use often in the following is due to Siméon-Denis Poisson (1781–1842) and is called a *Poisson formula*. It is valid if the magnitude A is constant.

In the general case in which the vector $\mathbf{A}$ varies both in direction and magnitude, its time derivative is immediately obtained by writing the vector as the product of its magnitude and its unitary vector

$$\frac{d\mathbf{A}}{dt} = \frac{d(A\mathbf{u}_A)}{dt} = \frac{dA}{dt}\mathbf{u}_A + A\frac{d\mathbf{u}_A}{dt}.$$

But the vector $\mathbf{u}_A$ is constant in magnitude, being unitary, and we can use the Poisson formula for its derivative. We get

$$\frac{d\mathbf{A}}{dt} = \frac{dA}{dt}\mathbf{u}_A + A\omega \cdot \mathbf{u}_p = \frac{dA}{dt}\mathbf{u}_A + \boldsymbol{\omega} \times \mathbf{A}, \tag{1.60}$$

which is an important result that we shall use often in the following.

1.14 Motion on the Plane

We now consider a general motion in a plane. We indicate with $\mathbf{u}_t$ the unit vector tangent to the trajectory in its generic point in the direction of the velocity in that point. In general $\mathbf{u}_t$ varies in time. Figure 1.20 shows the situation in two consecutive instants.

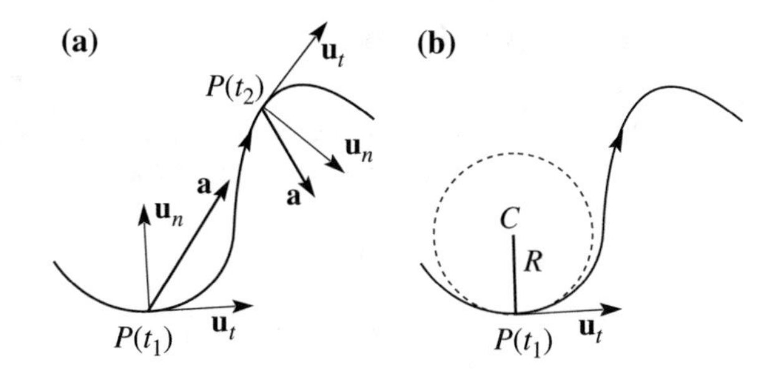

Fig. 1.20 **a** The acceleration vector in two different points of the trajectory; **b** the osculating circle

In every instant, i.e. in every point of the trajectory, in general the velocity is different. We indicate with $\mathbf{u}_n$ the unit vector normal to the trajectory. Its positive direction is the direction obtained by rotating $\mathbf{u}_t$ by 90° in the direction of the instantaneous rotation of the velocity vector. This geometrically means that $\mathbf{u}_n$ is directed towards the curvature center. The latter may lie on the left or the right of the trajectory depending on the case. To obtain the acceleration we take the derivative of the velocity expressed as the product of magnitude times unit vector, $\mathbf{v} = \upsilon\ \mathbf{u}_t$.

$$\frac{d\mathbf{v}}{dt} = \frac{d\upsilon}{dt}\mathbf{u}_t + \upsilon\omega \cdot \mathbf{u}_n = a_t\mathbf{u}_t + a_n\mathbf{u}_n. \tag{1.61}$$

As anticipated, the acceleration has two components. One is tangent to the trajectory and equal to the time derivative of the magnitude of velocity. It is null if the motion is uniform, positive if it is accelerated, negative if decelerated. The other component is normal to the trajectory in any case towards the "interior" of the curve. It is zero when the direction of the velocity does not vary, even if instantaneously, as in the flex points of the trajectory.

We can express the normal component of the acceleration in terms of the *curvature radius* of the trajectory in the point P under consideration. Figure 1.20b shows the situation. Consider all the circles tangent to the curve in P having radiuses between 0 and infinity. One of these gives locally the best approximation of the curve. It is called an *osculating circle*, from the Latin word *osculum*, meaning kiss. Its radius R is called the *curvature radius* of the curve in the point P. Its reciprocal is the *curvature*. In an inflexion point the curvature radius is infinite and the curvature is null.

Now we can approximate the small curve segment around P with the arc of the osculating circle and think of the point as moving on that arc with angular velocity $\omega = \upsilon/R$. In conclusion, the two components of the acceleration are

$$a_t = \frac{d\upsilon}{dt}, a_n = \frac{\upsilon^2}{R}. \tag{1.62}$$

We see that the normal component of the acceleration is proportional to the curvature and to the square of the velocity.

1.15 From Acceleration to Motion

Figure 1.21 represents the trajectory of a material point P in a given Cartesian reference frame, its position vector $\mathbf{r}(t)$, its velocity $\mathbf{v}(t)$ and its acceleration $\mathbf{a}(t)$, that are all functions of time.

We recall their expressions

$$\mathbf{r}(t) = x(t)\mathbf{i} + y(t)\mathbf{j} + z(t)\mathbf{k}, \tag{1.63}$$

$$\mathbf{v}(t) = \upsilon_x(t)\mathbf{i} + \upsilon_y(t)\mathbf{j} + \upsilon_z(t)\mathbf{k} = \frac{d\mathbf{r}}{dt} = \frac{dx}{dt}\mathbf{i} + \frac{dy}{dt}\mathbf{j} + \frac{dz}{dt}\mathbf{k}, \tag{1.64}$$

$$\mathbf{a}(t) = a_x(t)\mathbf{i} + a_y(t)\mathbf{j} + a_z(t)\mathbf{k} = \frac{d\mathbf{v}}{dt} = \frac{d\upsilon_x}{dt}\mathbf{i} + \frac{d\upsilon_y}{dt}\mathbf{j} + \frac{d\upsilon_z}{dt}\mathbf{k}, \tag{1.65}$$

$$\mathbf{a}(t) = \frac{d^2\mathbf{r}}{dt^2} = \frac{d^2x}{dt^2}\mathbf{i} + \frac{d^2y}{dt^2}\mathbf{j} + \frac{d^2z}{dt^2}\mathbf{k}. \tag{1.66}$$

In words, the velocity is the time derivative of the position vector and the acceleration is the time derivative of the velocity or the second time derivative of the position vector. We shall see in the next chapter that acceleration is proportional to the force.

We consider now the inverse problem, namely to find the velocity and the law of motion once the acceleration $\mathbf{a}(t)$ is given. As the velocity is the time derivative of the position vector, the latter is given by the integral of the velocity on time from the initial instant t_0 to the time t considered, namely

$$\mathbf{r}(t) - \mathbf{r}(t_0) = \int_{t_0}^{t} \mathbf{v}(t)dt.$$

In general, we want to know the position of P at the time t and rewrite the expression as

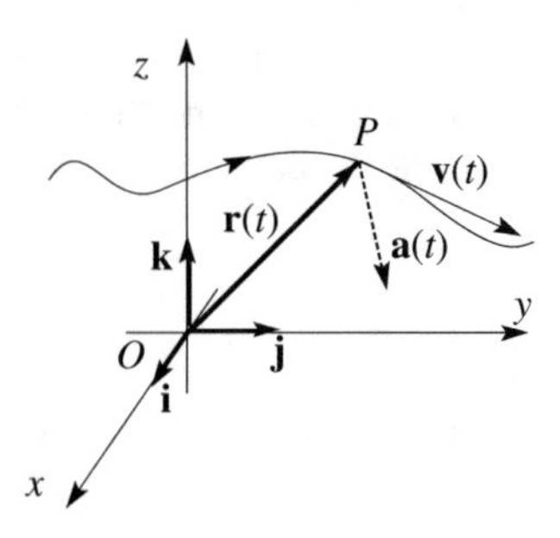

Fig. 1.21 Trajectory, position vector, velocity and acceleration

$$\mathbf{r}(t) = \mathbf{r}(t_0) + \int_{t_0}^{t} \mathbf{v}(t)dt. \tag{1.67}$$

We see that knowledge of the velocity $\mathbf{v}(t)$ is not sufficient. We need also to know the position of the body at a certain instant t_0. This instant can be any, but generally we know how the motion began, namely we know the *initial position*. It is customary to choose that instant as the origin and $t_0 = 0$.

To a question like “A car has been travelling at a constant speed of 100 km/h. Where is it after 2 h?” We can only answer it has travelled 200 km. We can know its position only if we know from were it started.

Equation (1.67) corresponds to three integrals

$$x(t) = x(t_0) + \int_{t_0}^{t} \upsilon_x(t)dt,\ y(t) = y(t_0) + \int_{t_0}^{t} \upsilon_y(t)dt,$$
$$z(t) = z(t_0) + \int_{t_0}^{t} \upsilon_z(t)dt. \tag{1.68}$$

If we want to know the velocity for a given acceleration, we proceed in the same manner by integrating

$$\mathbf{v}(t) = \mathbf{v}(t_0) + \int_{t_0}^{t} \mathbf{a}(t)dt \tag{1.69}$$

or, in terms of the components

$$\upsilon_x(t) = \upsilon_x(t_0) + \int_{t_0}^{t} a_x(t)dt$$
$$\upsilon_y(t) = \upsilon_y(t_0) + \int_{t_0}^{t} a_y(t)dt \tag{1.70}$$
$$\upsilon_z(t) = \upsilon_z(t_0) + \int_{t_0}^{t} a_z(t)dt.$$

Again, we need to determine the integration constants, namely the velocity at a certain instant, which is usually the initial one.

Once the velocity is known we need to integrate again to have the law of motion. For that we need to know both the initial position and the initial velocity.

1.16 Free Fall Motion

The study of the free fall motion of bodies near the surface of earth is an important example of the use of the just developed formalism. With “free fall” we mean an idealized situation in which the air resistance can be neglected and the bodies move only under the action of gravity We anticipate that under these conditions the vertical and horizontal motions are independent from one another, as we shall study in Sect. 3.7, and that any free body moves with a constant acceleration, **g**, which is vertically directed downwards and has a magnitude (approximately) $g = 9.8\ \text{m/s}^2$. We choose a reference frame with the *z-axis* vertical upward, and the x and y-axes in a horizontal plane, for example the ground. The acceleration of the body, that we shall consider point-like, P, has the components

$$\mathbf{a} = (0, 0, -g) = -g\mathbf{k}. \tag{1.71}$$

The motion of the body depends on the initial conditions. If for example we drop the body from a certain height with null velocity it will move vertically down with uniform acceleration. If we launch it vertically upwards it will gradually slow down, stop and then fall down. If we launch it at an angle with the horizontal it will describe a curved trajectory, etc. Let us study these motions.

Let us start from the simplest case. We drop the body at the height h above ground with null velocity at $t = 0$. The initial conditions are

$$x(0) = 0, \quad y(0) = 0, \quad z(0) = h; \quad \upsilon_x(0) = 0, \quad \upsilon_y(0) = 0, \quad \upsilon_z(0) = 0.$$

The x component of the velocity at the generic time t is

$$\upsilon_x(t) = \upsilon_x(0) + \int_0^t a_x(t)dt = 0 + 0.$$

The x component of the velocity is identically zero (i.e. is zero at every instant of time) because the x components of both acceleration and initial velocity are zero. A similar argument leads immediately to conclude that also $x(t) = 0$. The same is true for the y components of velocity and position vectors. Notice that the initial conditions $x(0) = 0$ and $y(0) = 0$ depend on the reference frame. Its origin has been chosen in such a way to have the point from which we drop the particle on the z-axis. A different choice would have led to the initial conditions, say, $x(0) = a$, $y(0) = b$. The two co-ordinates as functions of time would have been $x(t) = a$, $y(t) = b$. The motion obviously is the same.

We have found that the motion is along the z-axis. As the acceleration is constant, it is uniformly accelerated (acceleration may have both signs, if we want to be specific we can say accelerated, if the acceleration is positive, delayed if it is negative). Let us now find the velocity in the z direction.

$$\upsilon_z(t) = \upsilon_z(0) + \int_0^t a_z(t)dt = 0 - \int_0^t gdt = -gt. \quad (1.72)$$

Velocity is always negative. Indeed the body moves always in the z direction we have chosen as negative. We now integrate once more to find the position as a function of time

$$z(t) = z(0) + \int_0^t \upsilon_z(t)dt = h - \int_0^t gtdt = h - \frac{1}{2}gt^2 \quad (1.73)$$

which is the law of the motion. Knowing completely the motion, we can look for interesting properties, for example the time taken to reach the ground. This is the instant in which z = 0, hence $t_f = \sqrt{2h/g}$ and the velocity in that instant

$$\upsilon_f = \upsilon(t_f) = \sqrt{2gh}. \quad (1.74)$$

Consider now the same initial conditions with the difference that the initial velocity has a nonzero vertical value υ_0. With the same arguments as before, we obtain

$$\upsilon_z(t) = \upsilon_0 - gt, \quad (1.75)$$

$$z(t) = h + \upsilon_0 t - \frac{1}{2}gt^2. \quad (1.76)$$

We should now distinguish the two cases of positive (downwards) and negative (upwards) initial velocity.

If $\upsilon_0 < 0$, the velocity is always negative. To find the instant t in which the body is at the height z we solve Eq. (1.76), obtaining

$$t(z) = \frac{\upsilon_0 \pm \sqrt{\upsilon_0^2 + 2g(h-z)}}{g}.$$

We have two solutions because Eq. (1.76) is of second degree in t. However, in the case we are considering, one of them, the one with the negative sign, is always negative and consequently does not have physical meaning. We must choose the solution with a positive sign, because the motion starts at t = 0.

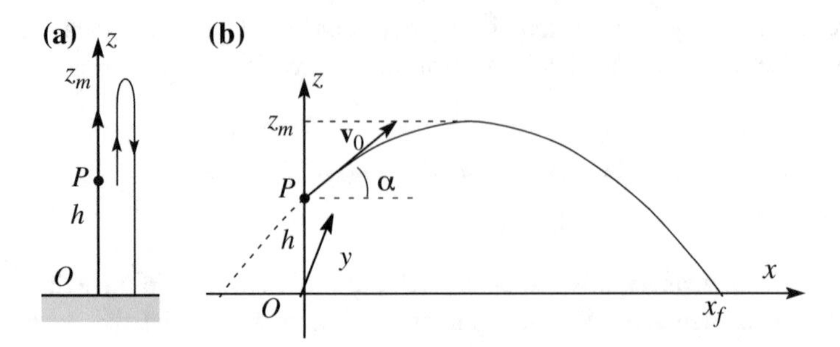

Fig. 1.22 Free fall trajectories with initial velocity **a** vertical upward, **b** at an angle α with the horizontal

The time of arrival at ground, the duration of the fall, is the time at which $z = 0$, namely

$$t_f = t(z = 0) = \frac{\upsilon_0 + \sqrt{\upsilon_0^2 + 2gh}}{g},$$

which is shorter than in the case of null initial velocity. Obviously the expressions found in the latter case are particular cases.

If $\upsilon_0 > 0$, from Eq. (1.75) we see that the velocity is positive, namely upwards, for a while, but it diminishes with increasing time. It is zero in the instant $t_m = \upsilon_0/g$, and negative in later times. Indeed, the body reaches the maximum height at t_m, namely $z_m = z(t_m) = h + \upsilon_0^2/(2g)$ (see Fig. 1.22a). In this case both roots for $t(z)$ have physical meanings provided $t \geq 0$. Indeed, the body goes twice through the same height, if it is $z \geq h$, first going up later going down. If $z < h$ one solution is negative and again does not have physical meaning.

Why may it happen that a mathematical solution should be discarded on physical grounds? The reason is that the equations stating the "initial conditions" do not give information on the system before the "initial" instant. In this case the body stood still, say in our hand. But it would have been possible that it was moving upwards in such a way as to reach $z = h$ at $t = 0$ with velocity equal to υ_0. The discarded solution would have made sense.

We now suppose that the initial position is again at the height h above ground, but that the velocity $\mathbf{v}_0$ is at an angle α with the horizontal. This is what happens when shooting with a cannon from the top of a tower. We choose the z vertical upwards as before and the x horizontal in the plane of z and of the initial velocity. The initial conditions are

$$x(0) = y(0) = 0, \quad z(0) = h; \quad \upsilon_x(0) = \upsilon_0 \cos\alpha, \quad \upsilon_y(0) = 0, \quad \upsilon_z(0) = \upsilon_0 \sin\alpha.$$

The motion is in the plane xz, as show in Fig. 1.22b. We find, as usual, the velocity using Eq. (1.69) and the initial conditions.

$$\mathbf{v}(t) = \mathbf{i}\upsilon_0 \cos\alpha + \mathbf{k}(\upsilon_0 \sin\alpha - gt). \tag{1.77}$$

We see that the horizontal, x, component of the velocity is constant and equal to its initial value and that the vertical one, z, decreases linearly in time, exactly as in the case we have considered.

We integrate once more and use the initial conditions to obtain the law of motion, finding

$$\mathbf{r}(t) = \mathbf{i}(\upsilon_0 \cos\alpha)t + \mathbf{k}\left[(\upsilon_0 \sin\alpha)t - \frac{1}{2}gt^2 + h\right] \tag{1.78}$$

or

$$x(t) = (\upsilon_0 \cos\alpha)t, \quad z(t) = (\upsilon_0 \sin\alpha)t - \frac{1}{2}gt^2 + h. \tag{1.79}$$

We now know completely the motion. If, for example, we want to know the shape of the trajectory we must eliminate t from the equations for the co-ordinates. From the first one we have $t = x/(\upsilon_0 \cos\alpha)$, which, substituted in the second equation, gives

$$z = x\tan\alpha - x^2 \frac{g}{2\upsilon_0^2 \cos^2\alpha} + h, \tag{1.80}$$

which is the equation of a parabola. The distance x_f at which the body touches the ground, namely the range of the weapon, is the value of x corresponding to $z = 0$. We then put this value in Eq. (1.80) and solve for x. We find

$$x_f = \frac{\upsilon_0^2}{g}\sin\alpha\cos\alpha\left(1 \pm \sqrt{1 + \frac{2gh}{\upsilon_0^2 \sin^2\alpha}}\right). \tag{1.81}$$

The negative root solution is for $t < 0$ and corresponds to the intersection of the parabola on the left of the tower. It is shown dotted in Fig. 1.22b and should be discarded. The positive root is the solution for which we searched.

We now find the duration of the shot, which is the time t_f at which the body touches ground. With $x = x_f$ the first of the (1.79) solved for t gives

$$t_f = \frac{x_f}{\upsilon_0 \cos\alpha} = \frac{\upsilon_0}{g}\sin\alpha\left(1 + \sqrt{1 + \frac{2gh}{\upsilon_0^2 \sin^2\alpha}}\right). \tag{1.82}$$

We now find the maximum height z_m reached by the body. This can be done in different ways. One is noticing that this is the height at which $\upsilon_z = 0$. From

Eq. (1.78) we see that a happening at $t_m = \upsilon_0 \sin\alpha/g$, which was substituted in the second Eq. (1.79), gives

$$z_m = \frac{\upsilon_0^2 \sin^2\alpha}{2g} + h.$$

The same result can be reached finding the maximum of the second Eq. (1.79).

It is interesting to consider the special case $\alpha = 0$. We want the time t_f taken by the bullet to reach ground. Equation (1.79) become

$$x(t) = \upsilon_0 t, \quad z(t) = -\frac{1}{2}gt^2 + h.$$

The bullet hits the ground in the instant $t_f = \sqrt{2h/g}$, which, as we see, is independent of υ_0. This implies that for whatever initial velocity, even if enormous, the time taken to fall from the height h is always the same and is then equal to the free vertical (the special case $\upsilon_0 = 0$). In other words, the vertical and horizontal motions are independent.

The law of independence of (the components of) motion was discovered by G. Galilei. In the "Dialogue concerning the two Chief World Systems" he writes (translation by the author):

> …suppose having on the top of a tower a horizontally arranged culverin (*a relatively light cannon*) and firing point-blank shots, namely parallel to the horizon; then for little or much gunpowder charge given to it, such that the cannonball would fall at a distance of either one thousands arms, or four thousand, or six thousand, or ten thousand, etc., all these shots would take place in times equal to each other, and each equal to the time the ball would take to fall from the cannon's mouth to earth, when dropped, without any other impulse, for a simple vertical fall. Indeed it looks really wonderful that in the same short time of the vertical fall from a height, for example, of one hundred arms, could the same ball travel either four hundred, or one thousand, or four thousand, or even ten thousand arms, in such a way that in all the point-blank (*horizontal*) shots it would be in the air for equal times.

A little later Galilei specifies that that would be true

> …when there were no accidental impediments by the air…

1.17 Scalars, Pseudoscalars, Vectors and Pseudovectors

In Sect. 1.4 we have defined the vector as an ordered triple of real numbers that under rotations of the reference frame transforms in the same way as the triplet representing the position vector.

In Sect. 1.6 we have met a scalar quantity, the dot product of two vectors. We have seen that it is the same in two reference frames differing for a rotation of the axes. Indeed, in general, a quantity is, by definition, a scalar if it is invariant under change of the reference frame. For example, the x component of a vector is a single number but is not, properly speaking, a scalar, because it is not invariant under rotations of the axes.

Hence, both vector and scalar properties are expressed in terms of transformations between reference frames. We shall now consider the behaviors of these quantities under the inversion of the axes. It is called *parity* operation. It leads from a, say, left-handed frame to a right-handed one.

We now consider the transformation properties of physical quantities.

A quantity can be scalar or pseudoscalar. Both are invariant under rotations but the former is invariant under parity operation, the latter changes sign, while keeping its absolute value.

The dot product of two vectors is a scalar; the "scalar" triple product is a pseudoscalar. This is immediately evident considering that under inversion of the axes all the three vector factors change sign.

A quantity can be a vector or a pseudovector (also called an axial vector). Both transform in the same way under rotations, but the components of the former change sign under inversion of the axes, as the position vector does, while the components of the latter do not change sign.

The cross product of two vectors is a pseudovector, because both the vector factors change sign and their product does not. We met both types of physical quantities. Position vector, velocity and acceleration are (proper) vectors; angular velocity and moment of a vector are pseudovectors.

This type of properties of the physical quantities belong to a class generically called *symmetry properties*.

Problems

1.1. The vector $\mathbf{V}$ varies by $\Delta\mathbf{V}$, its absolute value varies by ΔV in the time interval Δt. (a) Can ΔV be larger than the magnitude of the variation, namely $|\Delta\mathbf{V}|$? Can they be equal?

1.2. The vector $\mathbf{V}$ changes its verse. Express $\Delta\mathbf{V}$, ΔV and $|\Delta\mathbf{V}|$?

1.3. At the instant t_1 the velocity of a body is, with certain units, $\mathbf{v}_1 = (1, 3, 2)$, at time t_2 is $\mathbf{v}_2 = (5, 3, 5)$. Find: (a) The variation of the velocity $\Delta\mathbf{v}$, (b) the magnitude of the variation of the velocity $|\Delta\mathbf{v}|$ and (c) the variation of the magnitude of velocity Δv.

1.4. A particle travels on a circle with velocity υ constant in magnitude. After a complete turn, (a) which is the mean value of υ? (b) which is the mean velocity $\langle\mathbf{v}\rangle$?

1.5. A particle moves with a position vector, in the given frame, $\mathbf{r}(t) = (2t\mathbf{i} + 3t^2\mathbf{j} + t\mathbf{k})$ m. Find: (a) velocity and acceleration as functions of time, (b) the velocity at $t = 2$ s.

1.6. A point moves uniformly on a plane curve trajectory with velocity υ. The magnitude of acceleration on a certain point of the trajectory is a. What is the curvature radius in that point?

1.7. The position vector of a point is $\mathbf{r}(t) = \mathbf{i}\cos(\omega t) + \mathbf{j}\sin(\omega t)$. (a) Find the velocity and acceleration vectors and their magnitudes. (b) Express the scalar product of **r** and **v**. What does the result mean? (c) Express the scalar product of **r** and **a**. (d) Find the trajectory of the point. (e) How would the motion change changing the sign of $y(t)$?

1.8. A cyclist travels at 10 km/h heading north. Wind blows with a speed (relative to ground) of 6 km/h from a direction between N and E. To the cyclist the wind appears to come from the direction at 15° from North to East. (a) Find the speed of the wind relative to the cyclist and the direction of the wind, relative to ground. When the cyclist goes back, which are velocity and apparent direction of the wind (wind did not vary).

1.9. We are on a ship travelling at 10 kn heading east. We see another ship, which we know moves at 20 kn to North, 6 miles distant in the South direction. What is the minimum distance the two ships will be (without changing their courses)? After how much time? Refer to Fig. 1.16. N.B. On the sea distances are measured in nautical miles and velocities in knots (1 kn = 1 mile/h). Assume for the mile the round figure of 1800 m.

1.10. Consider a flat platform rotating with angular velocity $\boldsymbol{\omega}_1 = K\,t^2\mathbf{k}$ where k is the unit vector of the z-axis directed vertically upwards. A body on the platform rotates with angular velocity, relative to it, $\boldsymbol{\omega}_2 = 2\,Kt^2\mathbf{i}$ (the x axis is horizontal). $K = 1$ rad/s^3. (a) Find the direction of the body relative to the ground. (b) Find the angle ϕ of which the body has rotated relative to ground at $t = 3$ s. (c) Does the magnitude of the resultant angular velocity vary in time? And its direction?

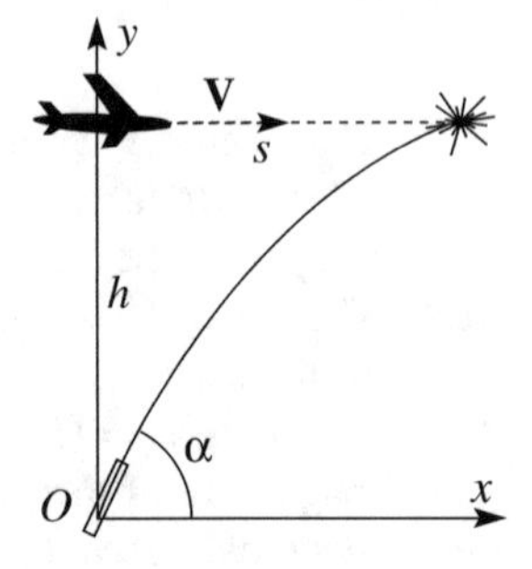

Fig. 1.23 The plane and the cannon of problem 11

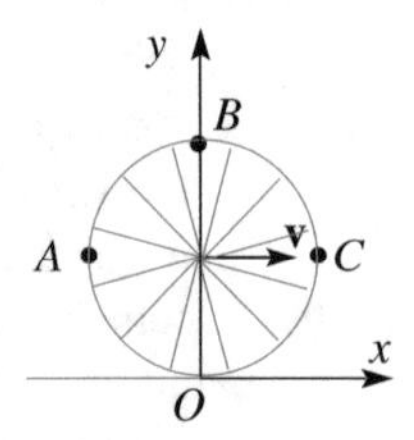

Fig. 1.24 The wheel of problem 12

1.11. An airplane is flying at constant velocity **V**, of horizontal direction and magnitude $V = 100$ m/s at the height $h = 5000$ m. A (super)cannon on earth shoots against it a ball at the moment in which the plane is just above the weapon (Fig. 1.23). The velocity of the ball is $v_0 = 500$ m/s. Neglecting the presence of air, find: (a) the angle α at which we must shoot to hit the plane; (b) the time of the collision (which of the two solution should be chosen?); (c) how much did the plane travel up to this moment.

1.12. The wheel shown in Fig. 1.24 rotates without slipping. Its axis moves forward at the velocity **v**. Find the velocities (namely their components on the two co-ordinate axes) of the points A, B, C.

Chapter 2
Dynamics of a Material Point

In this chapter we study the dynamics of a material point, namely the laws governing motion by its causes, which are the forces. We shall then start by defining and discussing the concept of force. The experimental method was introduced by Galileo Galilei at the end of the XVI century. He also discovered part of the laws of mechanics. The complete theory of mechanics was built by Isaac Newton, who published in 1686 the "Philosophiae Naturalis Principia Mathematica", known generally as simply "Principia".

The law of inertia was discovered by Galilei and assumed by Newton as the first law of mechanics. It will be studied in Sect. 2.3. The law states that a body in absence of forces acting on it moves naturally with constant velocity in a straight line, a rectilinear uniform motion. The second law was also discovered by Galilei and precisely formulated by Newton. It states that the rate of change of the momentum, a vector that we shall define, namely its time derivative, is equal to the force acting on the body. In an equivalent manner the acceleration is proportional to the force. This is the subject of Sect. 2.4. In the same section we shall discuss Newton's third law, the action-reaction law.

There are several types of force in Nature, as we shall see in the next chapter. In this one, however, in Sect. 2.5, we shall talk of weight, the force acting on all the bodies near the surface of the earth. A few examples will be discussed in Sects. 2.6 and 2.7.

In Sect. 2.8 we introduce two of the fundamental mechanical quantities (beyond momentum, or quantity of motion, already introduced in Sect. 2.4), the angular momentum and the moment of a force.

In Sect. 2.9 we shall study a simple but very important system, the pendulum and its harmonic motion. We shall also see how two concepts of mass, the inertial and the gravitational mass, are in fact only one.

After having introduced the concept of work made by a force and shown the theorem of energy conservation in Sect. 2.10, we shall describe an interesting experiment by Galilei. It establishes that the work done on a body by the weight force depends only on the difference between initial and final heights, not on the

A. Bettini, *A Course in Classical Physics 1—Mechanics*,
Undergraduate Lecture Notes in Physics, DOI 10.1007/978-3-319-29257-1_2

particular path followed. In modern language the experiment established that the weight force is conservative. This very important concept will be defined in Sect. 2.13. We then demonstrate the energy conservation theorem. Energy conservation is a fundamental law of all physics. We shall deal in this book only with mechanical energy, in its kinetic and potential forms, but we warn the reader that other important forms of energy exist, in particular thermal energy, as we shall discuss in the second volume of this course when dealing with thermodynamics.

The historical process leading to a precise definition of the concept of energy and to the establishment of the law of energy conservation took more than two centuries. Starting with Galilei, it came to maturity around mid XIX century, with the experiments of Mayer and Joule and enunciation of the energy conservation law by Mayer and Helmholtz. We shall give some hints in Sect. 2.14.

In Sect. 2.15 we shall discuss a particular type of force, the central forces. The gravitational attraction of the sun on a planet is an important example of this category.

In the last paragraph we introduce the concept of power, which is the work done by a force per unit of time.

2.1 Force, Operational Definition

The primitive concept of force is linked to muscular strain. If we lift a weight, push an object, we must exert a force with our hands and arms and we feel strain. Since ancient times humans developed simple mechanical devices to exert forces or amplify the muscular effect. The string of an archer's drawn bow exerts a force on the arrow, throwing it in the air; a lever can be used to lift big weights, etc. However, in physics the concept must be quantitative. For that, we must define force accurately enough to be able to measure it. This means that we must be able to compare two forces and establish when they are equal, when one is twice the other, etc. In other words we must be able to determine the ratio between two different forces.

A direct method to compare two forces is based on the lever rule, which was discovered by Archimedes of Syracuse (287–212 BC) more then two thousand years ago. The rule states that two equal forces balance when applied at equal distances on two sides of the pivot (Fig. 2.1a) and that two different forces F_1 and F_2 balance when applied at distances from pivot (l_1 and l_2 respectively) inversely proportional to the forces (Fig. 2.1b), i.e. such as

$$F_1 l_1 = F_2 l_2. \tag{2.1}$$

The first statement can be proven simply with symmetry arguments. If the two forces are equal and the two arms are equal, the system is symmetric. How could it choose on which side to bend? The second statement on the contrary, namely the validity of Eq. (2.1), must be experimentally verified.

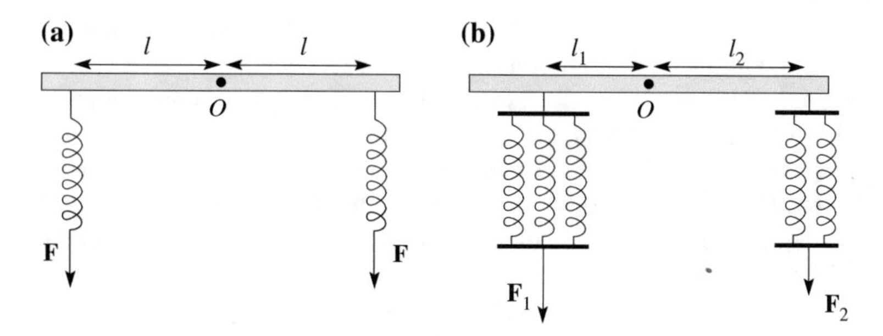

Fig. 2.1 Comparison of two forces

We know that a spring exerts a force when compressed or stretched relative to its natural length; we feel the muscular strain when we compress or pull it. We build a certain number of springs as equal as possible to each other. We can then verify that they exert equal forces when compressed (or stretched) in the same measure by applying those forces at equal distances from the pivot of a lever as in Fig. 2.1a. We can now define as unitary the force expanded of a specific length (N.B.: this is not the official definition).

We can then define the multiples of the unit force. If for example, we want a force of three units, we put three of our springs in parallel. We can experimentally verify the lever rule Eq. (2.1) as shown in Fig. 2.2b with different combinations of unit forces. Once we have stated that, we can use it to measure forces. As a matter of fact the method has been used in steelyards since very ancient times and is still used now in fruit or other goods markets to weigh a wide variety of goods. The weight to be measured is compared with the weight of a standard object seeking for equilibrium by changing the length of the lever arm of the latter.

In the operational definition of the force we have just chosen, we did not make any hypothesis on the relation between the force exerted by the spring and its length. However, this definition is not simple to use in practice. A handier device is the dynamometer (from the Greek *dynami* for force and *metro* for measure).

The dynamometer, shown schematically in Fig. 2.2, is made of a spring fixed at one extreme on a wood, or other material, plate and with a ring at its other extreme. The force to be measured is applied to the ring. A pointer moving on a scale gives a measurement of the dilation of the spring. Once we have built the device we must calibrate it. With the above described procedure we have built a number of springs, multiple and submultiples of the unit. We apply each of them to the ring and mark the position of the pointer on the table. In this way we build a scale on which we will read the values of unknown forces. In practice, we find that the scale is linear,

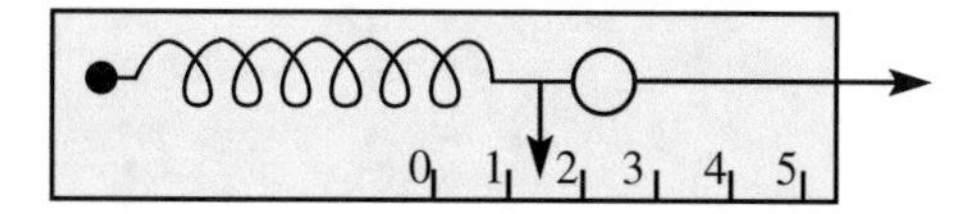

Fig. 2.2 The dynamometer

namely the stretch is proportional to the applied force, if the stretch is not too large. However, this property is comfortable, but not necessary.

The method we have described is used in practice, but does not allow a precise definition of force. In the SI the unit of force is a derived one, It is the force imparting the unit acceleration (1 m/s^2) to the unit mass (1 kg). It is called newton (N). To have an idea of the order of magnitude, think that the weight of one liter of water, 1 kg, is about 9.8 N. In other words one Newton is about the weight of the water filling a glass.

2.2 Force Is a Vector

In giving the operational definition of force in the previous section we have implicitly assumed, and we did that by definition, that two equal and opposite forces when applied to a point do not cause acceleration. Namely, the two forces are in equilibrium. Clearly, a force not only has a magnitude but also a direction. We can exert a force on a body applying one of our springs and pulling in different directions. We are led to think that force is a vector quantity. However, the conclusion cannot be reached by logic, rather it needs experimental verification. To be a vector, a quantity not only should have a magnitude and a direction, but also satisfy the rule of addition of vectors.

The experience with three forces was originally devised by Pierre Varignon (1654–1722), a contemporary of Newton. Its device is shown in Fig. 2.3. In the

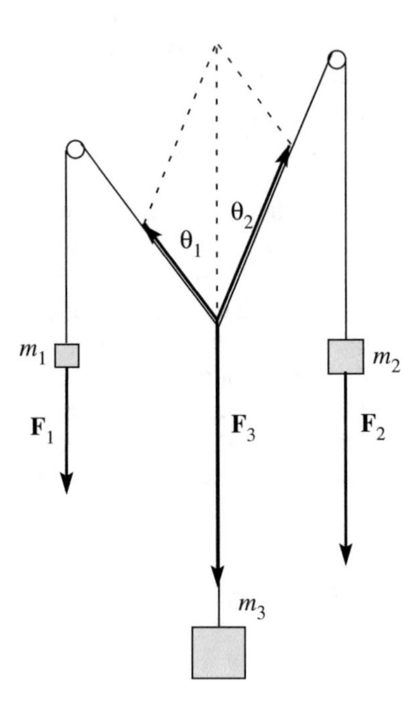

Fig. 2.3 Varignon experiment showing the composition of forces

plane of the figure, which is vertical, three pulleys are fixed. The three weights of masses m_1, m_2 and m_3, act by means of wires, drawn in the figure, joined in the point O. The forces exerted by the wires have magnitudes proportional to the weights and the directions of the wires. Once we have joined the three wires in O and let the system alone, the system moves until it reaches its equilibrium configuration, the one represented in the figure. We know the values of the weights, say F_1, F_2 and F_3, and measure the angles θ_1 and θ_2. We find that the following relations are satisfied:

$$F_1 \sin\theta_1 = F_2 \sin\theta_2, \quad F_1 \cos\theta_1 + F_2 \cos\theta_2 = F_3$$

or

$$\mathbf{F}_1 + \mathbf{F}_2 + \mathbf{F}_3 = 0.$$

The Varignon experiment and similar ones made afterwards verify the vector character of the force. The most precise tests, however, are indirect and come from the agreement of the experimental data with the predictions made under this hypothesis in the most different conditions.

Once we have established that forces add as vectors, we define as the *resultant* of the set of forces $\mathbf{F}_1$, $\mathbf{F}_2$, $\mathbf{F}_3$, ... and their vector sum

$$\mathbf{F} = \mathbf{F}_1 + \mathbf{F}_2 + \mathbf{F}_3 + \cdots. \tag{2.2}$$

Let us now think of some forces that we know from our everyday experience. We can distinguish two types. The just considered forces exerted by a spring, the force a table exerts on an object it supports, the force we exert with our hand pushing an object, are each exerted by contact. A body, the spring, the plane of the table and the hand each apply force to the object touching it. The everyday example of the second type of force is weight. Weight is the force with which earth gravitationally attracts all bodies. It is directed vertically down, towards the center of earth. This force is exerted at a distance i.e., it does not need contact.

2.3 The Law of Inertia

One of the most revolutionary discoveries of Galilei was the establishment of the behavior of a body not subject to forces. The problem lies in the fact that in practice it is impossible to eliminate all the forces. Weight is always present on earth. It cannot be eliminated, but it can be balanced. If we put a body on a horizontal plane, the latter will exert on the body a force equal and opposite to its weight. However, when the body moves, frictional forces due both to the contact between the surfaces of the plane and the body and the air are present. The effect of these "passive" forces is much more difficult to control and was not known before Galilei.

Consider the following experiment. We put a bronze sphere on a horizontal plane. We then give it a push. That is, we apply a force for a brief time interval, giving it a certain initial velocity on the plane. We observe the sphere's motion and see that its velocity gradually decreases and finally stops. To have the sphere moving at constant velocity we need to apply a force continuously. The conclusion seems to be that, when not acted on by forces, a body stands still. If it moves at constant velocity it is acted upon by a force proportional to its velocity. We now know that the conclusion, thought to be true for centuries, is actually false.

Galilei's argument can be summarized as follows. The fact that, when we apply a force to a body and then we cease to apply it, the body slows down and finally stops is obviously true. But the cause is not the absence of acting forces. On the contrary, the cause is the presence of forces that we do not apply, we do not see, yet exist (they are called passive) and we are unable to avoid, like friction and air drag,

Galilei could not prove his statement experimentally by eliminating all the passive resistive forces. He observed however that, when launching a solid polished sphere of brass or ivory on a horizontal guide, the distance travelled by the sphere before coming to rest was longer and longer when the surfaces of the guide and the spheres were smoother and smoother. Mentally going to the limit of infinite smoothness, he concluded that in those conditions the sphere would never stop, but would continue to move forever with the same velocity.

The conclusion is the *law of inertia*. In the words of Newton

> Every body preservs in its state of rest, or of uniform motion in a right line, unless it is compelled to change that state by impressed forces.

The law of inertia is not however valid in just any circumstance. Whether it is valid or not depends on the reference frame. Up to now we have made experiments in a reference fixed to earth. We now suppose that we want to build a laboratory on a carriage moving on straight rails at constant velocity, relative to earth. In our laboratory we have a smooth horizontal plane. We lay a bronze sphere on the table and observe that, as expected, it remains still. However, suddenly the sphere moves, accelerates and moves quickly forward, without any visible force acting on it. What did happen? It happened that the carriage suddenly started to slow down till coming to rest. Even if our laboratory is closed with no window to look out, we know that the carriage decelerates because we also experience a mysterious force pushing us forwards.

An observer on earth, namely in the frame we had been considering above, easily interprets the phenomenon. The sphere is free to move horizontally, the table being smooth. A force acted upon by brakes on its reels has slowed the carriage down. This force, however, does not act on the sphere, because the support plane is smooth. The resultant of the forces on the sphere is null. For the law of inertia it will continue in its motion with constant velocity. This is relative to the ground. But the observer on the carriage, which slows down relative to the ground, sees the sphere accelerating to reach the velocity that the carriage had before braking.

A reference frame in which the law of inertia is valid is called an *inertial frame*. We shall see that inertial frames have a privileged role in mechanics, and more generally in physics.

More precisely, the law of inertia can be stated as: *Reference frames do exist in which every body not subject to force indefinitely remains in its state of rest or uniform rectilinear motion.*

One might think that the law of inertia is a consequence of our definition of inertial frame, in other words that the argument is circular. But this is not true. Indeed, we can give arbitrarily any definition we like, but we can never establish by definition a law of nature, namely how she behaves. The existence of inertial frames is a law of nature not a definition by men.

We further observe that we have considered inertial any reference stationary on earth. The conclusion comes from the fact that, while doing experiments in such laboratories, we never observe objects suddenly moving when no force acts on them, nor do we feel as though we are being pushed in one direction or another. However, the conclusion is valid only in a first approximation. Accurate measurements show that frames that are stationary on earth are not exactly inertial. This is due to the fact that earth moves around the sun and rotates on its axis. We shall come back to that in Chap. 4. For the moment it will be enough to know that stationary reference frames on earth are close enough to be inertial for the vast majority of measurements carried out in laboratories and, on the other hand, procedures exist to define inertial reference systems with all the requested precision in case this is needed.

2.4 The Newton Laws of Motion

In the *Principia*, Newton begins by stating, as axioms induced from the experiments, the three fundamental laws from which the description of all the mechanical phenomena, both on earth and in the Universe can be deduced. The first law is the law of inertia we already discussed. The causes of any change of the state of rest or rectilinear uniform motion of a body are to be searched for in the bodies around it. For example the racket that hits it changes the state of motion of a tennis ball, the state of the compass needle is changed by the presence of a magnet, etc. The same hit imparted with a racket to a tennis or ping-pong ball produces different accelerations in the two bodies. By the term *inertial mass* we mean the characteristic of a given object that makes it more or less resistant to changing its state of motion under the action of a given force. Galilei had already proven with his experiments that a body under the action of a constant force, its weight or a component of its weight, moves with a constant acceleration in the direction of the force.

Let us study the phenomenon quantitatively. We have already built springs producing forces of different magnitudes. We have performed an analogous procedure for mass. We have built a number of blocks of the same material making them as equal as possible to each other. We can say that one block has unit inertial mass, two blocks inertial mass equal to two, etc.

We have also prepared a horizontal plane, the function of which is to equilibrate the weights of our blocks. In our experiments we shall put the blocks in motion sliding on the plane and we want to reduce as much as possible the friction forces between the plane and the blocks. We prepare the surface of the plane as smooth as possible. We can also play the following trick. We can build the blocks with a cavity inside and a series of holes between the cavity and the lower face. We fill the cavity with dry ice (frozen CO_2), which will sublimate pushing CO_2 gas through the holes. The thin layer of gas between the block and plane surfaces reduces friction to negligible values.

1. We attach one of our springs to one block, we give it a certain deformation, stretch or compression (Fig. 2.4a). We observe that the body moves with constant acceleration, say a_0, in the direction of the force, as long as we keep constant the force (i.e. the deformation)
2. We attach two springs (Fig. 2.4b) to the block and give them the same deformation as in the first experiment. We observe the body moving again with constant acceleration in the direction of the force. The acceleration is twice as large, 2 a_0.
3. We fix two blocks one on top of the other and attach one spring to which we give once more the same deformation. The acceleration is now one half as in the first experiment, $a_0/2$ (Fig. 2.4c).

Continuing with similar experiments changing the force on a body or the inertial mass, we come to the conclusion that its acceleration a is proportional to the force F and inversely proportional to its inertial mass m_i and we write $F = m_i a$.

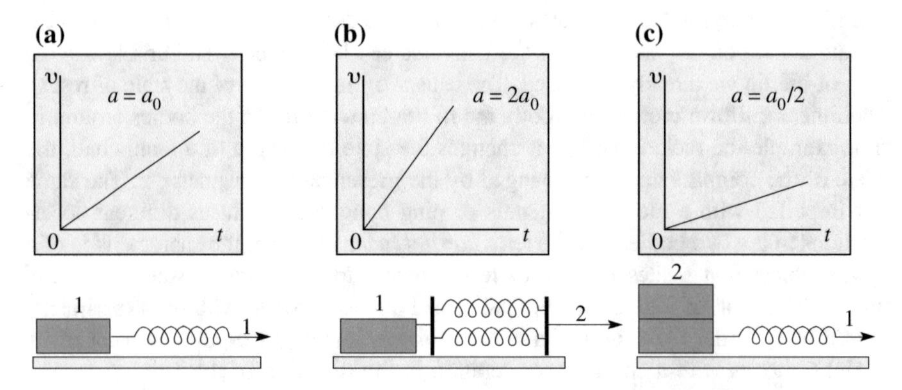

Fig. 2.4 Simple experiments to study the relation between force, acceleration and inertial mass

We can do better, because we have found that acceleration and force, which are two vectors, have the same direction. The second law states that

$$\mathbf{F} = m_i \mathbf{a} = m_i \frac{d\mathbf{v}}{dt} = m_i \frac{d^2\mathbf{r}}{dt^2}. \tag{2.3}$$

This is the form that is more often expressed. However, Newton stated it as

A change of motion is proportional to the motive force impressed, and takes place in the direction of the right line in which the force is impressed.

The quantity called by Newton "motion" is a fundamental vector quantity, $\mathbf{p}$, now called *quantity of motion*, or *momentum* (sometimes *linear momentum*). It is the velocity times the inertial mass

$$\mathbf{p} = m_i \mathbf{v} = m_i \frac{d\mathbf{r}}{dt}. \tag{2.4}$$

Two bodies of different masses can have the same quantity of motion if their velocities are in the inverse proportion of the masses. The second Newton law is

$$\mathbf{F} = \frac{d\mathbf{p}}{dt}. \tag{2.5}$$

In words, the rate of change of the momentum of a material point is equal to the force acting on it. Considering that m_i is a constant, and using Eq. (2.4) we have

$$\mathbf{F} = \frac{d\mathbf{p}}{dt} = m_i \frac{d\mathbf{v}}{dt}. \tag{2.6}$$

As for the law of inertia, the second law is not valid in every reference frame. Recall the example of the sphere in a laboratory on a carriage that starts suddenly to accelerate without any force being acting. Like the first law, the second Newton law is valid only in inertial frames.

Equation (2.3) says that acceleration has the same direction as the relevant force. This may appear to be obvious but it is not true in every circumstance. The equation also says that the acceleration due to a given force acting on a given body is independent of the velocity of the body. Experiments show that both of these, while true at common experience velocities, are not so for velocities close to the speed of light. In these conditions, called relativistic, Eq. (2.3) fails. However, even in these high velocities regimes, Eq. (2.5) remains valid, namely, as Newton stated, the force and the time derivative of momentum are equal. What needs to be changed is the relation between momentum and velocity.

We shall study relativistic mechanics in Chap. 6; we anticipate that in a relativistic regime, the concept of inertial mass remains exactly the same. Mass is a constant, independent of velocity, characteristic of the body. The concept of momentum however must be made more general. Its expression is

$$\mathbf{p} = m_i \gamma(\upsilon) \mathbf{v}, \tag{2.7}$$

where $\gamma(\upsilon)$ is a function of velocity, called the Lorentz factor, after Hendrik Lorentz (1853–1928), one of the fathers of relativistic mechanics. Its value is very close to 1 up to velocities close to that of light, $c \approx 3 \times 10^8$ m/s, but increases very rapidly when υ approaches c.[1]

For comparison, the speed of the earth relative to the sun is about 3×10^4 m/s, 10^{-4} of the speed of light, the speeds of the stars relative to their galaxies, including our sun, are an order of magnitude larger, but still 10^{-3} of the speed of light. For the latter, the Lorentz factor differs from 1 only by 0.5×10^{-6}.

A second limit of validity of the Newton laws is at very small dimensions. Indeed, classical physics ceases to be valid and must be modified in quantum physics, at atomic scales. These however are very small compared to the objects of everyday experience, e.g., atomic radiuses are typically 30–300 pm.

The Newton law gives the acceleration once the forces are known. Consequently, in the analysis of any motion we deal with the position vector, the velocity, which is its first time derivative, and the acceleration, its second time derivative. We do not need higher derivatives. For these reasons we did not go beyond the second derivative of the position vector when we studied kinematics. We recall on purpose that to know the motion of a particle we need to know not only the acting forces, but also the initial position and velocity.

Let us now look at another aspect. The second law can be used in three main ways:

1. If we know the inertial mass of a body and all the forces acting on it, and the initial conditions, we can calculate its motion
2. If we know the motion of a body and its inertial mass, we can infer the forces acting on it.
   Distinguishing the two points of view is not as trivial as it may look. The first point of view is deductive. The laws of mechanics are used to calculate the motion of bodies in all possible circumstances. In this way physicists and engineers design mechanical devices and engines. The second point of view is inductive and is the point of view taken to make progress in physics. The challenge of the physics research is to understand from the study of motion the fundamental nature of the forces that cause it. This is the way followed by Newton to discover universal gravitation from study of the motions of heavenly

[1]The reader is warned that one can still find books and articles calling the product $m_i\gamma(\upsilon)$ "relativistic mass" and m_i "rest mass". The former in a relativistic regime increases with increasing velocity. These concepts were introduced in the last years of the 19th century and the first ones of the 20th when relativity theory was being developed and things were not yet completely clear. They are misleading concepts (what varies with velocity is the Lorentz factor, not the mass, which is invariant) and should be avoided. We shall treat relativity in Chap. 6.

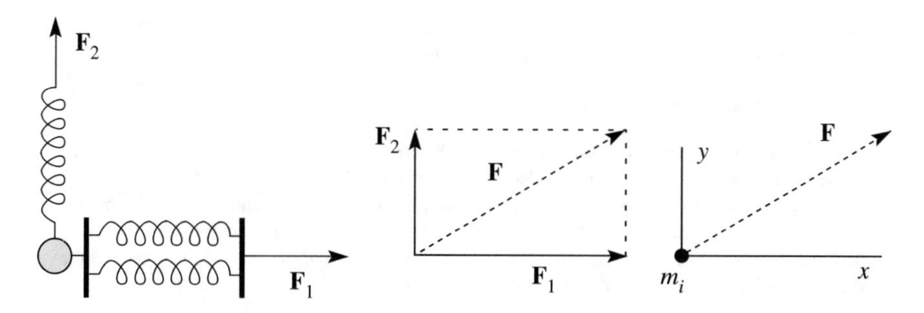

Fig. 2.5 Two forces acting at the same time

bodies. This is the way in which Ernest Rutherford (1871–1937) discovered the atomic nucleus in 1911 when studying the scattering of energetic alpha particle by a thin gold sheet. This is the way followed today to study the properties of atomic nuclei and elementary particles.

We can state that the success of the Newton law is just as follows. It substantially tells us: if you see a body that does not move in a uniform rectilinear motion, a force should act. Search for it and search for the physical agent to which it is due. You will find a force, the mathematical expression of which will be *simple* and, as a consequence, you will be able to lay down a simple theory. From this point of view the Newton law is a research program. We shall see in Chap. 3 that, indeed, the various forces of nature have simple expressions in terms of the co-ordinates and characteristics of the system. The program is successful.

3. A third possibility is that, if we know both forces and motion we can deduce the inertial mass of the body. To know the mass of the proton for example, we can measure how its momentum and energy vary under the action of a known force.

The law of composition of forces. If more than one force act at the same time on the material point we are discussing, their effect is the same as if only one force were acting, equal to the resultant of those forces. Consider for example that two forces are applied as in Fig. 2.5. The first spring exerts the force $\mathbf{F}_1$ in the x direction. When acting alone it produces the acceleration F_1/m_i along x. The second spring exerts the force $\mathbf{F}_2$ in the y direction. When acting alone it produces the acceleration F_2/m_i along y. To know what happens if the two forces act contemporarily is something that cannot be found by logic, rather it has to be found experimentally. Indeed, what experiments show is that the acceleration is just what one calculates assuming that only one force were acting, equal to the resultant $\mathbf{F}$ of $\mathbf{F}_1$ and $\mathbf{F}_2$. In other words, the observed acceleration is $\mathbf{a} = \mathbf{F}/m_i$.

The third Newton law is the *law of action-reaction*.

If a body exerts a force (an action) on a second body, the second always exerts on the first a force (a reaction) that is equal and opposite on the same line of action.

Given its importance, we reproduce how it is stated, in an equivalent manner, by Newton.

> To every action there is always opposed an equal reaction: or, the mutual actions of two bodies upon each other are always equal, and directed to contrary parts.

Newton gives then a few examples.

> Whatever draws or presses another is as much drawn or pressed by that other. If you press a stone with your finger, the finger is also pressed by the stone. If a horse draws a stone tied to a rope, the horse (if I may so say) will be equally drawn back towards the stone; for the distended rope, by the same endeavor to relax or unbend itself will draw the horse as much towards the stone as it does the stone towards the horse and will obstruct the progress of the one as much as it advances that of the other.

We notice that, differently from the first two, the third law deals with two, rather than one, bodies. It tells us that isolated forces (actions) do not exist, only **inter**actions do exist.

Pay attention to the fact that action and reactions are applied in different points, one on one body, the other on the other body. If we push a stone with a finger, the action of the finger is applied in a point of the stone; the reaction of the stone is on the tip of our finger. The force exerted by the horse drawing the stone is exerted on the stone through the rope, the reaction acts, again through the rope, in the point of the horse at the end of the rope. Every object whether it is falling or laying on a support, weighs, meaning that the weight force is applied on it. Weight is the force with which the earth attracts all bodies. As a reaction, each body attracts the earth with an equal and opposite force. The reaction is applied to a point of the earth, its center.

The action-reaction principle, as all physical laws, must be experimentally verified. Direct verifications are based on the fact that in a collision between two bodies the total quantity of motion, namely the vector sum of the two, is conserved, meaning that its values before and after the collision are equal (while each of the two vary).

The vectors we have met so far, position vector, velocity and acceleration depend, as we have seen, on the reference frame. On the contrary, force does not.

2.5 Weight

We know from every day experience that all the bodies on earth are subject to a force, vertically directed downwards, called the weight. We can measure the weight of a body, for example, attaching it to a dynamometer vertically positioned and reading on its scale the position of the pointer, namely the stretch of the spring. If we repeat the measurement in different points of our laboratory we find that it does not vary. However, if we repeat the measurement at much larger distances, for example at the Equator and at 45° latitude, or at different altitudes, for example at the sea level and at 2000 m altitude, we notice small differences (of the order of a few per mille) between them. As we shall discuss in Sect. 5.7, these small variations are due to the rotation of the Erath. Apart from these small corrections, the weight is the gravitational attraction exerted by the earth on the body. This is universal; it is the same force with which the earth attracts the moon. We shall discuss this fundamental force in Chap. 4. We anticipate that the gravitational attraction decreases as the reciprocal of the distance squared. This is one of the reasons (the other is the rotation motion of earth) why the weight of an object is a bit smaller on a mountain than at the sea level.

Different objects, in the same place, may have different weights. This means that the force with which earth attracts a body depends on a characteristic of the body. We state that the gravitational force on a body is proportional to its *gravitational mass*, which we denote with m_g. This is similar to the electric attraction. A charged body A at a certain distance from another body that is also charged, is subject to an electrical force. If in the place of A we put a body B with twice the charge, the force on it is double. Hence, the electric force on a body is proportional to its electric charge. In a similar way two massive bodies, for example two spheres, at a certain distance attract with the gravitational force that is proportional to the gravitational mass of each of them. This force, if between two objects of every day life is quite small, but can be measured with very delicate experiments, as we shall see in Sect. 4.7, but is large between Heavenly bodies. Considering that the gravitational mass is for the gravitational force the analogous of the electric charge for the electric forces, we might call it gravitational charge, but we shall soon see the reason why we call it mass.

The weight force $\mathbf{F}_W$ acting on a body of gravitational mass m_g is then

$$\mathbf{F}_w = m_g \mathbf{g}. \tag{2.8}$$

The vector quantity $\mathbf{g}$ does depend on the location, but in a given site it is equal for all bodies. If $\mathbf{r}$ is the position vector, the vector $\mathbf{g}(\mathbf{r})$ is the gravitational force at $\mathbf{r}$ per unit gravitational mass. It is called *gravity acceleration*. We shall see soon the reason for the name. We notice that the gravitational mass being a characteristic of a body is the same in any point, differently from its weight. If we measure the weights of two bodies in different points on the earth we find that each of them varies a bit, as already mentioned, but the ratio of the two remains rigorously equal. Even if we should do this experiment on the moon.

Operationally, the gravitational mass is the physical quantity measured by a balance. A balance, see Fig. 2.6, consists of a lever with pivot in O and two pans, which we shall consider, to make it simple, exactly at the same distance on the two

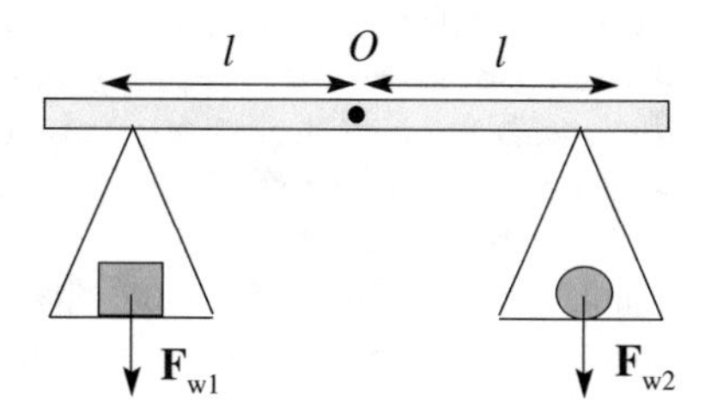

Fig. 2.6 Comparing the weights of two equal masses

sides of O. The balance compares the weights of the two objects on its pans. If they are equal the balance is in equilibrium. We have seen that, by definition, the weights of different objects in the same place are proportional to their gravitational mass. We can then the state that two objects have the same gravitational mass when, put on the pans of the balance, they are in equilibrium.

We now need a body having unit mass by definition. We put it on a pan. Another body has gravitational mass equal to one when, put on the other pan it is in equilibrium. A body has gravitational mass equal to 2, if put on a pan is in equilibrium with two of the unit masses on the other, etc.

Gravitational mass and inertial mass are two different properties of every body. The former is a measure of the strength of the gravitational attraction to which it is subject, the latter of how difficult it is to modify its quantity of motion. However, we know from every day experience, that heavier bodies are also more difficult to accelerate because they are more inert. To search for a mathematical relation, suppose to observe the free fall of two different bodies. Their inertial masses are m_{1i} and m_{2i} and their gravitational masses m_{1g} and m_{2g}. The weight of the first is, $\mathbf{F}_{1w} = m_{1g}\mathbf{g}$, the weight of the second $\mathbf{F}_{2w} = m_{2g}\mathbf{g}$. Calling $\mathbf{a}_1$ and $\mathbf{a}_2$ the two accelerations, we have:

$$m_{1g}\mathbf{g} = m_{1i}\mathbf{a}_1, \quad m_{2g}\mathbf{g} = m_{2i}\mathbf{a}_2,$$

which can be written as

$$\mathbf{a}_1 = \frac{m_{1g}}{m_{1i}}\mathbf{g}, \quad \mathbf{a}_2 = \frac{m_{2g}}{m_{2i}}\mathbf{g}. \tag{2.9}$$

We see that the free fall accelerations of different bodies in the same place are proportional to the ratios of their gravitational and inertial mass. Consequently, if this ratio is equal for all the bodies, light or heavy, all of them fall with the same acceleration. This fundamental property was experimentally shown to be true by G. Galilei.

It is often told that Galilei dropped contemporarily two balls, one made of lead, one of wood, from the Pisa tower and that he observed them reaching ground at the same instant, showing in this way that they fall with the same acceleration. The experiment was absolutely success and spectacularly carried out in 1971 by the NASA Apollo 15 astronaut D. Scott dropping a hammer and a feather on the moon. As a matter of fact Galilei never mentions having made his fundamental experiments in such a way. He new very well that it could not work, both for the perturbing effects of the atmosphere and due to the smallness of the fall times, a fact that did not allow him precise measurements. His very precise experiments were done with reduced, to say so, weight forces, with spheres on inclined planes and with pendulums. We shall discuss this in Sect. 2.9.

We can conclude that the free fall accelerations of all bodies in a given place are equal, action of the atmosphere apart. The ratio between gravitational and inertial mass is a universal constant, the same for all bodies. The value of the constant is arbitrary, because depends on the choice of the two units. Clearly, the most convenient choice is to have the ratio equal to one. With this choice gravitational and inertial mass are not only proportional, they are equal. The unit of both is the kilogram. From now on we shall indicate with the same symbol, for example m without any subscript, both quantities.

2.6 Examples

In this section we study a number of examples of application of the Newton laws. A good way to proceed is the following.

The first step is to identify all the bodies present in the problem. Next we identify for each of them all the forces acting on it. To do that it is convenient to wrap it, ideally in an envelope, in order to identify all the forces acting on the body from its exterior. To this aim it is often useful to draw each object separately, in its ideal envelope, and the acting forces and write down for each of them its type and its agent (for example: weight due to earth, normal force due to the constraint, friction due to the supporting surface). If the problem contains more than one body, we must identify the action and reaction pairs, and the bodies on which they act. Once all the forces are identified we must calculate the resultants on each of the bodies. To do that we choose a reference frame. The choice should be guided by any symmetry the problem might have. We must then calculate the Cartesian components of the resultant by summing the correspondent components of all the forces. The components divided by the mass of the body are the three components of the acceleration of the body. From the acceleration we find the law of motion with the procedures we studied in Sects. 1.15 and 1.16.

Example E 2.1. Place a block on a horizontal frictionless surface horizontally drawn by a rope.

Frictionless means a physical surface that does not exert forces parallel to it. It is an idealization. Friction always exists, but we can reduce it, for example with the dry ice trick of Sect. 2.4. We attach a rope to the block and draw it horizontally with the force $\mathbf{F}_r$. The situation is shown in Fig. 2.7.

Knowing $\mathbf{F}_r$ and the mass m of the block we want to know its motion, considering it as a point. We draw the body in its ideal envelope. We identify the forces

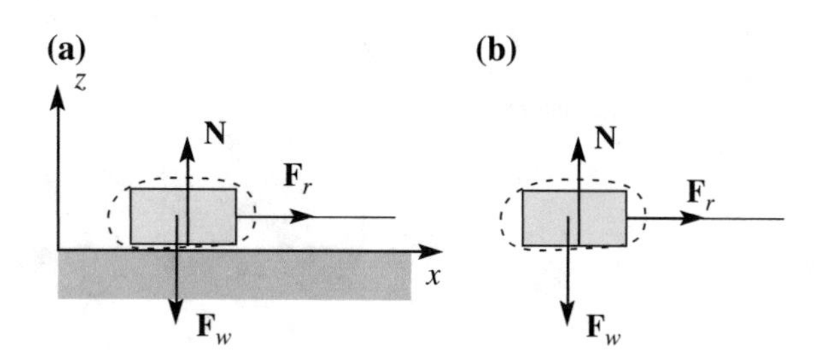

Fig. 2.7 **N** normal constraint force, $\mathbf{F}_r$ force exerted by the rope, $\mathbf{F}_w$ weight, due to earth

acting through the surface: (1) the weight of the block $\mathbf{F}_w$, due to earth, vertically directed downwards, (2) the constraint force exerted by the plane. As we have assumed it to be frictionless the force is normal to the surface, upwards and we call it $\mathbf{N}$, (3) the force (tension) exerted by the rope, $\mathbf{F}_r$. We have drawn all of that in Fig. 2.7b. As we are considering the block as a material point, all the forces are applied in the same point. One of the forces, $\mathbf{N}$, is not given. This is always the case of constraint forces. The body cannot penetrate the support plane because the molecules of the body and the plane repel each other. We know that the body has no vertical acceleration. We infer that the support develops the force that is exactly what is needed to keep it steady. We will find it by solving the equations.

All the forces of the problem lay in the same vertical plane. It is then convenient to choose a reference frame with one axis, say z, vertical upwards and a second one, say x, horizontal to the right in the figure. We do not need the third axis because there are neither forces nor motion in that direction. We now write the second Newton law and its two components

$$\mathbf{F}_r + \mathbf{N} + \mathbf{F}_w = m\mathbf{a}, \quad N - F_w = 0, \quad F_r = ma_x.$$

We conclude that the normal force exerted by the support plane has magnitude equal to the weight. Both forces are vertical and have opposite direction; hence their resultant is zero. The resultant of the forces is the tension of the rope, which causes a uniformly accelerated motion in the x direction.

Example E 2.2 A block moving on a horizontal frictionless surface drawn by a rope at an angle with the horizontal.

The situation is the same as in the previous example, but for the rope now pulling at an angle θ with the horizontal (see Fig. 2.8a). However, we still assume that the motion is on the plane, namely that there is no vertical acceleration. The forces are the same, but $\mathbf{F}_r$ has different components. We have

$$\mathbf{F}_r + \mathbf{N} + \mathbf{F}_w = m\mathbf{a}, \quad N - F_w + F_r \sin\theta = 0, \quad F_r \cos\theta = ma_x.$$

The equation for the z components gives again the normal constraint force, $N = F_w - F_r \sin\theta$. If $\theta > 0$ as in the figure, N is smaller than in the previous example because the rope helps in sustaining the block, the opposite if $\theta < 0$. The second equation gives horizontal acceleration.

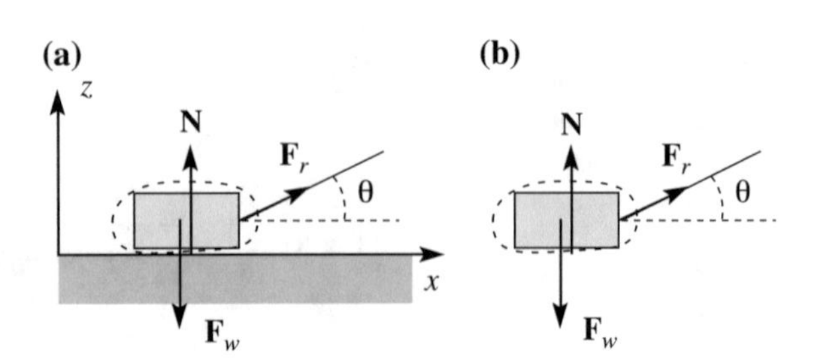

Fig. 2.8 **N** normal constraint force, $\mathbf{F}_r$ force exerted by the rope, $\mathbf{F}_w$ weight, due to earth

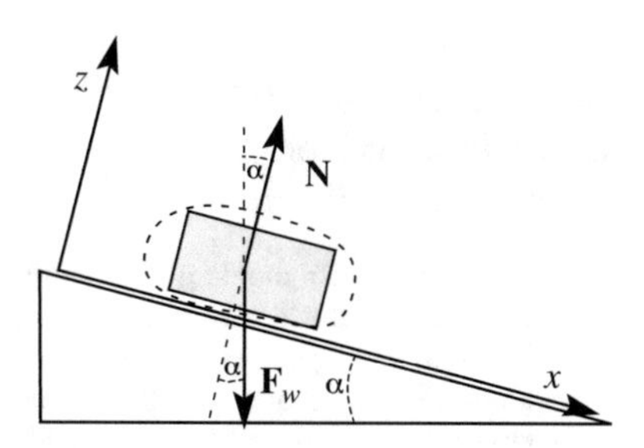

Fig. 2.9 A *block* on a frictionless incline

Notice that a physical limitation of this analysis exists. The normal force cannot be negative, because the support plane cannot attract the body (there is no glue). Hence, if $F_r \sin\theta > F_w$, the assumed conditions cannot be satisfied. Clearly, in this situation the block is lifted up and its acceleration has a vertical component.

Example E 2.3 Block on an inclined frictionless surface.

There are two forces acting on the body (Fig. 2.9), the weight $\mathbf{F}_w$ and the constraint force $\mathbf{N}$ perpendicular to the support plane, which is now inclined. The convenient choice of the axes is to take z perpendicular to the plane and x along the plane, downwards. Clearly, the body will slide accelerating downwards, namely in the x direction we have chosen.

The Newton equation and its components are

$$\mathbf{N}+\mathbf{F}_w = m\mathbf{a}, \quad F_w \sin\alpha = ma_x, \quad N - F_w \cos\alpha = 0.$$

The z component gives us the normal force $N = F_w \cos\alpha$. The x component gives the acceleration ($a = a_x$). Recalling that $F_w = mg$, we have that the notion on an inclined frictionless plane is uniformly accelerated with acceleration

$$a = g\sin\alpha. \tag{2.10}$$

We see that the motion on an incline is completely similar to the motion of free fall, as long as we can neglect the resistive forces. The difference is that the acceleration is smaller on the incline by a factor $\sin\alpha$. We can reduce acceleration by reducing the slope of the plane. If the motion starts from rest from the origin, the law of motion is obtained by integrating twice Eq. (2.10), obtaining

$$x(t) = \frac{1}{2}at^2 = \frac{1}{2}(g\sin\alpha)t^2. \tag{2.11}$$

In words: *the distances travelled are proportional to the squares of the times taken to travel them.*

The incline allows us to slow down the free fall motion and to study its laws over longer times, which can be measured with better precision.

As mentioned in Sect. 2.5 this is one of the great discoveries of Galilei. He did not have a modern chronometer, but invented an ingenious water chronometer, with which he was able to measure the times of the motion, a few seconds long, with a precision better than 0.1 s. He describes his experiments in the book "Dialogues and mathematical demonstrations concerning two new sciences" or "Two new sciences" published in 1638. He writes:

> A piece of wooden molding or scantling, about 12 cubits long, half a cubit wide, and three finger-breadths thick, was taken; on its edge was cut a channel a little more than one finger in breadth; having made this groove very straight, smooth, and polished, and having lined it with parchment, also as smooth and polished as possible, we rolled along it a hard, smooth, and very round bronze ball. Having placed this board in a sloping position, by lifting one end some one or two cubits above the other, we rolled the ball, as I was just saying, along the channel, noting, in a manner presently to be described, the time required to make the descent. We repeated this experiment more than once in order to measure the time with an accuracy such that the deviation between two observations never exceeded one-tenth of a pulse-beat. Having performed this operation and having assured ourselves of its reliability, we now rolled the ball only one-quarter the length of the channel; and having measured the time of its descent, we found it precisely one-half of the former. Next we tried other distances, comparing the time for the whole length with that for the half, or with that for two-thirds, or three-fourths, or indeed for any fraction; in such experiments, repeated a full hundred times, we always found that the spaces traversed were to each other as the squares of the times, and this was true for all inclinations of the plane, i.e., of the channel, along which we rolled the ball. We also observed that the times of descent, for various inclinations of the plane, bore to one another precisely that ratio which, as we shall see later, the Author had predicted and demonstrated for them.
>
> For the measurement of time, we employed a large vessel of water placed in an elevated position; to the bottom of this vessel was soldered a pipe of small diameter giving a thin jet of water, which we collected in a small glass during the time of each descent, whether for the whole length of the channel or for a part of its length; the water thus collected was weighed, after each descent, on a very accurate balance; the differences and ratios of these weights gave us the differences and ratios of the times, and this with such accuracy that although the operation was repeated many, many times, there was no appreciable discrepancy in the results.

Example E 2.4 A block at rest in a lift.

A block of mass m lies in a lift on a horizontal pan of a balance, one of those, for example, that are used to weigh people. What is the apparent weight of the block when the lift accelerates up or down?

As usual we imagine the block in an ideal envelope (Fig. 2.10). Two forces act on it, the weight $\mathbf{F}_w$ vertical down, and the normal constraint of the pan $\mathbf{N}$ upwards. The balance measures the reaction to $\mathbf{N}$, namely the force on it, which is $-\mathbf{N}$. Hence, N is the apparent weight of the block.

If the lift moves with acceleration a upward, the unknown N is given by the Newton law $N - F_w = ma$. Hence, the apparent weight is $N = F_w + ma = m(g + a)$, which is larger than the true weight. If the lift accelerates downwards, the apparent weight is $N = m(g - a)$, smaller than the real one. Notice that if the acceleration downwards is g the apparent weight is null. Indeed, the block is falling with the same acceleration of the lift.

If the lift moves uniformly both upwards and downwards the apparent weight is equal to the real one, as if it were standing. We feel an increase of our weight either

Fig. 2.10 A *block* in an accelerating lift

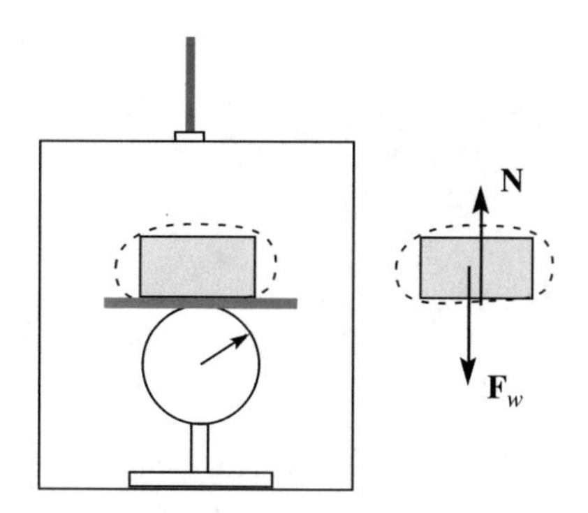

if the lift accelerates going up or if it decelerates going down. In both cases its acceleration is upwards. Similarly we feel a decrease of our weight when the lift slows down going up or accelerates going down.

Tension of the ropes and wires. In some of the examples we made we have used a stretched rope or wire to apply a force in a point of a body. This force is equal to the tension of the wire. We generally assume the wire to be inextensible, meaning that its length does not vary whichever the tension may be, and perfectly flexible, meaning that the tension is always parallel to the wire, and of negligible mass. Once more, these are idealizations.

Let us clarify the concept of tension. Consider a wire, stretched and steady as in Fig. 2.11a. We mentally isolate a small segment, enlarged in Fig. 2.11b. Two forces act on the segment (neglecting the weight), applied to its extremes and due to the contiguous elements of the wire. These are the tension forces. As the wire is at rest, the two forces are equal and opposite. Consequently, the tension is the same in every section of the wire.

Each of the extremes of the wire is not in contact with another element. As it does not accelerate, a force must act on it from outside equal in magnitude to the tension and directed outwards, as in Fig, 2.11a. The forces on the extremes are equal and opposite and have the magnitude of the tension.

Consider now the case in which the wire moves. As an example, suppose that one extreme is fixed to a block of mass M lying on a horizontal plane of negligible friction. We draw the block applying to the free extreme of the wire a force $\mathbf{F}_1$ obtaining an acceleration $\mathbf{a}$, as shown in Fig, 2.12a. We want to understand under which conditions we really can neglect the mass of the wire. To do that, let us start assuming the mass of the wire to be m.

We are now dealing with two bodies, the block and the wire. We ideally isolate each of them and draw the force diagrams on each of them, in Fig. 2.12b, c.

Fig. 2.11 **a** The tension forces on a wire and, **b** on a segment

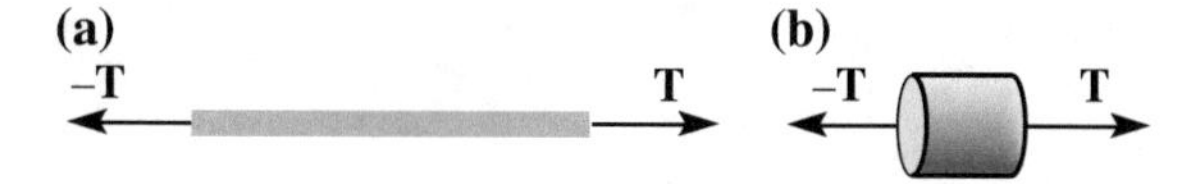

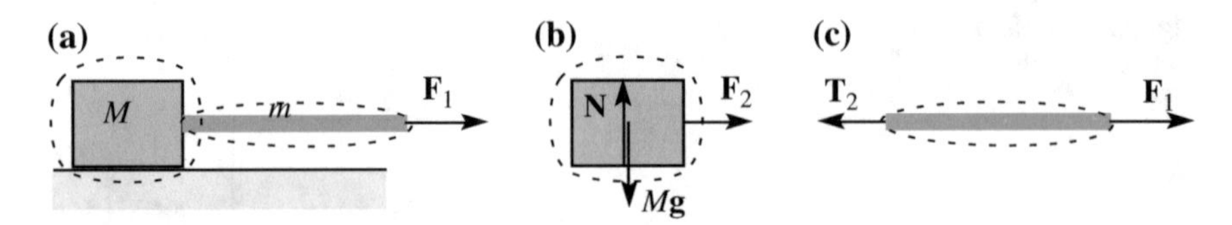

Fig. 2.12 **a** Accelerated motion of a *block* drawn by a rope, **b** **N** normal constraint force, $M\mathbf{g}$ weight due to earth, $\mathbf{F}_2$ force due to the wire, **c** $\mathbf{T}_2$ force on the wire due to the *block*, $\mathbf{F}_1$ force pulling the wire

We next identify the action reaction pairs. There is one such pair, consisting of the forces $\mathbf{F}_2$ applied to the block and $\mathbf{T}_2$ applied to the left extreme of the wire. They are equal and opposite. The force $\mathbf{F}_1$ applied to the right extreme of the wire is its tension and we can call it $\mathbf{T}_1$. The Newton equations for the two bodies are

$$\mathbf{F}_2 = -\mathbf{T}_2 = M\mathbf{a}, \quad \mathbf{T}_1 + \mathbf{T}_2 = m\mathbf{a},$$

hence, for the magnitudes, $T_1 = (M + m)a$ and $T_2 = Ma$. We see that the tensions at the two extremes are different. Indeed $T_1 > T_2$ because T_1 must accelerate wire and block, T_2 only the block. Let us consider their ratio

$$\frac{T_1}{T_2} = \frac{M + m}{M} = 1 + \frac{m}{M},$$

which becomes unity for $m/M \to 0$. We can then state that the tensions at the extremes can be considered equal if the mass of the wire is negligible compared to the mass of the block. When we speak of massless ropes or wires we mean of negligible mass compared to the masses of the other objects.

Notice that we can arrange a stretched wire, or rope, to have forces at its extremes of equal magnitude but different directions, by using pulleys. We did so already discussing the Varignon experiment (Fig. 2.3). Notice that in these cases, if the motion is accelerated, the magnitudes of the tensions at the extremes can be considered equal only if also the mass of the pulley is negligible and if it can rotate with negligible friction on the pivot (Fig. 2.13).

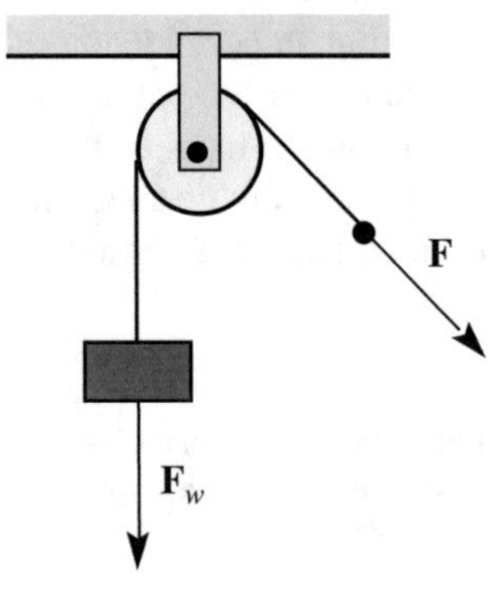

Fig. 2.13 With a pulley, the direction of the force exerted by a wire can be changed

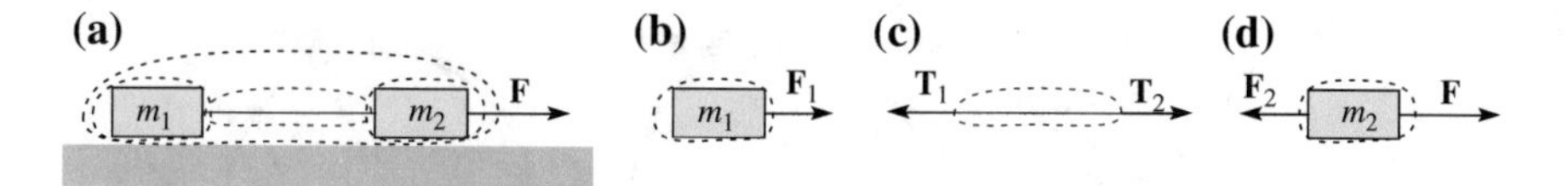

Fig. 2.14 **a** Two *blocks* connected by a wire, **b** force on m_1, **c** forces on the wire, **d** forces on m_2

Example E 2.5 Two blocks linked by a rope of negligible mass.

Figure 2.14a shows two blocks of masses m_1 and m_2 lying on a horizontal frictionless plane, connected by an inextensible wire of negligible mass. To the second block, at the right, a horizontal force **F** is applied. The motion is on the support plane. To know it, we do not need to analyze the vertical forces, which have zero resultants (Fig. 2.14a, b, c, d).

We start by considering the whole system, thinking of it as a unique ideal envelope. The only force acting on this surface is **F**. Hence we have $F = (m_1 + m_2)a$. which gives an acceleration a equal for the two bodies.

We now isolate each of the bodies. The block on the left (Fig. 2.14b) is attached to an extreme of the wire. This exerts on the block the horizontal force $\mathbf{F}_1$. For the action-reaction law the block exerts on the extreme of the wire an equal and opposite force, which is the tension of the wire at that extreme ($\mathbf{F}_1 = -\mathbf{T}_1$). Two other forces act on the block, the external force **F** and the force $\mathbf{F}_2$ due to the right extreme of the wire (Fig. 2.14d). Again, for the action-reaction law the block exerts, on the right extreme of the wire, a force equal and opposite to $\mathbf{F}_2$ that is the tension $\mathbf{T}_2$ at that extreme ($\mathbf{F}_2 = -\mathbf{T}_2$). As we have discussed above, the magnitude of the tension is the same in all points of the wire. Taking into account the directions we have $\mathbf{T}_1 = -\mathbf{T}_2$ (Fig. 2.14c). Calling T the magnitude of the tension we can write the Newton equations as $T = m_1 a,\ F - T = m_2 a.$

The sum of the two equations gives the acceleration of the system $a = F/(m_1 + m_2)$. If we want the value for tension, we substitute a in the first equation obtaining

$$T = \frac{m_1}{m_1 + m_2} F.$$

We see, in particular, that $T < F$, namely the tension is smaller than the force with which we pull.

2.7 Curvilinear Motion

In the previous section we have studied a few examples in which the forces were known, a part of the constraint ones, and the motion that had to be found. In this action we shall consider the inverse problem, namely, the motion of a material point being known, find the resultant of the forces. The singular forces, in case more than one is present, cannot be found, because systems of forces with the same resultant produce the same motion in the case of material points.

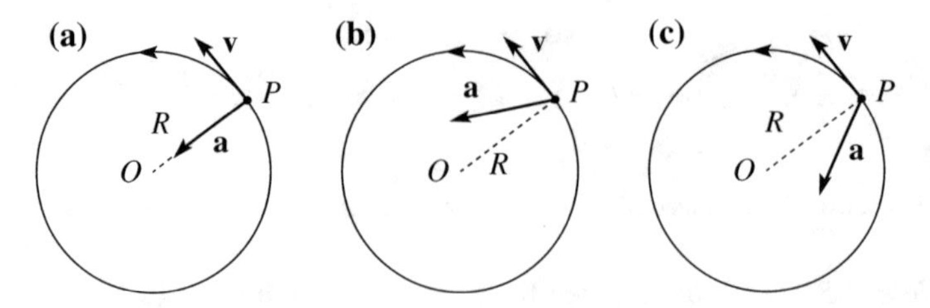

Fig. 2.15 Circular motion, **a** uniform, **b** increasing velocity, **c** decreasing velocity

Circular uniform motion
Consider the motion of a material point P with mass m constrained to move on a circumference of radius R. Suppose the motion to be uniform, namely the magnitude of its speed υ to be constant, as in Fig. 2.15. The motion is however accelerated, because the direction of the velocity varies. As we already found, the acceleration has a constant magnitude (Eq. 1.57) $a = \upsilon^2/R$ and is in every point directed to the center (centripetal acceleration). This acceleration must be given by a force of magnitude

$$F = ma = m\frac{\upsilon^2}{R}. \tag{2.12}$$

The corresponding force has the same direction as the acceleration and is called *centripetal force*. The adjective "centripetal, from the Latin "petere" for "point towards", recalls only its direction but does not specify at all its nature. It may be the tension of a wire, the normal force of a circular guide, the gravitational force of the earth on the moon, etc. We shall discuss a few examples in Sect. 3.4.

Variable speed motion.
If the magnitude of the velocity of a particle moving on a circle varies, its acceleration has two components. One component, a_n, is perpendicular to the trajectory, or, the latter being circular, directed to the center. It is again the variation of the direction of the velocity, namely the just discussed centripetal acceleration of value υ^2/R where υ, we must now specify, is the instantaneous velocity. The second component, a_t, is in the direction of the motion, i.e. tangent to the trajectory and expresses the variation in time of the magnitude of the velocity. We have

$$a_t = \frac{d\upsilon}{dt}, \quad a_n = \frac{\upsilon^2}{R}. \tag{2.13}$$

The acceleration vector, and the force, is directed at an angle with the radius that is forward if the velocity is increasing (Fig. 2.15b), backward if it is decreasing (Fig. 2.15c). The magnitude of the force is $F = ma = m\sqrt{a_n^2 + a_t^2}$.

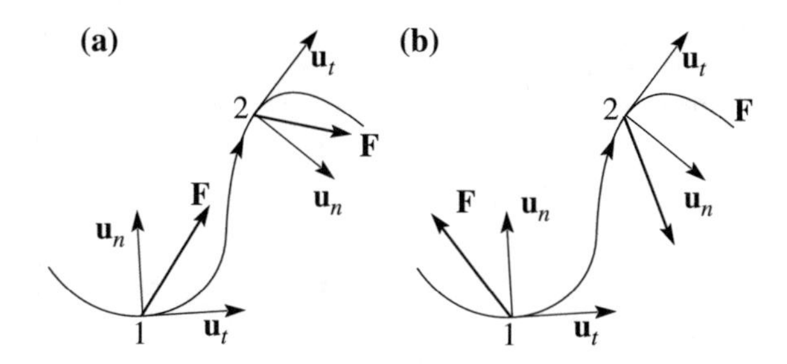

Fig. 2.16 General plane motion. **a** Increasing speed, **b** decreasing speed

As an example, consider a block lying on the platform of a merry go round, which is initially still. When the platform starts moving, gradually increasing its angular velocity, the acceleration of the block has two components, one centripetal and one tangential. The corresponding force, equal to the mass of the ball times this acceleration, is given by the friction on the platform. If the latter is not enough, the block slides towards the periphery of the platform.

As a second example consider the launch of the hammer. The athlete acting on the rope he holds in his hands puts the hammer in rotation with increasing speed. The force on the hammer must be adequate to keep it on a circular orbit (component $m\upsilon^2/R$ towards the center) and makes its speed increase (a component in the direction of the motion). The rope must then be directed forward, as in Fig. 2.15b

General plane motion.
We consider now a material point of mass m moving on a plane trajectory of arbitrary shape with velocity not necessarily constant in magnitude. We have already studied the kinematics of the problem in Sect. 1.14. Even in this case, the acceleration has two components, a tangential and a normal one, as in Eq. (1.62). They are given by Eq. (2.13).

The only difference from the circular case is that now R is the local curvature radius, which is not fixed but varies along the trajectory. The second Newton law tells us that the resultant of the forces acting on the point must be its acceleration times its mass.

If we know only the trajectory, but nothing of the velocity, we can still say that in every point of the trajectory in which the curvature is not zero, the resultant of the forces must be directed on the side of the curvature center, pointing forward from it (Fig. 2.16a) or backwards (Fig. 2.16b) depending on whether the motion is accelerated or delayed respectively.

2.8 Angular Momentum and Moment of a Force

Consider a material point P moving in an inertial frame as shown in Fig. 2.17. Let $\mathbf{p} = m\mathbf{v}$ be its momentum and $\mathbf{r}$ its position vector. Consider a generic point Ω, which may be at rest or moving relative to the frame. We shall now introduce the concepts of *angular momentum* and *moment of a force* about the *pole* Ω.

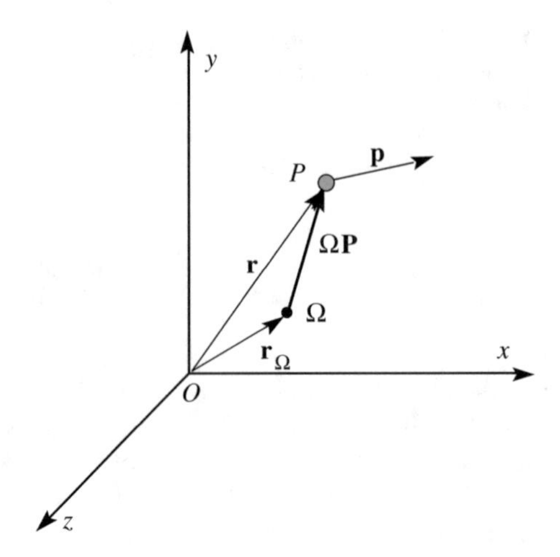

Fig. 2.17 The vectors relevant for angular momentum

We have already defined the moment of a bound vector in Sect. 1.8. The angular momentum is the moment of the linear momentum, considering it, for this purpose, as applied to the material point, as shown in Fig. 2.17.

Hence, the angular momentum of the point P about the pole Ω is the vector product of the vector from Ω to P and its quantity of motion (or momentum).

$$\mathbf{l}_\Omega = \mathbf{\Omega P} \times \mathbf{p}. \tag{2.14}$$

Consider the force $\mathbf{F}$ applied to P. The moment of the force about the pole Ω is the vector product of the vector from Ω to P and $\mathbf{F}$

$$\tau_\Omega = \mathbf{\Omega P} \times \mathbf{F}. \tag{2.15}$$

Remember that the order of the factors matters in cross products. Notice also that the moments change if the reference frame changes.

Let us now see how the angular momentum changes in time. For that, we take the time derivative of Eq. (2.14) using the rule of the derivative of products, paying attention to the order of the factors

$$\frac{d\mathbf{l}_\Omega}{dt} = \frac{d\mathbf{\Omega P}}{dt} \times \mathbf{p} + \mathbf{\Omega P} \times \frac{d\mathbf{p}}{dt}. \tag{2.16}$$

To find the derivative of the vector $\mathbf{\Omega P}$ we notice that it is the difference of two vectors, both varying with time, $\mathbf{\Omega P} = \mathbf{r} - \mathbf{r}_\Omega$. Deriving we have.

$$\frac{d\mathbf{\Omega P}}{dt} = \mathbf{v} - \mathbf{v}_\Omega.$$

The meaning of this expression is clear: the derivative of a vector joining two moving points is the relative velocity of those points. We substitute this expression in Eq. (2.16) and also notice that the derivative of the momentum is equal to the resultant **F** of the forces acting on P, because the frame is inertial. We get

$$\frac{d\mathbf{l}_\Omega}{dt} = \mathbf{v} \times \mathbf{p} - \mathbf{v}_\Omega \times \mathbf{p} + \Omega\mathbf{P} \times \mathbf{F}.$$

The first term in the second member is zero, being the cross product of two parallel vectors; the last term is the moment of the resultant about the pole $\boldsymbol{\tau}_\Omega$. In conclusion

$$\frac{d\mathbf{l}_\Omega}{dt} = \boldsymbol{\tau}_\Omega - \mathbf{v}_\Omega \times \mathbf{p}. \tag{2.17}$$

This is a very important equation that we shall use often in the following. It becomes particularly simple if we choose a stationary pole in the reference frame. The equation becomes

$$\boldsymbol{\tau}_\Omega = \frac{d\mathbf{l}_\Omega}{dt}. \tag{2.18}$$

In words the equation is called the angular momentum theorem for a material point: *the time derivative of the angular momentum of a material point about a pole fixed in an inertial reference frame is equal to the moment of the resultant of the forces acting on it about the same pole.*

Notice that if the body is extended, as we shall discuss in the following chapter, the different forces acting on it, say $\mathbf{f}_1$, $\mathbf{f}_2$,…, may be applied in different points and the moment of their resultant $\mathbf{F} = \mathbf{f}_1 + \mathbf{f}_2 + \cdots$, $\boldsymbol{\tau}_\Omega = \Omega\mathbf{P} \times \mathbf{F}$ is in general different from the vector sum of their moments. In the case under study however, all the forces are applied in P and

$$\begin{aligned}\boldsymbol{\tau}_\Omega &= \Omega\mathbf{P} \times \mathbf{F} = \Omega\mathbf{P} \times (\mathbf{f}_1 + \mathbf{f}_2 + \cdots) = \Omega\mathbf{P} \times \mathbf{f}_1 + \Omega\mathbf{P} \times \mathbf{f}_2 + \cdots \\ &= \boldsymbol{\tau}_{\Omega_1} + \boldsymbol{\tau}_{\Omega_2} + \cdots.\end{aligned}$$

The resultant of the moments is equal to the moment of the resultant of the forces. We stress that this is true only if all the forces are applied at the same point.

2.9 The Simple Pendulum

The pendulum is a material point constrained to move on an arc of a circumference. It can be simply made by fixing a thin wire to a small sphere on an extreme and to a fixed point on the other, which we call Ω. The length l of the wire, or better the distance between the fixed point and the center of the sphere, is called the *length of the pendulum*. If we take the pendulum away from its equilibrium position O and abandon it with zero velocity, the body moves towards O under the action of two

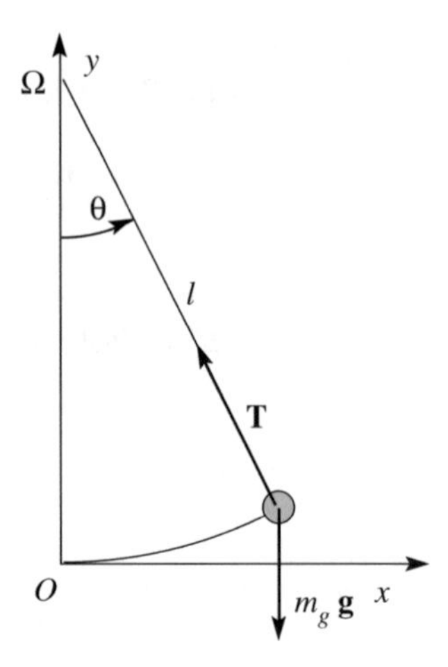

Fig. 2.18 The simple pendulum

forces, the weight, directed vertically down, and the tension of the wire (**T**), directed as the wire. The acceleration has the direction of the resultant of these two forces. Consequently it is always in the plane defined by the wire and the vertical. If the initial velocity is zero, the motion is on the plane. As the distance from Ω is kept fixed by the wire, which we assume inextensible, the trajectory is an arc of a circle of radius l.

As shown in Fig. 2.18, we take a reference system with the origin O in the rest position of the pendulum, the y-axis vertical upwards, the x-axis horizontally in the plane of motion and z such as to complete the triplet. The z-axis is normal to the figure towards the observer. We call θ the angle between the wire and the vertical, taking it positive if seen anticlockwise by the observer.

Historically, as we have already mentioned, the study of the motion of pendulums, with their periodic motion, made a fundamental contribution to the development of mechanics. Galilei discovered two important properties. The first one is the isochronism of small oscillations; if the amplitude is not too large (we shall be more precise in the following), the oscillation period is independent of the amplitude. This property allowed building of precise clocks. The second property is even more important; the oscillation periods of pendulums of the same lengths and different masses are identical. This proves, as we shall now see, that gravitational mass and inertial mass are equal. The property was later called *equivalence principle* and is at the basis of general relativity.

In our demonstration, we start by assuming that the two masses might be different. We call m_i the inertial mass of the pendulum, namely the proportionality constant between acceleration and force, and m_g its gravitational mass, the constant that appears in the weight, which is then $m_g\mathbf{g}$.

The tension is a constraint force, due to the wire, which we assume to be perfectly flexible and inextensible. The constraint develops a force, in general unknown a priori, automatically adjusted to make the motion happen, in our case, at a fixed distance from Ω. We do not know the intensity of the wire tension **T**, but we know its direction, which is along the wire.

In our study of the motion we shall use the angular momentum theorem. We choose the pole in the suspension point Ω, for reasons that will become clear soon. We use Eq. (2.18) with

$$\boldsymbol{\tau}_\Omega = \boldsymbol{\Omega}\mathbf{P} \times \left(\mathbf{T} + m_g\mathbf{g}\right) = \boldsymbol{\Omega}\mathbf{P} \times \mathbf{T} + \boldsymbol{\Omega}\mathbf{P} \times m_g\mathbf{g}.$$

We now see the reason for our choice of pole. The first term is always zero, being the vector product of two parallel vectors. Consequently we do not need to know the intensity of the tension. We have

$$\boldsymbol{\tau}_\Omega = \boldsymbol{\Omega}\mathbf{P} \times m_g\mathbf{g}. \tag{2.19}$$

The angular momentum about the same pole is

$$\mathbf{l}_\Omega = \boldsymbol{\Omega}\mathbf{P} \times m_i\mathbf{v}, \tag{2.20}$$

where the mass is the inertial one. Equation (2.18) gives

$$\boldsymbol{\Omega}\mathbf{P} \times m_g\mathbf{g} = \frac{d(\boldsymbol{\Omega}\mathbf{P} \times m_i\mathbf{g})}{dt}. \tag{2.21}$$

All the vectors in these equations, in any position of the pendulum, belong to the plane xy. Both vector products are consequently in z direction. The equation has only the z component. The z component of $\boldsymbol{\Omega}\mathbf{P} \times m_g\mathbf{g}$ is $-lm_g g \sin\theta$. The velocity is always perpendicular to $\boldsymbol{\Omega}\mathbf{P}$. As a consequence the z component of $\boldsymbol{\Omega}\mathbf{P} \times m_i\mathbf{v}$ is simply $lm_i\upsilon$, where $\upsilon = l\frac{d\theta}{dt}$. So, we have

$$-lm_g g \sin\theta = lm_i\frac{d\upsilon}{dt}.$$

And finally we can write

$$\frac{d^2\theta}{dt^2} + \frac{m_g g}{m_i l}\sin\theta = 0. \tag{2.22}$$

This is a differential equation, whose unknown is a function of time $\theta(t)$. Once it is solved, we know the motion of the pendulum, because if we know θ, we know its position. Equation (2.22) cannot be solved analytically. However, if the oscillations are “small”, we can approximate the sine with its argument and the equation becomes

$$\frac{d^2\theta}{dt^2} + \frac{m_g g}{m_i l}\theta = 0. \tag{2.23}$$

This is a well-known differential linear equation with constant coefficients, which we shall meet several times. We leave its study to calculus courses and directly give the general solution, which is

$$\theta(t) = \theta_0 \cos(\omega_0 t + \phi), \tag{2.24}$$

where

$$\omega_0 = \sqrt{\frac{m_g g}{m_i l}} \tag{2.25}$$

is called *proper angular frequency*. As one sees, it depends only on the characteristics of the pendulum, including its weight.

The reader can easily verify, with two derivatives, that this expression indeed satisfies Eq. (2.23), for whatever values of the constants θ_0 and ϕ. These constants do not depend on the characteristics of the pendulum but on how the motion has started. They should be found in each case on the basis of two initial conditions. We can use the position and velocity at the starting time that we shall take as $t = 0$. We immediately see that

$$\theta(0) = \theta_0 \cos\phi, \qquad \left(\frac{d\theta}{dt}\right)_{t=0} = -\theta_0\omega_0 \sin\phi.$$

The initial velocity being zero, the second equation gives $\phi = 0$ ($\theta_0 = 0$ is also a solution of Eq. (2.24) but identically null). The first condition says that θ_0 is just the initial angle, the angle at which we have let the pendulum go. In conclusion the motion of the pendulum is described by the equation

$$\theta(t) = \theta_0 \cos(\omega_0 t). \tag{2.26}$$

The motion is periodic, meaning that, for any instant of time t we can consider, both the position and the velocity become the same after a certain time interval T, called the *period*, namely at the instant $t + T$. From Eq. (2.26) we immediately see that the period is

$$T = \frac{2\pi}{\omega_0} = 2\pi\sqrt{\frac{m_i l}{m_g g}}, \tag{2.27}$$

where we used Eq. (2.25).

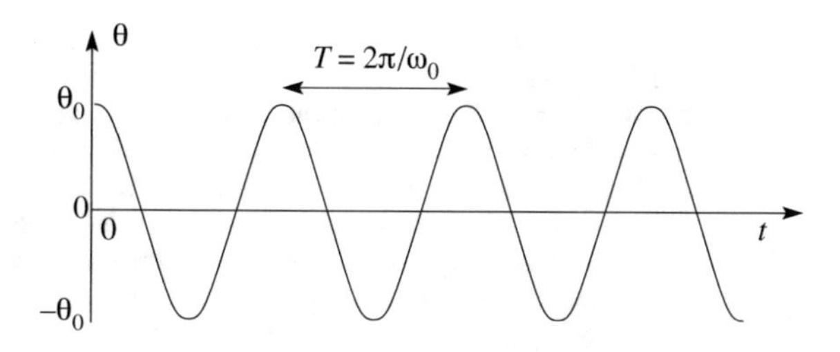

Fig. 2.19 Angular harmonic motion

The motion is represented in Fig. 2.19. This is the most common periodic motion in Nature. It is called harmonic motion. In the next chapter we shall study it in depth.

We now make an important observation on the expressions of proper angular frequency and period. Angular frequency and period depend on the length of the pendulum, but not on the oscillation amplitude: two pendulums of the same length (and in the same place, hence at the same g) are isochronous. On the other hand, angular frequency and period depend on the *ratio* between gravitational and inertial masses. If this ratio is the same for all bodies, independently of the substance they are made of and of their position, that ratio is a constant and angular frequency will be independent of the mass of the pendulum. If we want to experimentally test if gravitational and inertial masses are proportional or not, we can test whether pendulums of the same length, and different masses or made of different substances, do oscillate or not with the same period.

Galilei noticed that this method is much more accurate than others he knew. In principle, one could think to drop two spheres, e.g. one of wood and one of lead, from the top of a tower and check if they the reach the ground simultaneously. However, Galilei never mentions having done such an experiment, from the leaning tower of Pisa. This is a legend without any historical support. Indeed, Galilei observed, and wrote, that that method is not accurate enough, because is too fast and at high speeds the air resistance noticeably perturbs observations. Galilei used inclined slopes, as we have discussed, to slow down the motion, reduce the air drag, and increase the relative measurement accuracy due to the longer times to be measured. The use of pendulums allows even better accuracy. He used pairs of pendulums made of different materials and of exactly the same lengths, took them out of equilibrium and let them go at the same time. He found that they keep oscillating in phase for hundreds of periods. The air drag did act more effectively on the lighter pendulum gradually reducing its amplitude more than that of the heavier one. However this did not matter because the period is independent of amplitude.

G. Galilei describes accurately his progress toward a more precise experiment, gradually eliminating the spurious effects and the sources of errors in the "Dialogs concerning two new sciences" (1638) (translation from Italian by Henry Crew and Alfonso de Salvio). He established the proportionality of inertial and gravitational mass with an uncertainty of $2\text{–}3 \times 10^{-3}$.

The experiment made to ascertain whether two bodies, differing greatly in weight will fall from a given height with the same speed, offers some difficulty; because, if the height is considerable, the retarding effect of the medium, … will be greater in the case of the small momentum of the very light body than in the case of the great force of the heavy body; so that, in a long distance, the light body will be left behind; if the height be small, one may well doubt whether there is any difference; and if there be a difference it will be inappreciable. It occurred to me therefore to repeat many times the fall through a small height in such a way that I might accumulate all those small intervals of time that elapse between the arrival of the heavy and light bodies respectively at their common terminus, so that this sum makes an interval of time which is not only observable, but easily observable. In order to employ the slowest speeds possible and thus reduce the change which the resisting medium produces upon the simple effect of gravity it occurred to me to allow the bodies to fall along a plane slightly inclined to the horizontal. For in such a plane, just as well as in a vertical plane, one may discover how bodies of different weight behave: and besides this, I also wished to rid myself of the resistance which might arise from contact of the moving body with the aforesaid inclined plane. Accordingly I took two balls, one of lead and one of cork, the former more than a hundred times heavier than the latter, and suspended them by means of two equal fine threads, each four or five cubits long. Pulling each ball aside from the perpendicular, I let them go at the same instant, and they, falling along the circumferences of circles having these equal strings for semi-diameters, passed beyond the perpendicular and returned along the same path. This free vibration repeated a hundred times showed clearly that the heavy body maintains so nearly the period of the light body that neither in a hundred swings nor even in a thousand will the former anticipate the latter by as much as a single moment, so perfectly do they keep step. We can also observe the effect of the medium which, by the resistance which it offers to motion, diminishes the vibration of the cork more than that of the lead, but without altering the frequency of either.

In conclusion, Galilei experimentally demonstrated the equality of inertial and gravitational masses with an accuracy of about one per mille, namely that

$$\frac{m_i}{m_g} - 1 < 10^{-3}. \tag{2.28}$$

Newton repeated this later on the Galilei experiments. He writes in the "Principia":

It has been, now of a long time, observed by others, that all sorts of heavy bodies (allowance being made for the inequality of retardation which they suffer from a small power of resistance in the air) descend to the earth from equal heights in equal times; and that equality of times we may distinguish to a great accuracy, by the help of pendulums. I tried the thing in gold, silver, lead, glass, sand, common salt, wood, water, and wheat. I provided two wooden boxes, round and equal: I filled the one with wood, and suspended an equal weight of gold (as exactly as I could) in the center of oscillation of the other.

He concluded that:

By these experiments, in bodies of the same weight, I could manifestly have discovered a difference of matter (i.e. *inertial mass*) less than the thousandth part of the whole, had any such been.

Hence Newton confirmed what Galilei had discovered with a similar precision of 1×10^{-3}. After having found an expression of the gravitational force, Newton did also a check of the equivalence principle, on a solar system scale. He did that, in particular, on the system of Jupiter and its satellites. We shall see his argument in Sect. 4.4. Here we just say that the precision was, once more, of one per mille.

Having established the proportionality of the two types of mass, we can make them equal by choosing their units. With this choice Eqs. (2.25) and (2.27) become

$$\omega_0 = \sqrt{\frac{g}{l}}, \quad T = 2\pi\sqrt{\frac{l}{g}}. \tag{2.29}$$

To have a feeling of the orders of magnitude, we can easily calculate that a 1-m long pendulum has a period of about 2 s.

We now recall having approximated the sine of the angle with the angle (in radiants) itself. Let us verify when the approximation is good. For example, if $\theta = 30°$, or 0.52 rad, its sine is sin 30° = 0.50. The relative error is (0.52–0.50)/0.50 = 4 %, which is quite small. Even for $\theta = 60°$, or 1.05 rad, the error is not enormous, but already noticeable. Indeed, sin 60° = 0.87 and the corresponding error is 20 %. These are the relative errors making the sine equal to the angle, but the corresponding ones on the period are even smaller, as we now shall see.

The exact Eq. (2.22), as we said, cannot be solved analytically. However, it can be solved by successive approximations. In fact, the approximation we made is a series expansion stopped at the first term ($\sin\theta = \theta$); the next approximation we stop at the second term ($\sin\theta = \theta - \theta^3/6$). The resulting expression for the period with amplitude θ_0, calling T_o the period given by Eq. (2.28), is

$$T(\theta_0) = T_0\left[1 + \frac{1}{4}\sin^2\frac{\theta_0}{2}\right],$$

which, as it is seen, depends on the amplitude θ_0. The relative error made using the usual expression of the period is $\frac{1}{4}\sin^2\frac{\theta_0}{2}$. Going back to the above examples, we find that the relative error for $\theta = 30°$ is 1.6 %, the one for $\theta = 60°$ is 6.3 %. They are not large.

We make a last observation. If the oscillations are small, the pendulum moves substantially on the horizontal, namely on the x-axis in Fig. 2.18. Now $x = l\tan\theta$, that we can approximate with $x = l\theta$. We can then conclude that the motion, as represented by the x coordinate, has the equation

$$x(t) = x_0\cos(\omega t). \tag{2.30}$$

As expected it is a harmonic motion, of amplitude x_0.

2.10 The Work of a Force. The Kinetic Energy Theorem

In this section we introduce the concepts of work, done by a force, and kinetic energy, of a body. The meaning of "work" in physics is rather different from its meanings in everyday language and consequently from what intuition might suggest. For example, holding in one hand a heavy object even if we do not move it we still need to apply a force with our muscles and make some effort. However, we do not perform any work, in the language of physics. In physics, a force makes work only if its application point moves. In the example, the work done by the force we exert on the body is positive if we raise, negative if we lower it, but zero if we do not move it.

Consider the material point P moving in a reference frame with position vector $\mathbf{r}$, along a certain trajectory, the curve Γ. As shown in Fig. 2.20, consider the position vector in the instants t, $\mathbf{r}(t)$, and immediately after $t + dt$, $\mathbf{r}(t + dt)$. The displacement of P in the interval dt is the infinitesimal vector

$$d\mathbf{s} = \mathbf{r}(t+dt) - \mathbf{r}(t). \tag{2.31}$$

If $\mathbf{F}$ is a force acting on the point, its work for the infinitesimal displacement (2.31) is defined as

$$dW \equiv \mathbf{F}\cdot d\mathbf{s}. \tag{2.32}$$

The finite work having been done by the force, a finite displacement of the point, say from A to B along the trajectory Γ, is the line integral along the curve Γ from A to B

$$W_{AB;\Gamma} = \int_{A;\Gamma}^{B} dW = \int_{A;\Gamma}^{B} \mathbf{F}(\mathbf{r}) \cdot d\mathbf{s}, \tag{2.33}$$

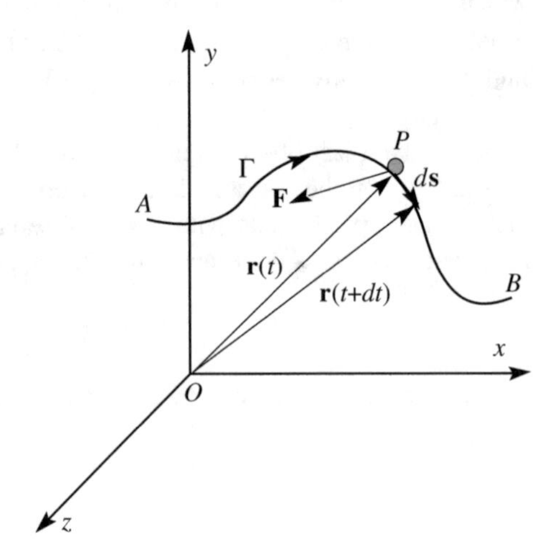

Fig. 2.20 The elements to define the work of force **F**

where $\mathbf{F}(\mathbf{r})$ is the force in the point of position $\mathbf{r}$. The line integral is the sum of all the elementary dot products $\mathbf{F}(\mathbf{r}) \cdot d\mathbf{s}$ on all the elements of the curve. Clearly, the integral does not depend only on the initial and final points A and B, but also on the specific path taken to go from the former to the latter. Indeed, if the path changes, also the force in the new points may change. To make this explicit in the notation we have included both A and B and Γ in the subscripts of W. The case in which the integral depends on the origin and the end but not on the path is however important and will be studied in Sect. 2.13.

Notice that more forces, call them $\mathbf{F}_i$, may act contemporarily on the point P, for example weight, friction, air resistance, etc. In this case, the total work made by all the forces is equal to the sum of the works each force would do if acting separately

$$W_{AB;\Gamma} = \sum_i \int_{A;\Gamma}^{B} \mathbf{F}_i(\mathbf{r}) \cdot d\mathbf{s} \tag{2.34}$$

Clearly, the elementary (meaning "infinitesimal") displacement of the application points of the forces $d\mathbf{s}$ is the same for all them. Considering that the sum of integrals is equal to the integral of the sum, which is in our case the resultant of the forces $\mathbf{R} = \sum_i \mathbf{F}_i$, we have

$$W_{AB;\Gamma} = \int_{A;\Gamma}^{B} \sum_i \mathbf{F}_i(\mathbf{r}) \cdot d\mathbf{s} = \int_{A;\Gamma}^{B} \mathbf{R} \cdot d\mathbf{s}. \tag{2.35}$$

Namely, the total work made by the acting forces is equal to the work made by their resultant. Notice, again, that this is true only if all forces are applied in the same point.

The physical dimension of the work is those of a force times a displacement. Its unit is the jule, with symbol J, which is the work done by the unit force, 1 N, when its application point moves one unit of length, 1 m, in the direction of the force. To appreciate the order of magnitude, a jule is roughly the work you do when you raise a glass of water by 1 m.

We now prove the work-kinetic energy theorem. Being a consequence of the second Newton law it is valid in inertial frames. Consider a material point and the resultant $\mathbf{R}$ of the forces acting on it. The Newton law says

$$\mathbf{R} = m\frac{d\mathbf{v}}{dt}.$$

We take the scalar product with the elementary displacement $d\mathbf{s} = \mathbf{v}\,dt$ of the two members

$$\mathbf{R} \cdot d\mathbf{s} = m\frac{d\mathbf{v}}{dt} \cdot \mathbf{v}dt = m\mathbf{v} \cdot d\mathbf{v}.$$

Now consider the dot product $\mathbf{v} \cdot d\mathbf{v}$. We recall that the square of a vector is the dot product of the vector by itself, in this case $\upsilon^2 = \mathbf{v} \cdot \mathbf{v}$. Differentiating this expression we have

$$d\left(\upsilon^2\right) = d(\mathbf{v} \cdot \mathbf{v}) = d\mathbf{v} \cdot \mathbf{v} + \mathbf{v} \cdot d\mathbf{v} = 2\mathbf{v} \cdot d\mathbf{v},$$

hence $\mathbf{R} \cdot d\mathbf{s} = \frac{1}{2}m(d\upsilon^2)$.

The work done by $\mathbf{R}$ when the point moves from A to B on the given trajectory is then

$$W_{AB;\Gamma} = \frac{1}{2}m\int_A^B d\left(\upsilon^2\right) = \frac{1}{2}m\upsilon_B^2 - \frac{1}{2}m\upsilon_A^2. \tag{2.36}$$

We then define the *kinetic energy* of the material point of mass m and velocity υ as

$$U_K = \frac{1}{2}m\upsilon^2, \tag{2.37}$$

which is independent of the position. The kinetic energy has the same physical dimension as the work and is measured in jule. We finally can write Eq. (2.36) as

$$W_{AB;\Gamma} = U_K(B) - U_K(A), \tag{2.38}$$

which is the work-kinetic energy theorem. In words: *when a material point moves on a certain trajectory from* A *to* B, *the work done by the forces acting on it is equal to the difference between the kinetic energy of the point has in* B *and that it had in* A.

It is sometimes useful to express kinetic energy in terms of momentum rather than velocity, namely

$$U_K = \frac{p^2}{2m}. \tag{2.39}$$

2.11 Calculating Work

In this section we shall see two examples of calculation of works, made respectively by weight and friction, when the application point P moves on its trajectory from the initial position A to the final one B. We shall see that in the former case the work

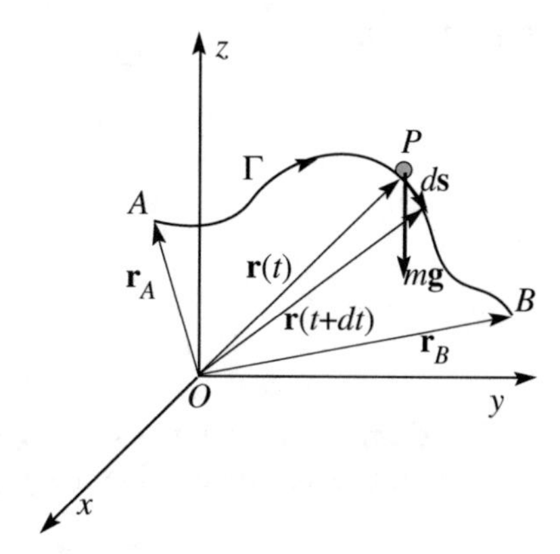

Fig. 2.21 Trajectory of the material point and its weight

depends only on the initial and final position, and not on the path taken between them, in the latter it depends on the path too.

Starting with weight, Fig. 2.21 shows the reference frame (not necessarily inertial) where we have chosen the z-axis to be vertical. Point P moves on the trajectory from the position A, with the position vector $\mathbf{r}_A = (x_A, y_A, z_A)$ to the position B, with the position vector $\mathbf{r}_B = (x_B, y_B, z_B)$. The figure shows also the position vector at the generic instant t and in the immediately following instant $t + dt$. The force acting on the point is its weight $m\mathbf{g}$, which is equal in all points. The elementary work done by the weight, which is vertically directed downwards is $dW = m\mathbf{g} \cdot d\mathbf{s} = -mgdz$. The total work is given by the integral

$$W_{AB;\Gamma} = \int_{A;\Gamma}^{B} m\mathbf{g} \cdot d\mathbf{s} = -\int_{A;\Gamma}^{B} mgdz = mgz(A) - mgz(B). \tag{2.40}$$

We see that in this relevant case the work is independent of the path, depending only on the final and initial position, even better, on their heights only. This conclusion was experimentally proven by Galilei with a simple experiment that we shall describe in the next section.

This is not the case of the second example, the friction force, which we shall study in Sect. 3.5.

Suppose we have an object, say a book or a brick, lying on a table. In real cases, the constraint does not apply to the body only the normal force, but also a friction that is tangent to the contact surface. If we want to move the body on the trajectory Γ in Fig. 2.22 at a constant speed, as we know from every day experience, we need to pull it, apply a force, parallel to the plane in the direction of the displacement. This means that the plane exerts on the body a force equal and opposite to our pull, because the velocity is constant in magnitude and then the resultant of the forces in the direction of the motion must be zero. Indeed, as we shall see in Sect. 3.5, the friction force, $\mathbf{F}_a$, is always parallel and opposite to the elementary displacement $d\mathbf{s}$. We now calculate the work of $\mathbf{F}_a$.

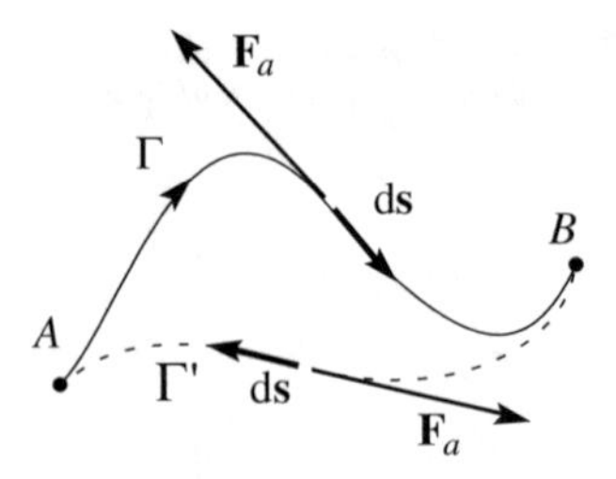

Fig. 2.22 Calculating the work of the friction force

The elementary work is $dW = \mathbf{F}_a \cdot d\mathbf{s} = -F_a ds$ which is always negative. The total work is given by the line integral on the trajectory

$$W_{AB;\Gamma} = \int_{A;\Gamma}^{B} \mathbf{F}_a \cdot d\mathbf{s} = -\int_{A;\Gamma}^{B} F_a ds = -F_a s_{AB}(\Gamma), \qquad (2.41)$$

where $s_{AB}(\Gamma)$ is the length of the trajectory Γ between A and B. The work is proportional to the length of the path, a quantity obviously depending on the path.

We conclude with an observation that we shall generalize in Sect. 2.13. We have seen that the work of the weight force for displacement A to B is $W_{AB} = -mg(z_B - z_A)$. Suppose now that the point goes back to A. The work of weight is $W_{BA} = -mg(z_A - z_B) = -W_{AB}$. Namely the total work of the weight on a closed path is zero. On the other hand, the work of the friction force to go from A to B on the curve Γ is $W_{AB,\Gamma} = -F_a s_{AB}(\Gamma)$. If we now go back on another curve, say Γ' in the Fig. 2.22, the work of the friction is $W_{BA,\Gamma} = -F_a s_{BA}(\Gamma)$, which is again negative. Consequently the work of the friction on a closed path is not zero, it is negative.

2.12 An Experiment of Galilei on Energy Conservation

One of the discoveries of G. Galilei was the fact, as we have mentioned, that the velocity of body descending under the action of is weight only, starting from rest, depends on the difference between the initial and final levels, and not on the followed path.

In the “Dialogue on Two new sciences” he states that the velocities of bodies descending on inclines of different slopes and the same height are equal. In his words (translations by the author):

> All contrasts and impediments removed… a heavy and perfectly round ball, descending through the lines CA, CD, CB would reach the final points A, D, C with the same moments

with reference to Fig. 2.23a reproduced from the book. Notice that, at the time, Galilei was searching for and developing the laws of mechanics and that several

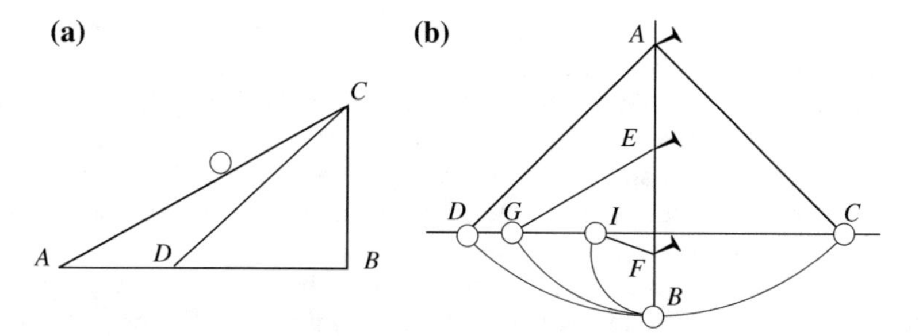

Fig. 2.23 **a** Ball falling on inclines of different slopes; **b** the pendulum and nail experiment

concepts had not yet been completely defined. In particular, impetus, momentum, kinetic energy were not well-separated concepts.

However, accurate measurements of those velocities were impossible to do. To prove the statement, he invented a simple and genial experiment, using a pendulum and a nail. Figure 2.23b is also reproduced from his book.

Salviati, the person who in the Dialogues represents Galilei, starts with the description:

> Suppose this sheet to be a vertical wall and to have a lead ball of one or two ounces hanging from a nail fixed in the wall, suspended to a thin wire AB, two or three arms long, perpendicular to the horizon… and about two finger far from the wall.

Then draw the vertical line AB and, perpendicular to it DC. Move the wire with the ball in AC and let it go. We shall see the ball

> descending first through the arc BCD, and going beyond point B as much as, sliding on the arc BD, almost reaching the drawn horizontal CD, failing to reach it by a very small gap, which has been taken away by the impediments of the air and the wire; from which we can likely conclude that the momentum (impetus) gained by the ball in B, in the descent on the arc CB, was so much to pull it back through the similar arc BD to the same height.

He continues with the request to repeat the experiments several times to check the result. Then

> I want we fix in the wall, grazing the vertical AB, a nail, like in E or in F, which should protrude out five or six fingers.

As before, the wire with the ball is moved to *AC* and let go. The ball will again move on the arc *CB*. But, when it is in *B*, the wire hits the nail, forcing the ball to move on the arc *BG*, having center in *E*.

> Now, my Lords, you will see with enjoyment the ball reaching the horizontal line in the point *G*, and the same to happen if the obstacle would be lower, as in *F*, where the ball would go through the arc *BI*, always finishing its ascent on the line *CD*.

Salviati concludes that the momentum acquired by a body descending from a certain height is just what is needed to bring it back to the same height, through whatever path. He observes that the momentum acquired in the descent on a given arc is equal to the momentum needed to rise through the same arc. He concludes that the momentum, and we can say also the velocity and kinetic energy in *B*, is the same whether it descends through *CB*, or *GB* or *IB* or any arc beginning on the horizontal *DC* and having its lowest point in *B*. On the other hand, the fall along an arc can be thought of as the fall on an "incline" of varying slope, proving the assumption.

The importance of the result of this experiment became clear in the following evolution of mechanics. In his experiment the kinetic energy of the ball in *B* is the same whatever the path starting from stillness from the same level. We now know that this energy is equal to the work done by the weight force. We conclude that the work done by the weight depends only on the difference of level and not on the particular path followed. We have already discussed this property in the previous section. Indeed, it is a fundamental one; it shows that there is a quantity, the energy, which is conserved, does not change in the motion under the action of weight. Weight is a conservative force, as we shall now see.

2.13 Conservative Forces

In general the work of a force on a point depends on the trajectory of the point. However, we have seen a case, the case of the weight force, in which the work depends only on the origin *A* and end *B* and not on the trajectory between them. Forces having these properties are said to be *conservative*. In the opposite case, as for the friction, they are said to be *non-conservative* or *dissipative*.

Let $\mathbf{r}$ be the position vector in the chosen reference frame and $\mathbf{F}(\mathbf{r})$ be a conservative force, a function of the position. The definition of conservative force states that, for whatever curve Γ with origin in *A* and end in *B*,

$$W_{AB;\Gamma} = \int_{A;\Gamma}^{B} \mathbf{F} \cdot d\mathbf{s} = f(\mathbf{r}_A, \mathbf{r}_B), \tag{2.42}$$

where f is a function of the co-ordinates of *A* and of *B*. It is easy to show that in this case it is always possible to find a function of the co-ordinates, which we shall indicate with $U_p(\mathbf{r})$, such as

$$W_{AB} = U_p(\mathbf{r}_A) - U_p(\mathbf{r}_B). \tag{2.43}$$

To show that, consider an arbitrarily chosen point o, as in Fig. 2.24. The work from *A* to o on whatever path is

$$W_{oA} = f(\mathbf{r}_o, \mathbf{r}_A) \tag{2.44}$$

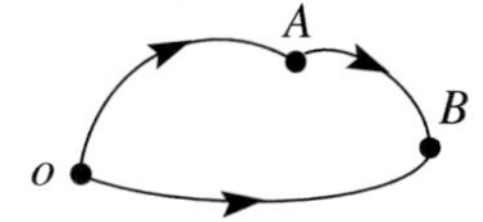

Fig. 2.24 Different paths

and similarly the work from o to B is

$$W_{oB} = f(\mathbf{r}_o, \mathbf{r}_B). \tag{2.45}$$

But, we can go from o to B also going from o to A and then from A to B. Considering that work is an additive quantity we can write $W_{oB} = W_{oA} + W_{AB}$. Hence

$$W_{oA} + W_{AB} = f(\mathbf{r}_o, \mathbf{r}_B). \tag{2.46}$$

By subtracting Eq. (2.43) from this expression we have

$$W_{AB} = f(\mathbf{r}_o, \mathbf{r}_B) - f(\mathbf{r}_o, \mathbf{r}_A). \tag{2.47}$$

We then reach the result by putting $U_p(\mathbf{r}) = f(\mathbf{r}_o, \mathbf{r})$. The function $U_p(\mathbf{r})$ is the *potential energy* of the force $\mathbf{F}(\mathbf{r})$ and is a function of the co-ordinates only. In conclusion the potential energy, or better its difference, is defined by the relation

$$U_p(\mathbf{r}_B) - U_p(\mathbf{r}_A) = -\int_A^B \mathbf{F} \cdot d\mathbf{s}. \tag{2.48}$$

In words: *the difference of potential energy of the force* **F** *in the point* B *and in the point* A *is equal to the opposite of the work done by the force when its application point moves from* A *to* B, *following any trajectory*.

The reason of the—sign, or the word "opposite", is the following. To be concrete, consider the weight. If we move a body of mass m from the level z_A to the higher level z_B, the displacement is opposite to the force and the work $-mg(z_B - z_A)$ is negative. The potential energy of the body is then larger when its level is higher. The work done by the weight force is equal and opposite to the gain of potential energy of the body. This energy can be given back as work by the body, taking it down to the original level. The higher the body, the greater is its potential to produce work.

We can conclude, and this is true in complete generality, by stating that *the potential energy difference between two states of a body is equal to the work we need to do against the force acting on the body to change it from the first to the second state.*

Notice again that a potential energy can be defined for a force only if its work is independent of the path. No potential energy exists, for example, for the friction forces.

Notice also that only differences of potential energy can be defined, not its absolute value. In other words, potential energy is defined up to an arbitrary

Fig. 2.25 The paths discussed in the text

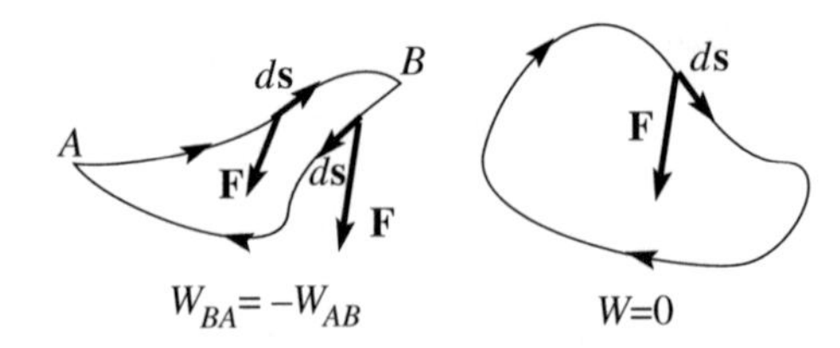

additive constant. In practice, we fix the constant choosing a reference position, say o, in which we define the potential energy to be zero ($U_p(o) = 0$), The potential energy in the arbitrary point P is then

$$U_p(P) = U_p(o) - \int_o^P \mathbf{F} \cdot d\mathbf{s} = - \int_o^P \mathbf{F} \cdot d\mathbf{s}.$$

For example for the weight, we arbitrarily fix a reference level at which the potential energy is zero by definition. This may be the ground level but some other level too. We take that level as the origin of the vertical upward directed z-axis and the potential energy is

$$U_p(z) = mgz. \tag{2.49}$$

We have stated that a force $\mathbf{F}$ is conservative if the work it does on a point when it moves from position A to B is independent of the path. There are two equivalent ways to state the same, which may be useful in certain circumstances.

1. A force is conservative if the work it does moving from A to B on any path is equal and opposite to the work done moving from B to A on any path (Fig. 2.25). This follows immediately from (2.48).
2. The work of a conservative force on any closed path is zero.
   In summary we can briefly say that the (equivalent) properties of conservative forces are: (1) its work does not depend on the path, (2) admits a potential energy, (3) the works going and going back are equal and opposite, (4) the work on a closed path is zero.

2.14 Energy Conservation

Consider a material point P of mass m moving from the position A to the position B on the trajectory Γ under the action of the (only) force $\mathbf{F}$. Whether the force is conservative or not its work is equal to the change of the kinetic energy of the point. Denoting with U_k the kinetic energy, we write

$$W_{AB,\Gamma} = U_k(B) - U_k(A). \tag{2.50}$$

If, and only if, **F** is conservative, the same work is also the opposite of the change of potential energy of the force

$$W_{AB,\Gamma} = U_p(A) - U_p(B). \tag{2.51}$$

It immediately follows that

$$U_p(B) + U_k(B) = U_p(A) + U_k(A). \tag{2.52}$$

Considering that the positions A and B are arbitrary, we conclude that the sum of the kinetic and potential energies is the same, i.e., is constant, in every position of the motion. The sum is the *total mechanical energy*, say U_{tot} of the material point. The conclusion is so important that it is often called a "principle". The principle, or law, of energy conservation states that

$$U_{tot} = U_p + U_k = \text{constant}. \tag{2.53}$$

If more than one force is acting on the point P and all of them are conservative, Eq. (2.53) is still valid, provided that U_p is the sum of the potential energies of all the acting forces, or, in an equivalent manner, if it is the potential energy of the resultant of those forces. Notice however, that the law is no longer valid even if only one of the forces is dissipative.

In words, the law of energy conservation states that *if a point moves under the action of conservative forces only, its total mechanical energy is conserved during its motion.*

Consider now the case, which is what happens in practice, that also dissipative forces are present. Consider for example the motion on an incline under the actions of weight and friction. The kinetic energy theorem is still valid. The work done by the forces for the displacement from A to B on the curve Γ, can be written as the sum of the work W_{AB}^C of the conservative forces and that $W_{AB,\Gamma}^D$ of the dissipative ones and we have

$$W_{AB}^C + W_{AB,\Gamma}^D = U_k(B) - U_k(A)$$

but $W_{AB}^C = U_p(A) - U_p(B)$, and in conclusion

$$U_{tot}(B) - U_{tot}(A) = W_{AB,\Gamma}^D. \tag{2.54}$$

We see that, if non-conservative forces are active, the total mechanical energy varies and its variation is equal to the work of the non-conservative forces. The work of these forces is negative, as we saw for friction. Hence the energy diminishes. This is the reason of the *dissipative* term.

Fig. 2.26 Fall on inclines or vertical

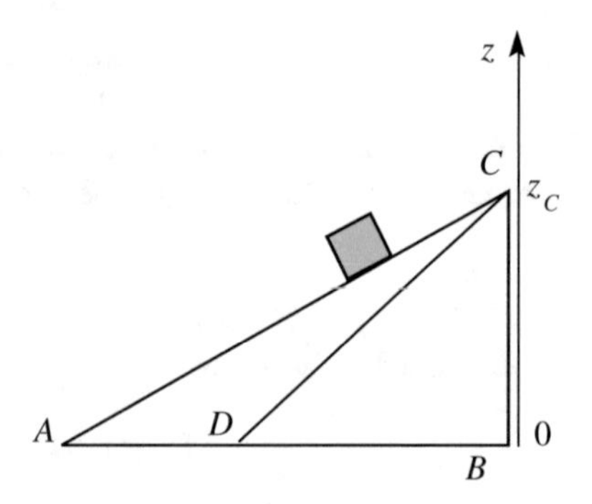

The physical dimension of kinetic, potential and total energies are the same as of the work. The measurement unit is consequently the jule.

Example E 2.1 Let us go back to the discussion made in Sect. 2.12 on the experiments by Galilei on inclined planes. Figure 2.26 shows a body of mass m, which can fall, starting from rest from point C, on inclines of different slopes CA or CD or vertically on CB. Take a vertical upwards axis z, and denote by z_C the height of C (that is the height of the inclined plane).

Consider the motion on CA. If friction is negligible the force exerted by the constraint is normal and does not make work. The other acting force is the weight $m\mathbf{g}$.

The energy conservation principle applied to the displacement CA from C, where the velocity is zero, to A, where $z = 0$, gives

$$mgz_C = \frac{1}{2}mv_A^2 \tag{2.55}$$

or

$$v_A = \sqrt{mgz_C}. \tag{2.56}$$

We see that the final velocity depends only on the difference in level not on the inclination.

If the friction is not negligible, the final energy is less than we have just calculated. We can obtain it with Eq. (2.54) calculating the work of friction. The latter does depend on the inclination for two reasons: the lengths of the paths are different and the body pushes with different forces on the plane. To do the calculation, however, we need to know something more on friction. We shall do that in the next chapter.

We finally observe that the above arguments are valid if the body can be considered a material point. If the body also rotates, like balls do, there is also kinetic energy associated to the latter that should be considered. We shall discuss this point in Sect. 8.16.

As we have just seen, in the presence of dissipative forces, the total mechanical energy, namely the sum of kinetic and potential energy, is not conserved. However, these are only two of many forms of energy. As a matter of fact the law of energy conservation is one of the basic laws of physics. The law is universally valid, without any exception, provided all the forms of energy are included in the balance. Other forms of energy are chemical energy, thermal energy, electric energy, nuclear energy, etc. Every time energy seems not to be conserved, it is because we have failed to include one of its forms. The issue is one of the main objects of thermodynamics, which will be discussed in the second volume of this course. The historic process that led to clarification of the concept of energy and to the establishment of the universal law of energy conservation was very long. Starting, as we have seen, already with Galilei, the process came to maturity only in the middle of the XIX century. It was then established with the first law of thermodynamics, mainly by Julius von Mayer (1814–1878) and James Prescott Joule (1818–1889). Energy is conserved also in the presence of dissipative forces if internal thermal energy is included in addition to macroscopic mechanical energy.

2.15 A Theorem Concerning Central Forces

A region of space in which a force that is a function only of the point, and possibly of time, is called a *force field*. If the force does not depend on time, the field is said to be *stationary*; if it does not depend on the position, it is said to be *uniform*.

The most common example of a uniform stationary field is weight, which is constant in time and space (at least within the limits of a laboratory). On the contrary, the viscous drag, the resistance of air to the motion, say, of a car or an airplane, is an (increasing) function of speed and consequently is not a force field.

A force field is said to be *central* if in every point P the force is directed as the line between P and a fixed point, called the *center of the forces*. The situation is sketched in Fig. 2.27, where C is the center of the forces.

It is clearly convenient to choose the center of the forces as the origin of the reference frame. We shall employ polar co-ordinates in which $\mathbf{r} = (r,\theta,\phi)$ is the

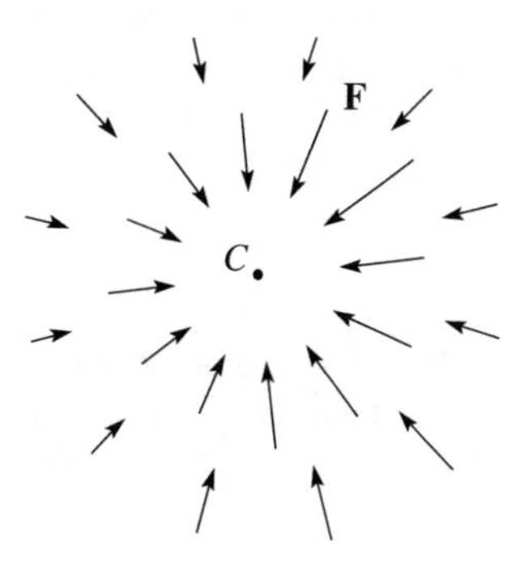

Fig. 2.27 A central field of forces

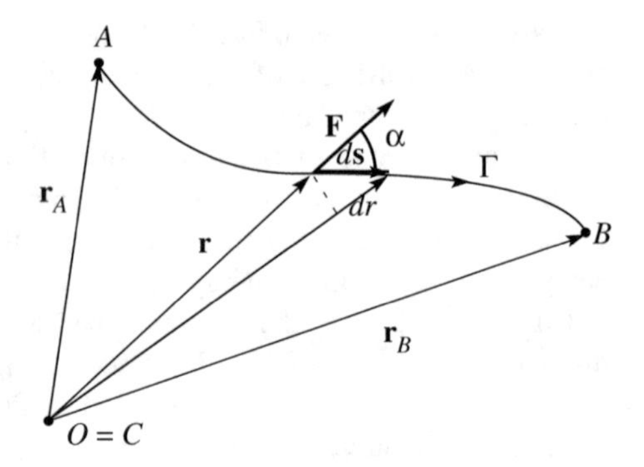

Fig. 2.28 Work by a central force

position vector. Let $\mathbf{F}(\mathbf{r})$ be the force under consideration. Saying that the force is central means that the two vectors $\mathbf{F}$ and $\mathbf{r}$ are everywhere parallel. They may have the same or opposite directions. The component of the force on the position vector, the radial component, is its magnitude in the former case, the opposite in the latter. This quantity may depend on the three coordinated, the two angles and the distance r from the center. If the force depends only on r, the field is said to have a *spherical symmetry*. On the other hand, a central force may be conservative or not. We shall now prove that these two properties are correlated: if a field of central forces has spherical symmetry, the force is conservative and, vice versa, if a central field of force is conservative it possesses spherical symmetry.

We start with the first statement. The radial component of the force, say $F_r(r)$, is by hypothesis a function of the distance from the center r only. Given any two points like A and B in Fig. 2.28, let us calculate the work done by the force on an arbitrary curve Γ, having A as origin and B as end. We shall proof that it is independent of the chosen curve. We indicate with $d\mathbf{s}$ the generic element of the curve. The work corresponding to this elementary displacement is

$$dW = \mathbf{F}(r) \cdot d\mathbf{s} = F_r(r) ds \cos \alpha, \tag{2.57}$$

where α is the angle between $\mathbf{F}$ and $d\mathbf{s}$, which is also the angle between the directions of $\mathbf{r}$ of $d\mathbf{s}$. Hence, $ds \cos \alpha$ is the projection of $d\mathbf{s}$ on the direction of $\mathbf{r}$, namely simply dr, i.e. the elementary variation of the distance from center. N.B. Pay attention! This notation is universally employed, but is ambiguous. The designation dr means the variation of the magnitude of the vector $\mathbf{r}$, namely $d|\mathbf{r}|$, not the magnitude of the vector variation of $\mathbf{r}$, namely $|d\mathbf{r}|$.

Anyway we have

$$dW = F_r(r) dr. \tag{2.58}$$

Notice that this elementary work may be positive or negative depending on F_r and dr having the same or opposite sign. The total work on the curve Γ is

$$W = \int_A^B F_r(r)dr \tag{2.59}$$

which is independent of the chosen curve, proving that the force is conservative.

What we have proven is valid for whatever dependence on r. A particularly important case is the gravitational force exerted by a mass M, which we shall consider to be point-like, on another mass m, point-like too. We shall study the gravitational force in Chap. 5. We anticipate here that such a force acting on m is in any point directed towards the position of M; namely it is central. Its magnitude is proportional to the product of the two masses and inversely to the square of their distance r. Indicating by G_N the proportionality constant, the force is

$$F_r(r) = -G_N \frac{Mm}{r^2}, \tag{2.60}$$

where the minus sign indicates that the force is always in the direction opposite to $\mathbf{r}$, namely is attractive. The work done on a displacement from A to B is

$$W = \int_A^B F_r(r)dr = \int_A^B -G_N \frac{Mm}{r^2} dr = G_N \frac{Mm}{r_B} - G_N \frac{Mm}{r_A}. \tag{2.61}$$

As expected, it is independent of path. We can then define the potential energy of the gravitational force. The potential energy difference between the point of position vector $\mathbf{r}_B$ and the point of position vector $\mathbf{r}_A$ is the opposite of the work Eq. (2.61), namely

$$U_p(\mathbf{r}_B) - U_p(\mathbf{r}_A) = -G_N \frac{Mm}{r_B} + G_N \frac{Mm}{r_A}. \tag{2.62}$$

As always, the potential energy is defined up to an arbitrary additive constant, namely

$$U_p(\mathbf{r}) = -G_N \frac{Mm}{r} + \text{constant}. \tag{2.63}$$

The constant is fixed choosing a point in which the potential energy is zero by definition. In this case it is obviously convenient (but not at all necessary) to choose this point at infinite distance, obtaining

$$U_p(\mathbf{r}) = -G_N \frac{Mm}{r}. \tag{2.64}$$

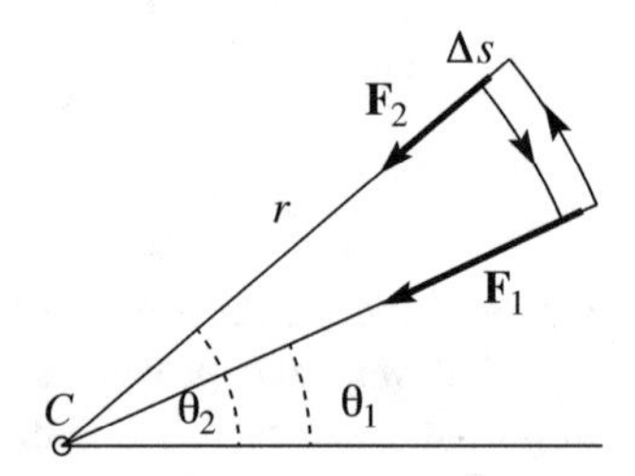

Fig. 2.29 The closed path used in the demonstration

This is the *potential energy* of a point-like mass m (the earth for example) in the gravitational field of the point-like mass M (the sun). Notice that, in fact, this is the energy of the pair of masses m and M (see Chap. 7).

We now prove the second of the above stated properties. We assume the force to be central and conservative and show that its component (magnitude with sign) on the position vector cannot depend on angles.

Let us consider for simplicity displacements on a plane. Consider a closed path, as in Fig. 2.29, composed of two circular arcs centered on the center of forces C, and two radial segments joining their extremes, at the angles θ_1 and θ_2 respectively. Take the radial segments of a very short length Δs. Assume by contradiction that the magnitude of the force F would depend not only on r but also on the angle θ. Under this hypothesis F_r has different values on the two radial sides that are at different angles, say F_{r1} and F_{r2}. Let us calculate the work of the force on this path. The contributions of the arcs are zero because on them the force is perpendicular to displacement. The contributions of the radial segments are $-F_{r1}\Delta s$ and $F_{r2}\Delta s$. The total work is then $W = (F_{r2} - F_{r1})\Delta s \neq 0$, in contradiction with the hypothesis that the force is conservative.

2.16 Power

In physics, power is defined as the work done per unit time. For a given delivered work, the power is larger for shorter delivery times. The simplest case is the work done by a force, say **F**, on a material point, say P. Consider the elementary displacement $d\mathbf{s}$ of the point, taking place between the instants t and $t + dt$. The work done by the force is $dW = \mathbf{F} \cdot d\mathbf{s}$. The *power* w given by the force the work divided by the corresponding time interval, that is

$$w = \frac{dW}{dt} = \mathbf{F} \cdot \frac{d\mathbf{s}}{dt} = \mathbf{F} \cdot \mathbf{v}. \tag{2.65}$$

In words: the power delivered by the force **F** acting on a material point moving at the velocity **v** in a given instant is equal to the dot product of the force and the velocity of the point in that instant. If the force is a function of the position, it must be obviously evaluated in the position of the point.

The physical dimensions of the power are those of a work divided by a time. Its unit is the watt, after James Watt (1736–1819) One watt is the power developed by a force delivering the work of one joule in one second (1 W = 1 J/1 s). To have an idea of the order of magnitude, you develop about 1 W if you raise a glass of water by 1 m in one second.

Problems

2.1 A person is sitting on a chair supported by a horizontal ground. Draw the diagrams of the forces for the person, the chair, and the earth. Describe each of the forces, identifying the body that produces them and the body on which they act. Identify the action reaction pairs.

2.2 A block hangs from the ceiling through a rope. A second rope is attached to the bottom of the block. It hangs vertically and you draw it with your hands downwards. Draw the diagrams of the forces for the block, each of the ropes, your body, the ceiling and the earth. Describe each of the forces, identifying the body that produce them and the body on which they act. Identify the action reaction pairs

2.3 Fig. 2.30 represents two blocks of masses m_1 and m_2 on frictionless planes. The plane of the first block is horizontal; the plane of the second is at an angle θ. The two blocks are tied by a mass less inextensible wire that can slide over a pulley without friction. (a) mentally insulate each block and draw the force diagrams; then write three equations of motion, (b) find the tension of the wire and the acceleration of m_2.

2.4 A body of mass $m = 1$ kg moves in a circular uniform motion on a circle of radius $R = 0.1$ m. What is the value of the centripetal force?

2.5 The system represented in Fig. 2.31 is in a vertical plane. $M > m$. Letting it free, M goes down and m goes up. Neglecting the frictions, draw the diagrams of the forces and determine the accelerations of M and of m.

2.6 With a hammer of mass $m = 0.1$ kg we beat on a nail, which is already partially stuck in a piece of wood, with a speed of $\upsilon = 1$ m/s. The nail advances a distance of $s = 1$ cm. Find the force exerted by the hammer.

2.7 Two people pull a rope, each on one end, each with a force of magnitude F. What is the tension? F or $2F$? Why?

2.8 Two ropes hang from the ceiling. Two spheres of different masses hang at the two ends. With both your hands you apply to the two spheres the same force

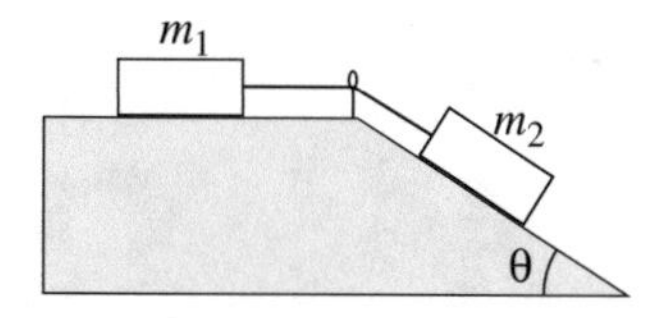

Fig. 2.30 The two blocks of problem 2.3

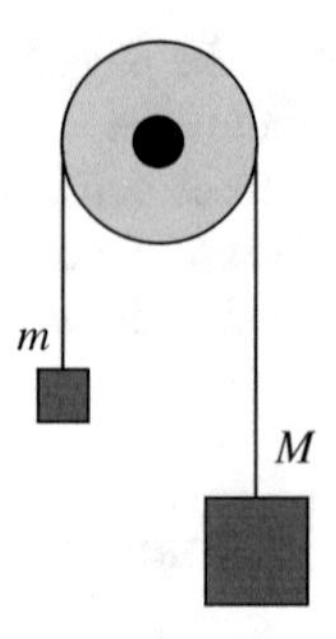

Fig. 2.31 The two blocks of problem 2.5

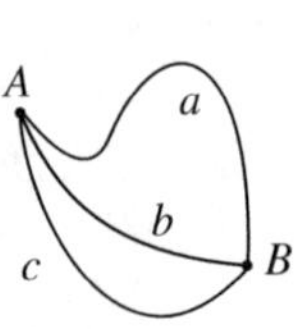

Fig. 2.32 The three guides of problem 2.9 in a vertical plane

F, which is not necessarily in the direction of the rope. What are the forces on each hand?

2.9 The three curves in Fig. 2.32 represent three rigid guides in a vertical plane. Three rings of different masses slide without friction, one on each of them. The three rings start from A at the same time with null velocity. State for each of the following statements if it is true or false. 1. The rings reach B contemporarily. 2. The rings reach B with velocities equal in magnitude.

2.10 A man of mass $m = 80$ kg jumps from a platform at the eight $h = 0.5$ m above ground. Reaching the soil he forgets to fold his legs. Fortunately the ground is quite soft and stops the motion in a distance $s = 2$ cm. What is the average force exerted on his bones during the stoppage?

2.11 Give an approximate evaluation of the height reached by a pole vaulter athlete able to reach in his run the speed of $\upsilon = 10$ m/s.

2.12 Fig. 2.33 shows three blocks of equal weight F_p. The pulley is frictionless. If we gradually increase all the weights, keeping them equal to each other, which rope will break?

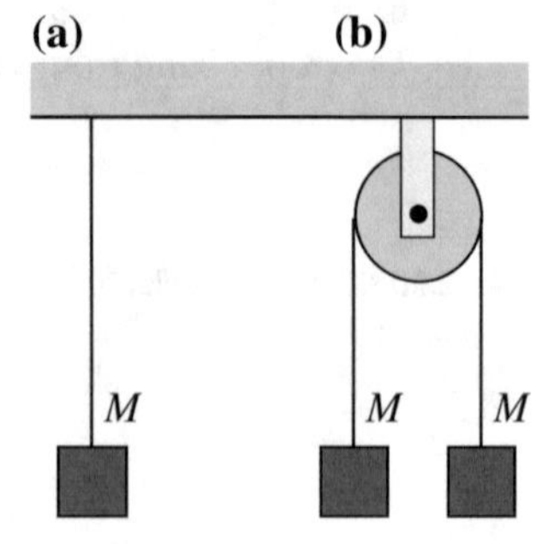

Fig. 2.33 The system of problem 2.12

2.13 Two spheres, one with mass double that of the other, are launched upwards with the same initial momentum p_0. If the resistance of air can be neglected, what is the ratio of the heights they reach.

2.14 A particle of mass $m = 2$ kg oscillates on the x-axis. The equation of its motion is $x = 0.2\sin(5t - \pi/6)$, with x in meters and t in seconds. (a) What is the magnitude of the force acting on the particle at time $t = 0$? What is the maximum value of the force?

2.15 A twine of length l can hold a maximum tension T. It is employed to rotate a mass m on a circle. Find the maximum velocity the body can rotate if (a) the rotation is in a horizontal plane, (b) in a vertical plane. Draw in each case the diagram of the forces.

Chapter 3
The Forces

In 1686, in his Preface to the First Edition of the Principia, Newton wrote

> … the whole burden of philosophy seems to consist in this – from the phenomena of motion to investigate the forces of nature, and then from these forces to demonstrate the other phenomena.

The second law of motion states that the time derivative of the momentum of a body is equal to the force acting on it. The law is not complete as long as the forces are not known. As a matter of fact, the forces present in nature have simple expressions. There are four fundamental forces: the gravitational force, the electromagnetic force, the strong nuclear force and the weak nuclear force. The two latter forces explain how matter behaves at a fundamental level. They appear at nuclear and subnuclear scales, at which quantum physics is valid, and do not directly appear in everyday macroscopic phenomena (even if, for example, the weak force is responsible for nuclear fusion processes in the Sun that give us light and energy). In the next chapter we shall study in some detail the gravitational force and related phenomena. The electromagnetic force is the object of the 3rd Volume of this course.

We have experience of several other forces. Apart from weight, which is (mainly) due to the gravitational attraction of earth, all the other forces are macroscopic effects of electromagnetic nature at microscopic level. Such are the elastic force, the normal force of constraints, friction and viscous drag in a fluid, both gas and liquid. These forces are not fundamental but are extremely important for the study of every day phenomena.

We shall study these forces and the corresponding phenomena in this chapter. The gravitational force will be treated in Chap. 4, fully dedicated to it. Further study of the viscous drag will be done in the second volume of this course.

The elastic force is met in a wide variety of circumstances. It gives rise to the most important periodic motion, the harmonic oscillations. Harmonic oscillations and the connected resonance phenomena, of which the mechanical ones are prototypes, are present with very similar characteristics in all the branches of physics, electromagnetism, optics, atomic an nuclear physics. Also, the vast majority of the strongly interacting particles, which are called hadrons, are extremely unstable,

A. Bettini, *A Course in Classical Physics 1—Mechanics*,
Undergraduate Lecture Notes in Physics, DOI 10.1007/978-3-319-29257-1_3

living only a few yoctoseconds. They are detected as resonances. We shall study the harmonic oscillator in Sects. 3.8 and 3.9 and, at a deeper level, in Volume 4 of this course.

In the last two sections we shall discuss the information that we can gather on the motion of the bodies starting from the potential energy, rather then from the force, which is possible if the forces are conservative. We shall introduce energy diagrams in Sect. 3.10 and employ them in three important cases, elastic force, pendulum and molecular forces, in Sect. 3.11.

3.1 Elastic Force

The solid objects of everyday experience have a definite shape. However, if stressed by a force, or a system of forces, they deform. Consider the geometrically simple situation of Fig. 3.1, showing a cylindrical metal bar attached on one face to a fixed support. If we apply a force, parallel to its axis, on the other face, the bar shortens if we push and lengthens if we pull. If the force is perpendicular to the axis, the bar flexes.

As another example consider a rubber band. If we pull on it at one end while keeping the other end fixed, the band lengthens according to the force exerted. When the force is removed, the ribbon returns to its natural length.

Many more examples can be cited. Their study shows that if a force is applied to a body and then removed, the body resumes the shape it had before being deformed, provided the deformation has not been too big. In these conditions we speak of *elastic deformation.*

For larger deformations, the original size is not completely recovered, rather some deformation is left permanently. This regime is called of *plastic deformation.* The transition between elastic and plastic regimes is smooth. It takes place at stress values that strongly depend on the chosen material. The bar in Fig. 3.1, for example, will be permanently deformed by more feeble forces if it is made of wax rather than of steel.

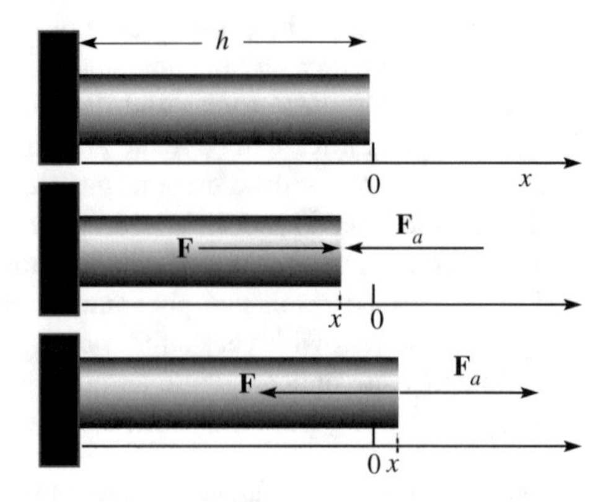

Fig. 3.1 A metal bar and its deformations under an applied force $\mathbf{F}_a$

Let us study the phenomenon with reference to the metal bar of Fig. 3.1. When it is not stressed by a force, its length is called *rest length* or *natural length* is h. We choose a reference axis parallel to the bar directed outside with the origin in its free end. When we apply a force $\mathbf{F}_a$ (a for applied) the extreme end moves and a new equilibrium state is reached. This means that the bar has reacted by developing a force, say $\mathbf{F}$, equal and opposite to $\mathbf{F}_a$. The deformation, namely the difference between the actual and original shapes is, in this, case, a change in the length of the bar. With the chosen co-ordinate, the infinitesimal change dx, is positive for lengthening, negative shortening. The x component of the force developed by the bar is in the positive x direction in case of compression, negative in case of stretching. The magnitude of this force increases with the deformation and is experimentally found to be, for not too large deformations, *proportional* to the deformation x, namely

$$F_x(x) = -kx. \tag{3.1}$$

The constant k is called *elastic constant* or *spring constant*. It is a property of the material characterizing its stiffness. Its physical dimensions are a force divided by a length, its units are the newton per meter (N/m).

The proportionality between force and lengthening was experimentally discovered by Robert Hooke (1635–1703) in 1676 and Eq. (3.1) is called *Hooke's law*. He made it public in a curious way. Initially he challenged his colleagues with the anagram “ceiiinosssttuv”. Two years later, considering that nobody had solved the quiz, he gave the solution: “ut tensio, sic vis” (as is the stretch so is the force).

The Hooke's law is very simple and very useful. However, it is not exact, but approximate. Let us study the phenomenon more precisely. We apply to the extreme of the bar a force of increasing and known values of intensity. At equilibrium, these are equal to the magnitude of the force developed by the bar. For each value we measure the deformation, both for extension (positive deformation) and compression (negative deformation). Plotting the results in a diagram we usually find the behavior of Fig. 3.2. For small enough values the dependence of the force on the deformation is linear, Hook's law holds. If the force is too large however, the deformation, in the case of metals we are considering, is larger than foreseen by the Hook law in compression, smaller in extension (we shall understand

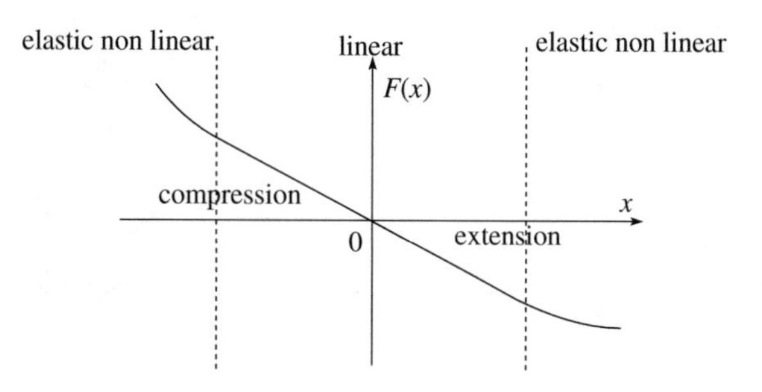

Fig. 3.2 Force versus deformation in the elastic and non-elastic regimes

the reason in Sect. 3.11). The non-linearity starts at values smaller than those at which the deformation is permanent. The regime in which the body returns to its original shape when the stress is removed is called *elastic*, both if the deformation is linear or not. The former is called a *linear regime*.

The transition between linear and non-linear behavior is smooth and is found at different values from metal to metal. It is smaller, for example, for lead than for steel.

Suppose that we now keep increasing the force further, for example in compression. The dependence of the deformation on the form is shown in Fig. 3.3, curve (a). Let us now suppose that, having reached point Q, we start decreasing the force, always measuring the deformation. We find that the representative point in the diagram does not go back on the curve (a) but on (b). Namely, for the same value of the force, the deformation is larger, in absolute value, when we start from a deformed state. In particular, when the external force, and the force of the bar with it, is back to zero, the deformation has a value, x_r, different from zero. It is called *permanent deformation*. We have deformed the bar so much that we went out of the elastic regime and entered the *plastic regime*.

Figure 3.3 shows, for one value of the deformation, two values of the force. In fact, the values are not only two, but a full range between a minimum and maximum. If we perform the same process, changing the point Q at which we invert somewhat further or somewhat sooner, the return branch is no longer (b), but a similar one lower or higher in the diagram, but always below the curve (a). In conclusion, the force does not depend only on the deformation but also on the past elastic history of the body. The phenomenon is called *elastic hysteresis*.

For a given material we can define the *elastic limit*, which we indicate with L. It is the maximum value of the deforming force (and of the force developed by the body) divided by the section of the bar to remain in the elastic regime. It is measured in newton per square meter (N/m^2).

As all the forces that depend only on distance, as the elastic force (within the elastic limit), are conservative. With reference to the co-ordinate in Fig. 3.1, we now express the work W of $\mathbf{F}$ when the extreme of the bar moves from x_1 to x_2 in the linear regime. The work for the elementary displacement dx is, in this regime, $dW = F_x dx = -kxdx$, hence

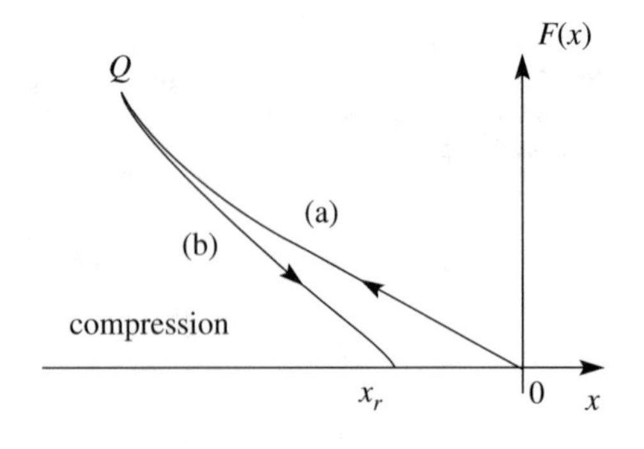

Fig. 3.3 The elastic hysteresis

$$W = \int_{x_1}^{x_2} F_x\, dx = -k \int_{x_1}^{x_2} x dx = -\left(\frac{1}{2}kx_2^2 - \frac{1}{2}kx_1^2\right)$$

and we can define the potential energy function of x or, better, its difference

$$U_p(x_2) - U_p(x_1) = \frac{1}{2}kx_2^2 - \frac{1}{2}kx_1^2.$$

As always, to define its absolute value we need to choose arbitrarily a point in which the potential energy is zero by definition. In this case it is quite obvious, but not necessary, to chose the point $x = 0$ (that is zero deformation). With this choice we have

$$U_p(x) = \frac{1}{2}kx^2. \tag{3.2}$$

This expression is valid within the linear regime. In the elastic non-linear regime the force is still conservative and a potential energy could be defined, but with a more complicated expression. In the plastic regime the force is dissipative and no potential energy can be defined. Indeed, to be very rigorous, small dissipative effects exist also in the elastic regime, but they can be neglected for many practical purposes.

A deeper study of the elastic force shows that it is the resultant of an enormous number of microscopic forces acting between the molecules of the material. These are ultimately electromagnetic forces. The elastic force and the Hook's law are a macroscopic description of a very complex situation, which depends on the specific microscopic structure of the matter of the body under consideration.

Let us go back to the linear regime. If the body has a simple geometry, a cylinder, a parallelepiped, a wire or a band we can define its length, say h, and its section, say S. In these cases it is found with a good approximation that the elastic constant of a body is directly proportional to its section and inversely to its length

$$k = E\frac{S}{h}. \tag{3.3}$$

The coefficient E, which depends on the material (and its temperature), is called *Young modulus* after Thomas Young (1773–1829) Using Eq. (3.1) in absolute value, we can express the Young modulus as

$$E = \frac{|F_x|}{S} \bigg/ \frac{|x|}{h}. \tag{3.4}$$

Table 3.1 Elastic characteristics of some materials in N/m^2

	Young module $10^{-10} \times E$	Elastic limit $10^{-7} \times L$	Fracture strength $10^{-7} \times \sigma_f$
Iron	20	20	35
Steel	22	30	50–200
Copper	10	10	20–40
Lead	1.5	1	1
Glass	6	2.5	3–9
Rubber	10^{-4}	10^{-4}	3×10^{-4}

Namely it is the ratio of the deforming force per unit section, which is called *stress*, and the deformation per unit initial length, called the *strain*. The stress is a pure number, the strain and the Young module are forces per unit area and are measured in N/m^2.

It is useful to appreciate the orders of magnitude. The Young modulus values of the metals range in the order of 10^{11} N/m^2 ($E = 2\times10^{11}$ N/m^2 for steels, $E = 10^{11}$ N/m^2 for Cu, etc.). The elastic limits are around 10^8 N/m^2 ($L = 3\times10^8$ N/m^2 for steel, $L = 10^8$ N/m^2 for Cu). A third quantity is the *fracture strength* σ_f, which is the stress under which the bar breaks. For the metals the values are two or three times larger than the elastic limits. Once the plastic regime is entered, the fracture is nearing. The issue of the resistance under stress is very important for engineering and the definition of safety limits is a much more complex issue than the definition we have given.

Typical values of the three quantities for some substances are given in Table 3.1.

Going back to the orders of magnitude, consider a steel wire $S = 1$ mm^2 in section and $h = 1$ m in length. We fix it at one extreme and pull the other with the force $\mathbf{F}_a$. Its stretch is

$$x = \frac{1}{E}\frac{h}{S}F_a = \frac{1}{2\times10^{11}}\frac{1}{10^{-6}}F_a = 5\times10^{-6}F_a.$$

The maximum force in the elastic regime is $F_{\max} = LS = 3\times10^8 10^{-6} = 300$ N (For example if the wire supports a 30 kg weight). The corresponding elongation is $x_{\max} = 5\times10^{-6}F_{\max} = 1.5$ mm and the stress is quite small, around 1.5 per mille. A stress a few times larger would break the wire.

Much larger stress without fracture can be obtained with other materials, like rubbers. Typical values are around $E = 10^6$ N/m^2, $L = 10^6$ N/m^2 and fracture strength of 3×10^6 N/m^2. Consider a rubber wire of the same geometry of the steel one we considered above, namely with $S = 1$ mm^2 and $h = 1$ m. Under the action of the force $\mathbf{F}_a$, the elongation is $x = \frac{1}{E}\frac{h}{S}F_a = \frac{1}{10^6}\frac{1}{10^{-6}}F_a = 1\times F_a$. The maximum force in the elastic regime is $F_{\max} = LS = 10^6 10^{-6} = 1$ N, which is quite small, as expected for rubber.

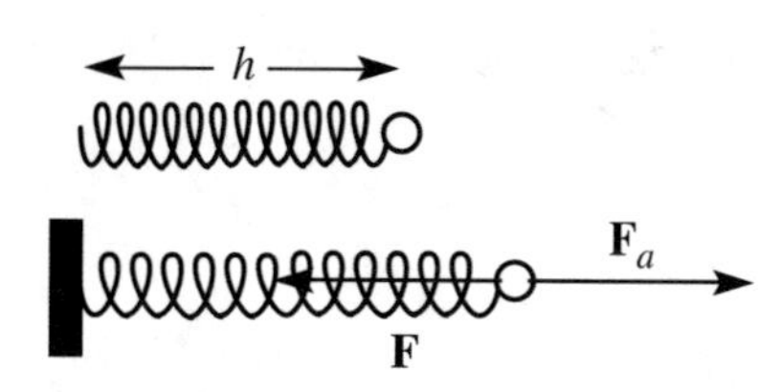

Fig. 3.4 Helical steel spring

The corresponding elongation is $x_{\max} = 1 \times F_{\max} = 1$ m. Hence, a rubber band can be stretched to twice its original length without reaching the elastic limit.

The reason for such different behavior of the metals and the rubbers is explained by the molecular structure of the materials. The metals are made of microscopic crystals. In each of these microcrystals the atoms are arranged at the nodes of a regular lattice. The distances between the atoms are such that the intermolecular forces are in equilibrium. When we try to deform a crystal we are attempting to change those distances by acting against intermolecular forces that are quite strong. Consequently the system is stiff, difficult to deform. On the other hand, rubber is composed of very long, spaghetti-like, molecules. These molecular spaghetti form a sort of tangled skein. The molecules interact amongst one another like a pasta that has been cooked too much and became sticky. When we pull the rubber we make the molecules straighter, but we do not change their length. Consequently, the process is much softer than for a crystal lattice and is reversible within much wider limits. We mention that when heated a metal expands, a piece of rubber contracts. With increasing temperature the equilibrium intermolecular distances in a crystal increase, while in the rubber the increased rate of collisions between molecules increases their entanglement.

In summary, the metal wires can be loaded with rather large stress and their strain is small. The rubber bands can have large strains, but do not bear large loads. If we need to work both with rather intense forces and relevant elongations we can use a steel helical spring, as in Fig. 3.4. When we pull the spring, its turns flex but the wire does not change appreciably its length. The elastic force is proportional to the tilt angle, hence to the elongation (or contraction) and the Hook's law holds.

3.2 Harmonic Motion

We have already discussed the motion, i.e. the oscillations, of the simple pendulum. This type of motion is important in every branch of physics and in physics based technologies. We shall now study the motion in its details.

To be concrete, let us consider the system shown in Fig. 3.5. The block, of mass m, lies on a horizontal plane, which we assume to be frictionless (we can use the trick explained in Sect. 2.4). In these conditions the resultant of the vertical forces is zero, the weight being equilibrated by the normal force of the constraint. A spring is

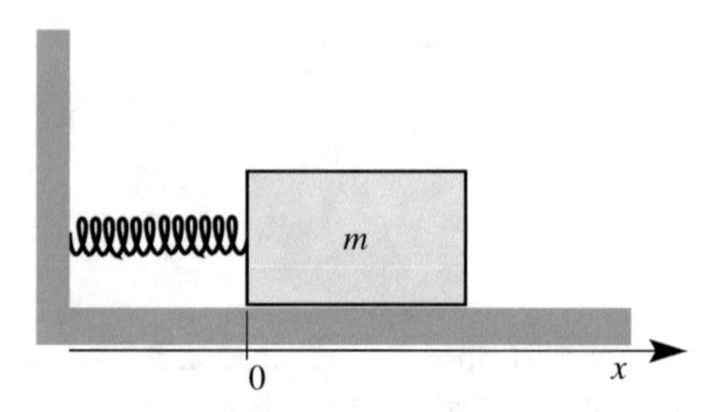

Fig. 3.5 A mechanical oscillator

connected to the block at one end and to a fixed point on the other. We take a co-ordinate axis, x, horizontally in the direction of the spring elongations and with the origin at the point in which the spring is attached to the block when it is at rest. In this way, x will measure the deformation of the spring. We assume we are in the range of validity of the Hook law. The force acting on the block is then

$$F_x(x) = -kx, \tag{3.5}$$

which is a *restoring force proportional to the displacement.*

The equation of motion is

$$-kx(t) = m\frac{d^2x}{dt^2}$$

which we write in the canonical form

$$\frac{d^2x}{dt^2} + \frac{k}{m}x(t) = 0. \tag{3.6}$$

We now introduce the positive quantity

$$\omega_0^2 = k/m. \tag{3.7}$$

This has a very important dynamical meaning. ω_0^2 is *the restoring force per unit displacement and per unit mass*. It depends on the characteristics of the system. We can then write Eq. (3.6) as

$$\frac{d^2x}{dt^2} + \omega_0^2 x(t) = 0. \tag{3.8}$$

We have already met it (with a different expression, Eq. (2.29) for ω_0) when discussing the pendulum. This very important differential equation describes the motion of many systems, including pendulums, near their stable equilibrium position, when subjected to a return force proportional to the displacement. The general solution, as learned by calculus, is

$$x(t) = a\cos\omega_0 t + b\sin\omega_0 t, \tag{3.9}$$

where the constants a and b must be determined from the initial conditions of the motion. They are two in number because the differential equation is of the second order.

The general solution can also be expressed in the, often more convenient, form

$$x(t) = A\cos(\omega_0 t + \phi), \tag{3.10}$$

where now the constants to be determined from the initial conditions are A and ϕ.

To find the relations between two pairs of constants, we start from

$$A\cos(\omega_0 t + \phi) = A\cos\phi\cos\omega_0 t - A\sin\phi\sin\omega_0 t.$$

Hence

$$a = A\cos\phi, \quad b = -A\sin\phi \tag{3.11}$$

and reciprocally

$$A = \sqrt{a^2 + b^2}, \quad \phi = -\arctan(b/a). \tag{3.12}$$

We now introduce the terms used when dealing with this type of motion. To do that in a general way, consider the expression (with a generic ω)

$$x(t) = A\cos(\omega t + \phi). \tag{3.13}$$

The motion is not only periodic, but its time dependence is given by a circular function. Such motions are said to be *harmonic*. A is called the *oscillation amplitude*, the argument of the cosine, $\omega t + \phi$, is called the *phase* (or *instantaneous phase* in case of ambiguity) and the constant ϕ is called the *initial phase* (indeed, it is the value of the phase at $t = 0$). The quantity ω, which has the physical dimensions of the inverse of time, is called *angular frequency* and also *pulsation*. Its kinematic physical meaning is to be *the rate of the variation of the phase with time* and, notice, is independent of the initial conditions of the motion. In the specific case we have considered above, the harmonic motion is the spontaneous motion of the system (in Sects. 3.8 and 3.9 we shall study motions under the action of external forces) and the angular frequency, ω_o, as in Eq. (3.10), is called *proper* angular frequency.

The motion is periodic with period

$$T = 2\pi/\omega. \tag{3.14}$$

The number of oscillations per unit time is called the frequency, ν. Obviously it is linked to the period and to the angular frequency by

$$\nu = \frac{1}{T} = \frac{\omega}{2\pi}. \tag{3.15}$$

The period is measured in seconds, the frequency in hertz (1 Hz = 1 s^{-1}), the angular frequency in rad s^{-1} or simply in s^{-1}. The unit is named after Heinrich Rudolf Hertz (1857–1899).

The harmonic motion can be viewed from another point of view. Consider a circular disc and a small ball attached to a point of its rim. The disc can rotate in a horizontal plane around a vertical axis in its center. Suppose the disc is rotating with a constant angular velocity ω. If we look at the ball from above, in the direction of the axis, we see a circular motion, but if we look horizontally, with our eye in the plane of the rotation, we see the ball oscillating back and forth periodically. Indeed, the motion is not only periodic, it is harmonic, as we now show.

Figure 3.6 shows the material point P moving on a circumference of radius A with constant angular velocity ω. We call ϕ the angle between the position vector at $t = 0$ and the x-axis. The co-ordinates of P at the generic time t are

$$x(t) = A\cos(\omega t + \phi), \quad y(t) = A\sin(\omega t + \phi).$$

The projection of the motion on the axes, in particular on x, is harmonic.

The conclusion leads to the simple graphical representation of the harmonic phenomena shown in Fig. 3.7. To represent an harmonic motion of amplitude A, angular frequency ω and initial phase ϕ, we take a fixed reference axis x and a vector $\mathbf{A}$, of magnitude A, rotating around its origin in the plane of the figure at the

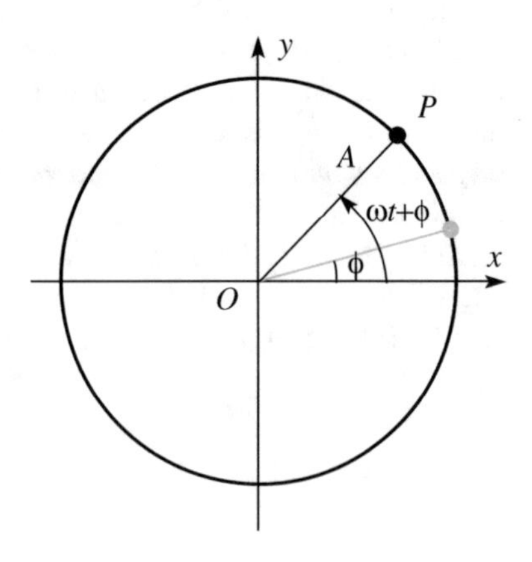

Fig. 3.6 A point P moving of circular uniform motion

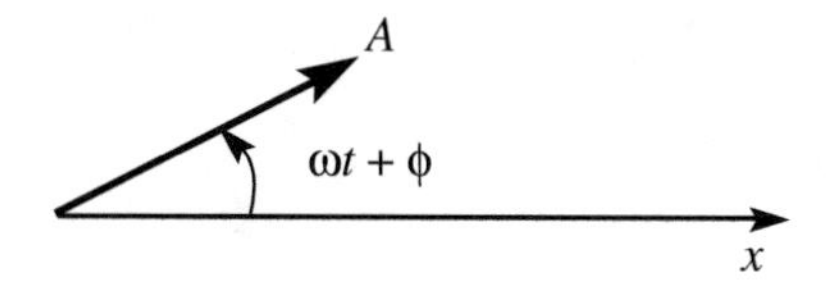

Fig. 3.7 Vector diagram for the harmonic motion

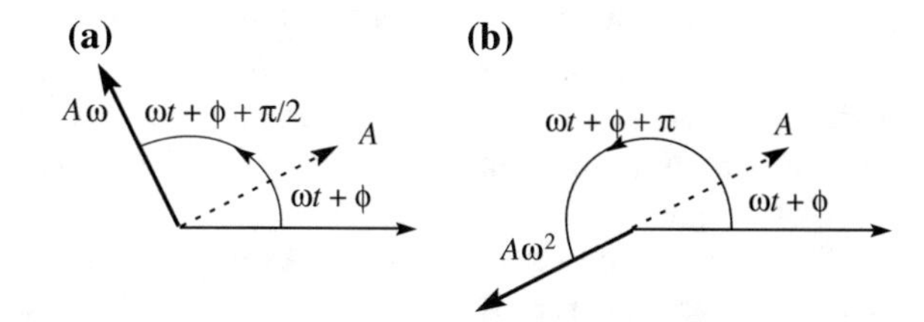

Fig. 3.8 Vector diagram for harmonic motion **a** velocity, **b** acceleration

constant angular frequency ω and forming with the x axis the angle ϕ at $t = 0$. The projection of **A** on the reference x axis is our harmonic motion.

We can use this representation also for velocity and acceleration of the harmonic motion. The derivative of Eq. (3.13) gives

$$\frac{dx}{dt} = -A\omega \sin(\omega t + \phi) = A\omega \cos\left(\omega t + \phi + \frac{\pi}{2}\right). \tag{3.16}$$

As written in the last side, the velocity is seen to vary in a harmonic way too, with a phase that is forward of $\pi/2$ radians to the displacement. This is shown in Fig. 3.8a.

Differentiating once more we have the acceleration

$$\frac{d^2x}{dt^2} = -A\omega^2 \cos(\omega t + \phi) = A\omega \cos(\omega t + \phi + \pi). \tag{3.17}$$

The acceleration is proportional to the displacement with the negative proportionality constant $-\omega^2$, or, as seen in the last side, its phase is at π radians to the displacement, or in *phase opposition* with it.

We now go back to the oscillator in the linear regime and consider its potential, kinetic and total energies. The former one is the potential energy of the elastic force, which we have already expressed in Eq. (3.2). We can use now Eq. (3.7) and write directly for the total energy

$$U_{\text{tot}} = U_k(t) + U_p(t) = \frac{1}{2}m\left(\frac{dx}{dt}\right)^2 + \frac{1}{2}kx^2 = \frac{1}{2}m\left[\left(\frac{dx}{dt}\right)^2 + \omega_0^2 x^2\right]$$

where $x(t)$ is given by Eq. (3.10), and we obtain

$$U_{tot} = U_k(t) + U_p(t) = \frac{1}{2} m\omega_0^2 A^2 \left[\sin^2(\omega_0 t + \phi) + \cos^2(\omega_0 t + \phi)\right] = \frac{1}{2} m\omega_0^2 A^2. \tag{3.18}$$

We see that neither the kinetic nor the potential energy are constant in time, rather, they vary as $\sin^2(\omega_0 t + \phi)$ and $\cos^2(\omega_0 t + \phi)$ respectively, but their sum, the total energy is, as we expected, constant. Notice also that kinetic, potential and total energies are all proportional to the square of the amplitude and to the square of the angular frequency.

The mean value of a quantity in a given time interval is the integral of that quantity on that interval, divided by the interval. It is immediate to calculate that the mean values of both functions $\cos^2$ and $\sin^2$ over a period are equal to ½ (the period of the square of a circular function is half the period of that function). Consequently the mean values of both potential and kinetic energy over a period are one half of the total energy.

$$\langle U_k \rangle = \langle U_p \rangle = \frac{1}{4} m\omega_0^2 A^2 = \frac{1}{2} U_{tot}. \tag{3.19}$$

3.3 Intermolecular Forces

All bodies are composed of very small particles that we call "molecules". These molecules combine to form gases, liquids and solids. Molecules are composed of atoms, a different type for each chemical element. Atoms are also composite objects. Each one has a positively charged central nucleus composed of protons and neutrons, while electrons form what may be thought of as a cloud surrounding the nucleus. Electrons and protons are equal in number, the *atomic number*, so that each atom is globally neutral. Different elements have different atomic numbers. Quantum mechanics, not classical mechanics, correctly describes the molecular and atomic phenomena. We can however give here a few semi-quantitative pieces of information, with a classical language, that are consistent with the prediction of quantum mechanics.

An electron inside an atom cannot be thought of as moving on a well-defined trajectory similar to the orbit of a planet (as was assumed in the early stages of the development of the theory, i.e., the Bohr model). We must instead consider the probability of finding an electron at a particular location around the nucleus. This probability is a known function of position different for each different atom. It, in particular, vanishes at a certain distance from the nucleus. It is the "cloud" we have mentioned above. In order of magnitude, the radiuses are tenths of nanometers (or 10^{-10} m). Nuclei are much smaller, 1–10 fm (10^{-15}–10^{-14} m). If we magnified a

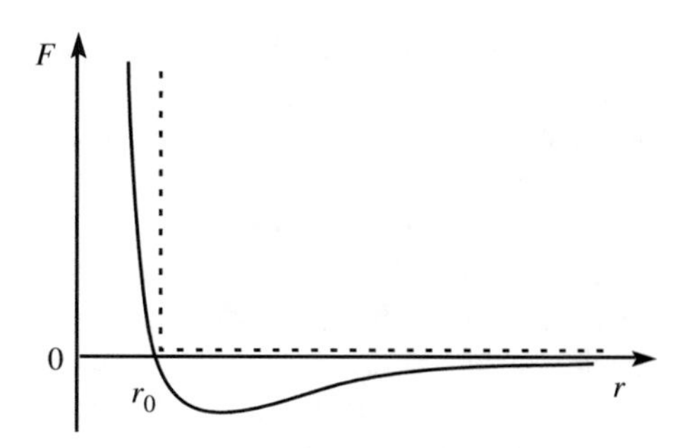

Fig. 3.9 The force between two molecules as a function of the distance between their centers

nucleus to the size of the dot on an "i" of this page, the diameter of the atom would be of the order of meters.

The elementary constituents of chemical substances are molecules. For example, a water molecule is made of two hydrogen and one oxygen atoms, nitrogen one of two nitrogen atoms, etc. Atoms are bound in a molecule by electric forces that are described by quantum mechanics. As a matter of fact, electric forces are very strong inside molecules, but, the molecules being globally neutral, are almost null outside the "cloud" of their electrons. Not completely however, as two molecules when they are close enough do interact with a force, much weaker than those inside the cloud, called *van der Waals force*, after Johannes Diderik van der Waals (1837–1923)

The van der Waals force between two molecules as a function of the distance between their centers r is shown in Fig. 3.9. It is repulsive at small distances, attractive at larger ones. In the diagram we adopted the convention of having positive repulsive forces.

When the centers of the molecules are at the distance r_0, at which the van der Waals force is zero, they are in equilibrium. At smaller distances the force is repulsive and becomes quickly enormous. In a very rough approximation we can consider them as rigid spheres of radius r_0. The dotted line in Fig. 3.9 is for an idealized rigid body, which would be non-deformable. The force would be repulsive and infinite when trying to squeeze it and null at distances larger than r_0 where it is not touched.

3.4 Contact Forces. Constraint Forces

If we put a heavy body, for example a brick, on a horizontal plane, it does not accelerate, it is in equilibrium. This implies that the plane, in general the constraint, has developed a force, called normal because it is perpendicular to the plane, which is exactly equal and opposite to the force that the body, the brick, exerts on the plane. The latter may be the weight, as in the example, or not. If we push with our hand on the brick, the magnitude of the normal force is equal to the sum of the weight and our push. Similarly, if we push a wall with a hand, it does not move. The normal force made by the wall is equal and opposite to the push.

The normal force is a *contact force*. If we raise the brick or take back our hand from the wall, even at very small distances, the force disappears. Contact force is the resultant of the forces between the molecules of the constraint and the molecules of the body. When the two are in contact, molecules on the two surfaces are at distances between their centers equal to molecular diameters. The applied force tends to bring the molecules of the body and of the constraint nearer to each other, namely to reduce their radii. This is opposed by the van der Waals force, which, as we have seen, quickly becomes enormous. For this reason two solid bodies cannot penetrate into each other. We have also seen that the intermolecular force goes quickly to zero at distances larger than the equilibrium position. This explains why the force disappears if we separate the surfaces even by very small distances. Already at a few molecular diameters the surfaces no longer interact.

Contact forces are used in practice when we want to constrain a body to move on a certain trajectory. For example, we have repeatedly used a horizontal plane to force a block to move in that plane; the rails force the train to move on a certain path, etc. The physical systems used for this purpose, the support plane, the rails, etc., are called *mechanical constraints*, because they constrain the motion. The constraints may inhibit motion on one side only or both, being named respectively *unilateral* and *bilateral*. A support plane is unilateral because it does not inhibit a body from rising above its surface. The rails of a train are unilateral but those of a roller coaster are bilateral, the coaster cannot detach from the rail.

Usually, forces produced by the mechanical constraints are not known a priori. They depend on the motion of the body, hence on other forces acting on it. For example the force exerted by the rail on the wheel of a train in a given curve depends on the curvature, but also on the speed of the train and on the mass of the wagon. Indeed, the rail develops a force that is exactly the centripetal force needed to have the wagon moving at that speed on that curvature with its mass. The forces exerted by the constraints are said to be *passive*, the other ones, which are usually under control, are called *active*.

We can however, calculate the passive forces if we know the motion of the body and all the active forces acting on it. Let us look at two examples.

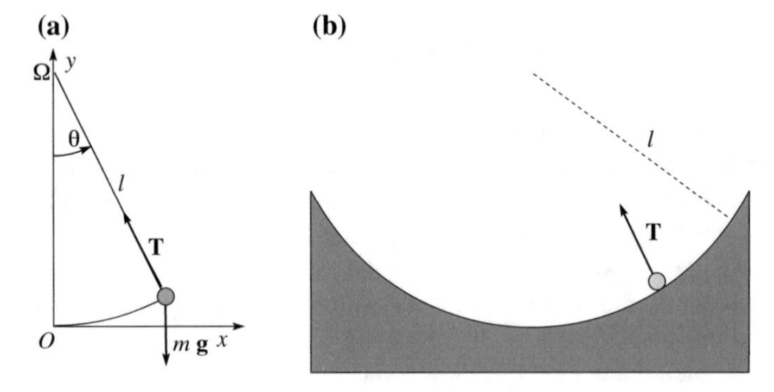

Fig. 3.10 Two different mechanical constraints for the same motion. **a** simple pendulum, **b** solid guide

Example E 4.1 We have already studied the pendulum in Sect. 2.9. We recall that the simple pendulum is a material body, of mass m, constrained to move on a circular arc of radius l. The easiest way to implement a mechanical constraint is like in Fig. 3.10a, with an inextensible wire fixed in Ω that exerts the tension $\mathbf{T}$ on the material point. Clearly, the constraint is unilateral, because the wire can fold. We could make it bilateral by using a light bar instead of the wire. In Fig. 3.10b the constraint is implemented with a wooden or plastic guide shaped as an arc of a circle of radius l, in which the body can slide. Assuming friction to be negligible, the guide will develop a normal force. We represent it with the same symbol as the tension of the wire, namely $\mathbf{T}$.

In both cases, the second law gives: $\mathbf{T} + m\mathbf{g} = m\mathbf{a}$. We already know the motion and are interested in the constraint force $\mathbf{T}$. We observe that in both cases $\mathbf{T}$ is directed always towards the center Ω. The radial component of the resultant of the forces must be the centripetal one, corresponding to the velocity υ of the body, namely $F = -m\upsilon^2/l$, where υ is the velocity at the considered instant and the minus sign means that the force is towards the center. The radial component of the resultant is $-T + mg\cos\theta$ and we have

$$-T + mg\cos\theta = -m\upsilon^2/l. \tag{3.20}$$

Clearly, T is not a constant, rather it depends on the position of the pendulum, which is defined by the angle θ. We could do that using the equation of motion we have found in Sect. 2.9. However, it is easier to employ energy conservation. The reason is the term $m\upsilon^2$ in the last expression, which is twice the kinetic energy. If the pendulum is abandoned from the initial position θ_0, corresponding to the height y_0, the energy conservation equation is $mgy_0 = mgy + m\upsilon^2/2$.

Hence $m\upsilon^2 = 2mg(y_0 - y)$. But, $y = l(1 - \cos\theta)$ and $y_0 = l(1 - \cos\theta_0)$, hence $y_0 - y = l(\cos\theta - \cos\theta_0)$, and we can write $m\upsilon^2 = 2mgl(\cos\theta - \cos\theta_0)$ and finally, substituting in Eq. (3.20), $T = mg(3\cos\theta - 2\cos\theta_0)$.

Example E 4.2 Consider, in a vertical plane, an inclined guide connected at its lower extreme with a circular guide, as shown in Fig. 3.11. We want to study the motion of a material point, a small rigid ball for example, on the circular rail, which

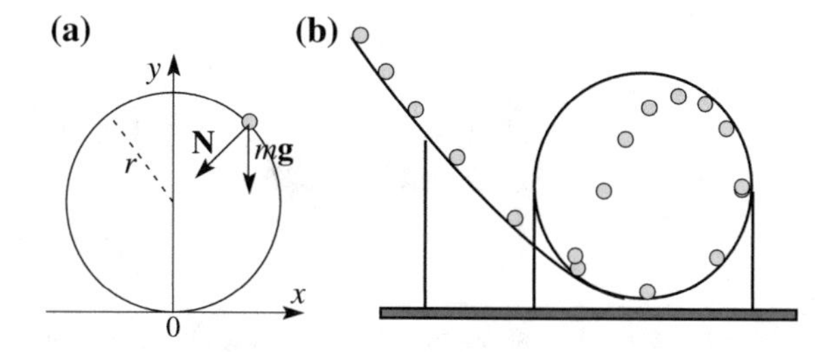

Fig. 3.11 **a** The forces on a ball moving on a vertical circular rail, **b** motion of the ball in case of detachment

is unilateral, of radius r. We use the incline to launch the ball with a certain initial velocity on that rail. More precisely, we want to find the minimum initial velocity in order that the ball would travel through the entire circle without detaching from the rail.

Two forces act on the ball, its weight $m\mathbf{g}$ and the force of the constraint, which we suppose to be normal, $\mathbf{N}$. The latter is directed as the radius, towards the center. The normal force cannot be directed outwards.

Again, the radial component of the resultant of the forces must be the centripetal force requested by the motion. This component is the sum of $\mathbf{N}$ and of the radial component of the weight. The latter is a maximum at the highest point of the guide.

To be sure that the ball does not detach, it is then sufficient to verify that in this point. Here, the weight and the constraint normal force are both directed vertically downwards. The condition of non-detachment is then $N + mg = m\upsilon^2/r$. Solving for the unknown N we have $N = m(\upsilon^2/r - g)$.

The condition of non-detachment is $N > 0$, hence the term $\upsilon^2 > gr$. If the velocity is smaller, the ball detaches following a trajectory as in Fig. 3.11b, which gives a sequence of images of the ball in its motion. We can think that in this situation the weight is providing a centripetal force too large for the radius of curvature of the guide, at that velocity. The motion must follow a trajectory with a smaller radius, and the ball detaches.

3.5 Friction

We have already seen several times that a physical rigid plane, when pushed by a body in contact with it, reacts with a normal force which is equal and opposite to the active force. In the example drawn in Fig. 3.12 the plane is horizontal and the active force, which is vertical, is simply the weight $\mathbf{F}_w$ of the block lying on the plane. The normal reaction $\mathbf{N}$ is vertical upwards.

We now apply to the block a force $\mathbf{F}$ *parallel* to the contact surface (horizontal in this particular case), by attaching a wire to the block and pulling. Suppose that we

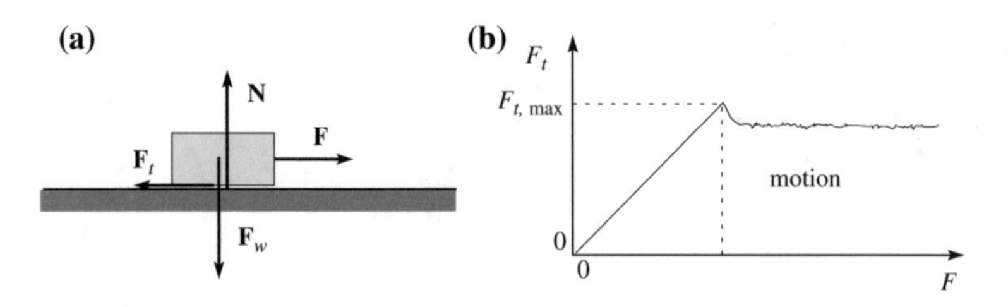

Fig. 3.12 **a** Active and constraint forces on a block, **b** friction force versus applied tangential force

gradually increase the tangential force starting from zero. We observe that initially, when F is not very strong, the block does not move, it is still in equilibrium. This implies that the resultant of the forces must still be zero, not only in the direction normal to the plane, where nothing is changed, but also in the tangent one, where now there is a force. The constraint must have developed also a force parallel to the contact surface, $\mathbf{F}_t$ equal and opposite to $\mathbf{F}$, namely

$$\mathbf{F}_t = -\mathbf{F}.$$

The force developed by the constraint parallel to the contact surface, when there is no motion, is called *static friction*.

If we continue to increase the tangential force on the block F, the tangential force by the constraint increases too, as long as the block does not move. This happens at a certain value of the active force, meaning that the friction force cannot be larger than a maximum value that we call $F_{t,\max}$.

This behavior is followed in all cases in which two dry surfaces are in contact. In these conditions, it is experimentally found that the maximum value of the static friction is proportional to the normal force, namely that

$$F_{t,\max} = \mu_s N. \tag{3.21}$$

The proportionality constant μ_s is called the *coefficient* of *static friction*, which is clearly a dimensionless quantity.

We now study the motion of the block when the tangential applied force is larger than $F_{t,\max}$. By measuring its acceleration, we infer that a tangential contact force $\mathbf{F}_t$ is present, which is in general somewhat smaller than $F_{t,\max}$ as shown in Fig. 3.12b. Also in the case of relative movements of the two contact surfaces, it is experimentally found that the tangential force by the constraint is proportional to the normal one. Its direction is always parallel and opposed to the velocity, namely

$$\mathbf{F}_t = -\mu_d N \mathbf{u}_\upsilon, \tag{3.22}$$

where $\mathbf{u}_\upsilon$ is the unit vector of the velocity. The dimensionless constant μ_d is called *coefficient of kinetic friction*.

Figure 3.12b shows schematically the tangential force of the constraint versus the applied tangential force. We see that F_t grows to be equal to the applied force up to $F_{t,\max}$. Then, when the motion is started, it diminishes somewhat, as we have already noticed, and then remains approximately, but not exactly, constant. Notice that in the majority of the cases $\mu_d < \mu_s$ but there are also opposite cases.

As a matter of fact, the static and dynamic friction forces are due to the interactions between the molecules on the surfaces of the two bodies. Consequently, Eqs. (3.21) and (3.22) are a macroscopic description of a complex microscopic situation. We observe that friction coefficients depend critically on the status of the surfaces in contact, on how they have been machined, on their cleanliness, etc. Notice carefully that the molecules on the surface of a body made of a certain

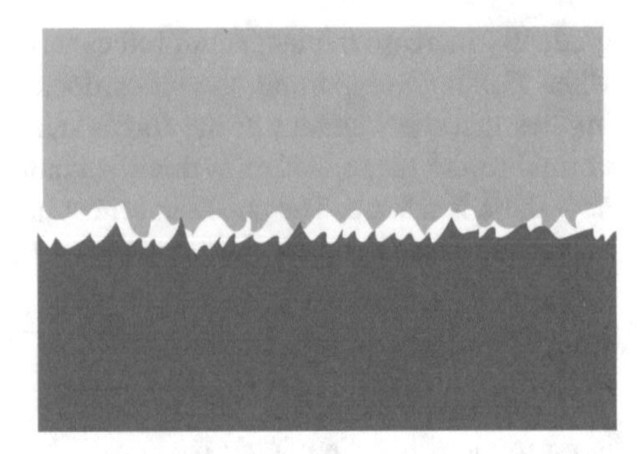

Fig. 3.13 Pictorial view of the contact surfaces between two metals, at nanometer scale

substance, for example copper or steel, are not only of that substance. Water is almost always present, oxidation too. One can find mentioned values of the friction coefficients between, say, copper and copper, copper and steel, etc. But, there is no single copper on copper, etc. friction coefficient, for the just mentioned reasons.

As a matter of fact, for example in the case of a piece of copper, it is possible to obtain surfaces populated by copper molecules only. The piece must be processed with ad hoc procedures under a vacuum, because in the presence of air, copper will oxidize and water molecules will be deposited on the surface immediately. Now suppose we have produced two such blocks in a vacuum and put their surface in contact. They immediately stick one onto the other and you will not be able to separate them. They became a unique copper bock. How are molecules supposed to know to which block they belong?

The first astronauts to land on the Moon observed this phenomenon. Putting two stones gathered from the soil in touch, they found them sticking together and difficult to separate, even if their surfaces were obviously irregular.

There is no universal mechanism at the origin of the friction between two contact surfaces. Consider the important case of two metal surfaces. Metallic surfaces can be worked to be extremely smooth. Even in these conditions, surfaces are not smooth if looked at nanometer scales. Figure 3.13 tries to show the surfaces as seen at a large magnification. The irregular patterns have a typical scale of 10 = 100 nm.

When two surfaces are, we think, in contact, the contact is indeed only between the "crests" on the two sides. Consequently the surface really in contact, say S_c is much smaller than the nominal surface S (typical values of S_c/S are between 10^{-4} and 10^{-5}). However, the larger is the normal force N pushing the two surfaces one against the other, the larger is the number of crests touching each other. We can then understand why the friction force is proportional to N. We can also understand why it is independent of the area of contact. Suppose we keep N constant and double the contact macroscopic surface S. The action of the normal force will distribute on a doubled area and its effect on the crests per unit area will halve. The number of contacts per unit surface will halve too, but they will cover a twice as large area. The total number of contact has not varied. In conclusion, S_c is proportional to N and independent of S.

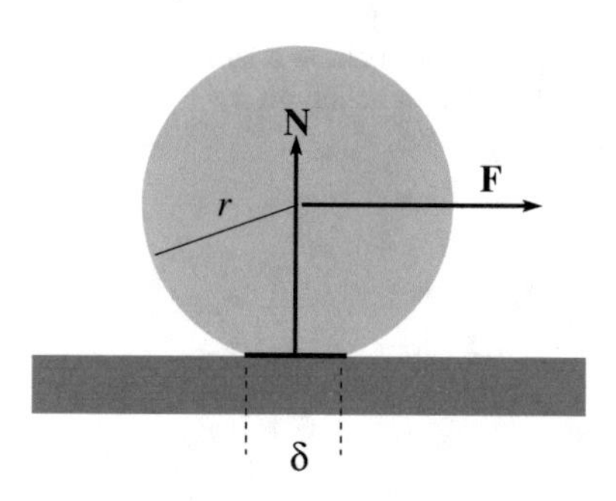

Fig. 3.14 Schematics of the rolling resistance

In the contact points the molecules of the two bodies interact strongly attracting each other and becoming, so to say, welded. To have one surface sliding on the other, these micro welding points must be broken. Again the necessary force is proportional to S_c and consequently to N and independent of S.

What we have just described is relative to dry surfaces between solid bodies and has nothing to do with the friction between lubricated surfaces. In this case, a film of liquid is present between solid surfaces, the molecules of which are far enough away from each other to have an interaction. In this case the friction is due to the viscosity of the lubricant (see Sect. 3.6).

The *rolling resistance* or *rolling friction* is the force resisting the motion developed by the constraint, for example the support surface, when a cylindrical or spherical body, such as a reel or a ball, rolls on the surface. Figure 3.14 represents in cross section such a cylinder, say a reel, of radius r. We apply a force $\mathbf{F}$ to the axis of the reel parallel to the support plane and normal to the axis. We assume that the reel does not slide on the plane due to the static friction force. This type of motion is called *pure rolling*. When the reel rolls, it does that about an instantaneous axis that is the contact generator in the considered instant. The moment of the applied force about the *instantaneous rotation axis* is $\tau = rF$. The moment τ necessary to have the rolling at a constant angular velocity is experimentally found to be proportional to the magnitude of the normal force N, namely

$$\tau = \gamma N, \tag{3.23}$$

where γ is the *rolling resistance coefficient*. Its physical dimension is a length, and is measured in meters. The applied moment is equal and opposite to the moment developed by the constraint.

The rolling resistance force is generally smaller than the dynamic friction. As a matter of fact it is due to quite complicated phenomena in the region of contact between the reel and the support plane. In Fig. 3.14 this region is shown as a flat area of longitudinal with δ. This is an idealization, because actually both the cylinder and the plane deform into shapes that are not forward-backwards symmetrical. We are here simplifying a lot. We can say that on the contact area a number of the above considered "crests" of both bodies are in contact. The difference is that now, to have movement, the microwelds are broken acting in a

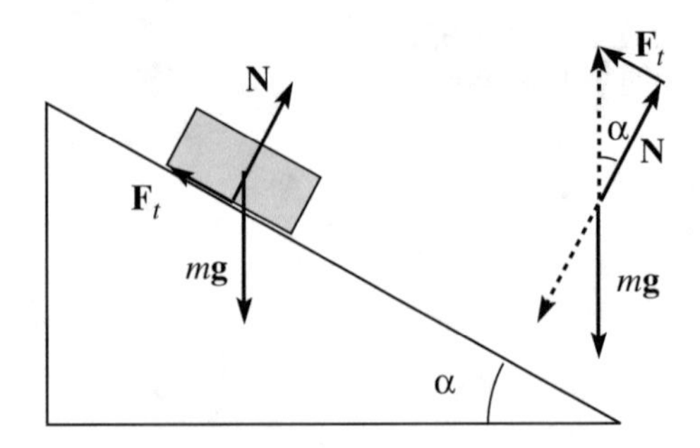

Fig. 3.15 A brick on a slide and the forces acting on it

direction normal, rather than parallel, to the surface. This requires, *caeteris paribus*, a smaller force.

Example E 4.3 Consider Fig. 3.15. A brick lies on an inclined surface, the inclination of which, α, can be varied. Given the coefficient of static friction μ_s, what is the maximum value of α at which the brick remains still?

The forces on the brick are its weight $m\mathbf{g}$ and the force exerted by the constraint. The latter can be decomposed in a normal, $\mathbf{N}$, and a tangential, $\mathbf{F}_t$, component, which is the friction. For equilibrium the components of the resultant must be zero. Namely, $F_t = -mg\sin\alpha$ and $N = -mg\cos\alpha$. Hence $F_t/N = \tan\alpha$. But, the static friction force cannot be larger than $\mu_s N$, and the no-slide condition is $\alpha \leq \arctan\mu_s$.

The maximum angle, say $\alpha_f \equiv \arctan\mu_s$ is called the *friction angle*. For example, the slopes of the piles of sand or of the screes in the mountains naturally settle on the corresponding friction angle.

We have seen in Sect. 2.11 that friction forces are dissipative, and that their work is negative when their application point moves, because they are always in a direction opposite to the motion, see Eq. (2.41). Indeed, the friction forces are always such as to oppose the relative motion of the two bodies. This does not imply that the friction acting on a body would always act to slow it down, on the contrary it can also accelerate it.

As an example, let us consider our brick, of mass m, ling on the horizontal platform of a cart. The latter moves straight forward with constant acceleration $\mathbf{a}$ (see Fig. 3.16) in the direction of its velocity $\mathbf{v}$. If the acceleration of the cart is not too large, the block remains still relative to the platform; its motion is accelerated with the same acceleration $\mathbf{a}$ as the cart. It must be acted upon by a force equal to $m\mathbf{a}$. But the only horizontal force acting on it is the friction $\mathbf{F}_t$. Hence, $\mathbf{F}_t = m\mathbf{a}$. The friction accelerates the brick. We know that F_t can be at most equal to $\mu_s N = \mu_s mg$.

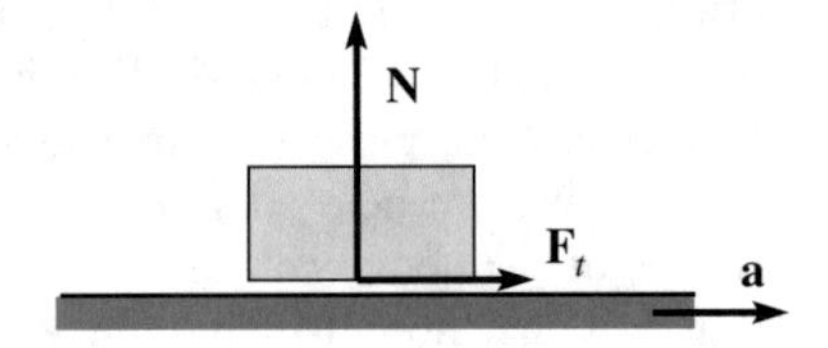

Fig. 3.16 A brick on an accelerating platform and the forces acting on it

Consequently the maximum acceleration of the cart at which the brick does not slide is $\mu_s g$.

Notice that in this case the friction has the direction of the velocity, namely of the displacement. Consequently its work is positive. In the same way, when we start running we are accelerated by the friction force between our shoe soles and the ground, when a car accelerates the accelerating force is the friction between its reel and the road. Notice however, that in these cases the work of the friction force is zero, because the point of application does not move.

3.6 Viscous Drag

A solid moving relative to a fluid, a liquid or a gas, is subject to a force, different from friction, but as friction opposing the relative motion of the body and the medium. It is called *viscous drag* or *viscous resistance*. Differently from friction, there is no drag when the relative velocity is zero, and an increasing function of the relative velocity. The direction of the drag force is always equal and opposite to the relative velocity.

The magnitude of a force depends on the magnitude of the relative velocity, on the shape of the body and on the fluid. Moving relative to the fluid, the body induces a number of effects that may perturb substantially its flow. Think for examples of vortices. Consequently, the dependence of the drag force on velocity is complicated. We shall study it in the second volume of this course, together with fluid dynamics. Here we anticipate only a few elements that are needed in our study of the motions of bodies.

The force depends on the shape of the body, for example it is different for a cylinder or a sphere, on its orientation, for example the case of a disc is different for its orientation parallel or perpendicular to the flow, and, for a given geometrical shape, on its size. We shall limit the discussion here to a spherical body, of radius a.

The force depends on two characteristics of the fluid, its density ρ (mass per unit volume) and the viscosity η. The latter will be discussed in the second volume. It suffices to know here that it characterizes the difficulty with which the fluid flows, so, for example, oil has larger viscosity, is more viscous, than water, but is less viscous than honey. For a given fluid, the viscosity depends on the temperature.

The physical units of viscosity are

$$[\eta] = \left[ML^{-1}T^{-1}\right] = \left[FL^{-2}T\right], \tag{3.24}$$

where, in the third member we have taken into account that the dimensions of the force are $[F] = [MLT^{-2}]$. Pressure has the dimensions of a force per unit surface (FL^{-2}) and its unit is the pascal (Pa), from Blaise Pascal (1623–1662). The unit for viscosity is then the pascal second (Pa s). For example, for some everyday fluids at ambient temperature, their viscosities are for oils $\eta \approx 0.5$–1.5 Pa s, for water $\eta \approx 10^{-3}$ Pa s, and for air $\eta \approx 1.8 \times 10^{-5}$ Pa s.

The *Reynolds number* is a parameter that gives relevant information on the regime of the motion, named after Osborne Reynolds (1842–1912). It is dimensionless, namely a pure number. The four quantities of the problem have the physical dimensions $[\rho] = [ML^{-3}]$, $[\eta] = [ML^{-1}T^{-1}]$, $[a] = [L]$ and $[\upsilon] = [LT^{-1}]$. They can be arranged in a dimensionless quantity as

$$Re = (\rho/\eta)\upsilon a, \tag{3.25}$$

which is the Reynolds number for a sphere. Its expressions for other shapes are similar.

Figure 3.17 shows schematically how the drag force on a body can be measured. The body is fixed to a thin bar and to the pointer of a dynamometer fixed on a support and is immersed in the fluid under study, which is moving at a known velocity υ, that we can vary in a known manner. Experiments of this type show that at small velocities the drag force can be written as the sum of a term proportional to the velocity and one proportional to its square

$$R = A\upsilon + B\upsilon^2, \tag{3.26}$$

where the coefficients A and B depend on the body and the fluid but, for not too large velocities, are independent of velocity. As the ratio between the second and the first term is proportional to the velocity, the first term dominates at small velocities, the second at larger ones. We define as critical velocity υ_c the velocity at which the two terms are equal. It corresponds to a quite small value of the Reynolds number

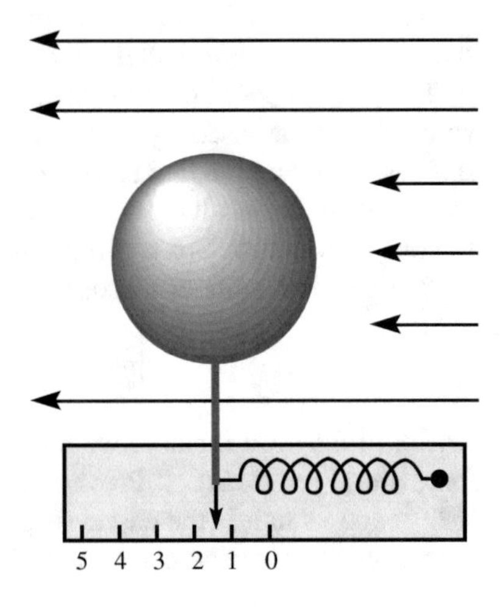

Fig. 3.17 Measuring the drag force

$$Re_c \approx 20 - 30. \tag{3.27}$$

Consider now the sphere moving in air, as pendulums or free falling bodies, at normal temperature and pressure conditions. The air density in these conditions is $\rho = 1.2\ \text{kg/m}^3$. With the value for viscosity already given, the Reynolds number is

$$Re(\text{air}) = 1.5 \times 10^{-5} \upsilon a \tag{3.28}$$

and the critical velocity, in a round number

$$\upsilon_c \approx 4 \times 10^{-4}/a\ \text{m/s}. \tag{3.29}$$

If for example $a = 1$ cm, the critical velocity is $\upsilon_c = 4$ cm/s. The time taken to reach it by a body freely (in a vacuum) falling from rest is $t = \upsilon/g = 4$ ms, which is very short indeed. In this time it would travel in vacuum $d = gt^2/2 = 80\ \mu\text{m}$. For larger dimensions bodies moving in the air the critical velocities are even smaller.

We conclude that only for very small velocities, smaller than υ_c, is the viscous drag proportional to the velocity. However, it becomes proportional to the square velocity very gradually, reaching that regime only at Reynolds numbers two orders of magnitude larger than in Eq. (3.27), corresponding to velocities of a few meters per second for a sphere of 1 cm radius.

As a second example consider the same sphere moving in water. With $\rho = 10^3\ \text{kg/m}^3$ and the viscosity given above, $\eta/\rho = 10^{-6}\ \text{kg/m}^3$, which is a value, notice, smaller than for air. The Reynolds number at velocity υ for $a = 1$ cm is $Re = 10^4\ \upsilon$. The critical velocity is only $\upsilon_c \sim 2.5$ mm/s.

In the elementary study of free fall, of the motion on an incline and of the pendulum, the viscous drag of air is usually neglected. Is this a good approximation? Let us control on a few typical cases. Consider a bronze (density $\rho = 8 \times 10^3\ \text{kg/m}^3$) ball of $a = 2$ cm radius and three cases: free fall from a $h = 20$ m tall tower, descent of an incline of elevation $h = 1$ m and oscillation of a pendulum abandoned at the height from the position at rest $h = 0.5$ m. The weight of the ball is $F_p = 2.7$ N. Neglecting the presence of the air, and the energy of the rotation in the second case, the velocities at the end of the fall would be in any case $\upsilon = \sqrt{2gh}$, hence $\upsilon_1 = 20$ m/s, $\upsilon_2 = 4.5$ m/s, $\upsilon_3 = 2$ m/s in the three cases respectively. In presence of air all velocities would be somewhat smaller, but larger than the critical velocity. The drag force is approximately proportional to the square velocity, but is not very large. For the just mentioned velocities its values are approximately $R_1 = 2.4 \times 10^{-2}$ N, $R_2 = 1.2 \times 10^{-2}$ N, $R_1 = 2.4 \times 10^{-3}$ N, which are in any case small compared to the weight. Neglecting the drag in these cases is not a bad approximation. However, the effect will be noticeable on much longer times.

Finally notice that, whatever its expression, the viscous drag is a dissipative force. As it is always directed opposite to velocity, its work is negative for any displacement of the application point.

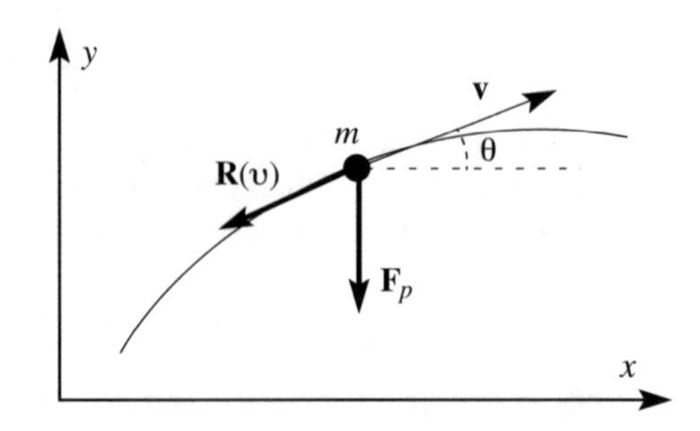

Fig. 3.18 The forces on a ball moving in air

3.7 Air Drag and Independence of Motions

In the study of the motion of a body under the action of one (or more) force, it is often convenient to decompose the motion in its components on the Cartesian axes. The component of the motion on an axis is due to the component of the force (or forces) on that axis. The component motions are independent of each other. This law of the independence of motions was discovered by Galilei, and assumed by Newton as a corollary of the second law. We have already quoted in Sect. 1.16 the following example from Galilei. Suppose we shoot a ball with a gun at the top of a tower, aiming horizontally. Simultaneously we drop a ball with zero velocity. The ball leaves the rifle barrel with a very high horizontal speed and, under the action of its weight, describes a parabola finally touching ground at a horizontal distance far away. The second ball falls vertically. Galilei established that both balls touch ground at the same instant, provided that the action of air is neglected, as he specifies.

We shall now analyze the motion in presence of the air and we shall see that the law of independence of motions is not always valid.

We refer the motion of the ball, of mass m, to a frame having the y-axis vertical and x horizontal in the plane of the motion as in Fig. 3.18. Two forces are acting on the ball, its weight $\mathbf{F}_p = m\mathbf{g}$ vertical downwards and the viscous drag

$$\mathbf{R} = -\left(A\upsilon + B\upsilon^2\right)\mathbf{u}_\upsilon \tag{3.30}$$

where υ is the velocity and $\mathbf{u}_\upsilon$ is its unitary vector. The Newton law gives

$$m\mathbf{g} - \left(A\upsilon + B\upsilon^2\right)\mathbf{u}_\upsilon = m\mathbf{a}. \tag{3.31}$$

The components on the axes of the equation, if θ is the angle of $\mathbf{v}$ with the horizontal, are

$$m\frac{d\upsilon_x}{dt} = -\left(A\upsilon + B\upsilon^2\right)\cos\theta, \quad m\frac{d\upsilon_y}{dt} = -mg - \left(A\upsilon + B\upsilon^2\right)\sin\theta. \tag{3.32}$$

This is a system of two non-linear differential equations, which cannot be easily solved. However, we are only interested here in knowing if and when the two motions are independent. To be so, only x and y components should appear in the first and second equation respectively. This is indeed the case for low velocities, when the term B can be neglected. In these conditions, considering that $\upsilon_x = \upsilon \cos\theta$ and $\upsilon_y = \upsilon \sin\theta$, Eq. (3.32) becomes

$$m\frac{d\upsilon_x}{dt} = -A\upsilon_x, \quad m\frac{d\upsilon_y}{dt} = -mg - A\upsilon_y. \tag{3.33}$$

The two motions are independent. However, if, as it is often the case, the drag is proportional to the square velocity, Eq. (3.32) become

$$m\frac{d\upsilon_x}{dt} = -B\left(\upsilon_x^2 + \upsilon_y^2\right)\cos\theta, \quad m\frac{d\upsilon_y}{dt} = -mg - B\left(\upsilon_x^2 + \upsilon_y^2\right)\sin\theta. \tag{3.34}$$

The motions are not independent. This is an obvious consequence of the proportionality of the drag force to the square of the velocity, which depends on both components. In the example of Galilei, the air resistance is larger for the gun ball than for the vertically falling one, because the velocity of the former is larger. The gun ball touches ground later than the ball falling from the tower if the effects of the air are not neglected.

3.8 Damped Oscillator

In Sect. 3.2 we discussed the motion of the harmonic oscillator. We then neglected the dissipative forces, which however, are always present. As we know these are basically of two types, friction and viscous drag of the air. We shall now include the viscous drag of the air, which we shall assume to be proportional to the velocity.

To be concrete, consider the system in Fig. 3.19, which is similar to that in Fig. 3.5, with the addition of an element providing the viscous force. We can think in terms of an absorber, like a piston moving in a fluid, but the element is meant to

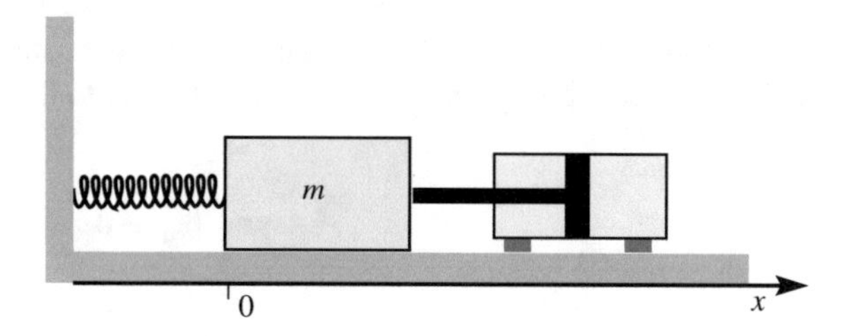

Fig. 3.19 A mechanical damped oscillator

represent all the viscous forces, including that due to the air. The viscous drag is proportional to velocity in magnitude and opposite to it in direction, namely

$$F_r = -\beta \frac{dx}{dt} \tag{3.35}$$

where β is a constant. We shall neglect the friction between the support plane and the block. The force (3.35) tends to slow down or damp the motion. Hence the oscillator is said to be *damped.* The second law gives

$$m \frac{d^2 x}{dt^2} = -\beta \frac{dx}{dt} - kx, \tag{3.36}$$

which we write, dividing by m and taking all the terms to the first member, in the "canonical" form

$$\frac{d^2 x}{dt^2} + \gamma \frac{dx}{dt} + \omega_0^2 x = 0. \tag{3.37}$$

In this form, the equation is valid for all harmonic damped oscillators. The two parameters depend on how the oscillator is built, the strength of the spring, the viscosity, etc. We have already met the first one while discussing the harmonic oscillator. It is the restoring force per unit displacement and per unit mass

$$\omega_0^2 = k/m. \tag{3.38}$$

The second, see Eq. (3.35) is the resistance force per unit velocity and unit mass

$$\gamma = \beta/m. \tag{3.39}$$

Notice that both constants have the dimension of the inverse of time. We already know that ω_0 is the angular frequency of the oscillator in absence of dissipative forces. The inverse of the second

$$\tau = 1/\gamma \tag{3.40}$$

is the time that characterizes the damping, as we shall now see.

The solution of the differential Eq. (3.37) is given by calculus. The rule to find it is as follows. First we write the algebraic equation obtained by substituting in the differential equation powers of the variable equal to the degree of the derivative. In our case it is

$$r^2 + \gamma r + \omega_0^2 = 0. \tag{3.41}$$

Then we solve it. The two roots are

$$r_{1}\,\mathit{underscore2} = -\frac{\gamma}{2} \pm \sqrt{\left(\frac{\gamma}{2}\right)^2 - \omega_0^2}. \tag{3.42}$$

The general solution of the differential equation is

$$x(t) = C_1 e^{r_1 t} + C_2 e^{r_2 t} \tag{3.43}$$

where C_1 and C_2 are integration constants that must be determined from the initial conditions.

Let us discuss the motion we have found. We observe that the effect of the dissipative force, which is to damp the motion, is larger for larger values of γ. Considering the two roots r_1 and r_2 of the algebraic Eq. (3.41), three cases should be distinguished called respectively: *under-damping* if $\gamma/2 < \omega_0$, the two roots are real and different, *over-damping* if $\gamma/2 > \omega_0$, the two roots are complex conjugate, and *critical damping* if $\gamma/2 = \omega_0$, the two roots are real and coincident. Let us analyze the three cases.

Over-damping. The two solutions, which are real, are both negative. The motion is the sum of two exponentials decreasing in time

$$x(t) = C_1 e^{-|r_1|t} + C_2 e^{-|r_2|t}. \tag{3.44}$$

The damping is so large that the system is not able to perform even a single oscillation. The displacement from the equilibrium position decreases monotonically. Mathematically speaking, Eq. (3.44) says that the time to reach that is infinite. In practice, after some time both addenda are so small, and so is the velocity, that other resistive forces that are always present, as the friction, stop the motion in the equilibrium position ($x = 0$). This happens in a time interval of a few times $1/|r_2|$ (which is larger than $1/|r_1|$).

Critical damping. The two roots coincide, $r = -\gamma/2 = -2/\tau$. In this particular case, Eq. (3.42) is not the solution. This is

$$x(t) = (C_1 + C_2 t)e^{-t/(2\gamma)}. \tag{3.45}$$

In this case too the system does not oscillate. The displacement reduces to zero, in practice, in a time interval of a few times 2τ. It can be shown that in the critical damping the time to reach equilibrium is minimum.

Under-damping. We can write the equation of motion in the form

$$x = C_1 e^{-(\gamma/2)t + i\omega_1 t} + C_2 e^{-(\gamma/2)t - i\omega_1 t}$$

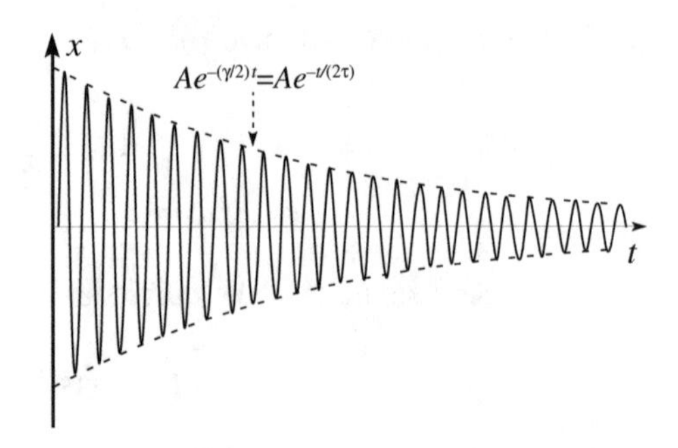

Fig. 3.20 Weakly damped oscillations

where

$$\omega_1 = \sqrt{\omega_0^2 - \left(\frac{\gamma}{2}\right)^2}. \tag{3.46}$$

We can now choose two different integration constants as $a = C_1 + C_2$ and $b = i(C_1 - C_2)$ and have the solution in the form

$$x(t) = e^{-(\gamma/2)t}(a\cos\omega_1 t + b\sin\omega_1 t). \tag{3.47}$$

For damping tending to zero ($\gamma \to 0$) the equation of motion becomes Eq. (3.9), as we expect since the oscillator is un-damped in these conditions. The solution can be written in a form analogous to Eq. (3.10)

$$x(t) = Ae^{-(\gamma/2)t}(\cos\omega_1 t + \phi) = Ae^{-t/(2\tau)}(\cos\omega_1 t + \phi) \tag{3.48}$$

where now the integration constants are A and ϕ. The motion is an oscillation similar to the harmonic motion with an amplitude, $Ae^{-t/(2\tau)}$, which is not constant but decreases exponentially in time with a decay time 2τ. The oscillations are damped. A weakly damped, namely with $\gamma \ll \omega_1$, motion is shown in Fig. 3.20. The oscillation amplitudes diminish gradually in a time long compared to the period. As a matter of fact, rigorously speaking, the motion is not periodic, because the displacement after every oscillation is somewhat smaller than before it. However, if the damping is small, $\gamma \ll \omega_1$, we can still identify a period

$$T = 2\pi/\omega_1. \tag{3.49}$$

The weak damping condition $\gamma \ll \omega_1$ can be written as $\tau \gg T$, in words, the decay time is much longer than the period.

Notice that the proper angular frequency ω_1 is smaller than the proper angular frequency of the free oscillator ω_0, but that for $\gamma \ll \omega_1$ the difference becomes infinitesimal of the second order compared to γ/ω_0.

We have seen in Sect. 3.2 that the total, kinetic plus potential, mechanical energy of the harmonic oscillator is constant in time. The difference now, even in the case of weak damping, is that a dissipative force is present. We expect that energy decreases. Without losing generality, we can assume the initial phase to be zero. The initial amplitude of oscillation is A. At every oscillation, the displacement reaches its maximum at, say, time t. The displacement is then

$$x_{\max}(t) = Ae^{-t/(2\tau)}. \tag{3.50}$$

In that instant the velocity is zero and the total energy is equal to the potential energy, which is proportional to the square of the amplitude

$$U_{\text{tot}}(t) \propto A^2 e^{-t/\tau}. \tag{3.51}$$

The total energy decreases exponentially in time, reducing to a value $1/e$ of the initial value in a time τ, which is one half of the time in which the amplitude reduces of the same factor. τ is called *decay time* of the oscillator.

An observation on the exponential function. The amplitude of a damped oscillation in the Eq. (3.48) and the energy of the damped oscillator, Eq. (3.50) are examples of physical quantities decreasing exponentially in time. This behavior is often met in physics. We make here a simple but important observation. Consider the function

$$f(t) \propto f_0 e^{-t/\tau}$$

and the ratio between its two values in two different instants t_1 and t_2 ($t_1 < t_2$). We immediately see that this ratio depends only on the interval $t_2 - t_1$ and not separately on the two times or the constant (the initial value) f_0. Indeed

$$f(t_2)/f(t_2) = e^{-(t_2 - t_1)/\tau}. \tag{3.52}$$

In particular, the function diminishes by a factor $1/e$ in every time interval $t_2 - t_1 = \tau$ and not only in the initial one.

In particular, we can reformulate the above statement in: "τ is the time interval in which energy reduces of a factor $1/e$".

3.9 Forced Oscillator. Resonance

Consider again the damped oscillator of the previous section and apply to the body a force in the direction of the x-axis that oscillates as a circular function of time with angular frequency ω and amplitude F_0. The component of the force on the

x axis (its magnitude or its opposite depending on the direction relative to x) is given by

$$F(t) = F_0 \cos(\omega t), \tag{3.53}$$

where we have chosen the origin of times and the instant in which the force is zero; its initial phase is then null. The second law is

$$m\frac{d^2x}{dt^2} = F_0 \cos \omega t - \beta\frac{dx}{dt} - kx, \tag{3.54}$$

which we write in the form

$$\frac{d^2x}{dt^2} + \gamma\frac{dx}{dt} + \omega_0^2 x = \frac{F_0}{m}\cos \omega t. \tag{3.55}$$

The left-hand side of this equation is that of the equation of the damped oscillation (3.37). But the right-hand side, which is zero for the latter, is now proportional to the external force. Equation (3.55) is a non-homogeneous differential equation and Eq. (3.37) is its *associated* homogeneous differential equation. A mathematical theorem states that the general solution of the former is the sum of the general solution of the associated homogeneous equation and of any particular solution of the non-homogeneous one.

We shall limit our discussion to the case of weak damping, as in Fig. 3.20. We can guess that a possible motion might be a harmonic oscillation at the angular frequency of the force; namely a particular solution might be

$$x(t) = B\cos(\omega t - \delta) \tag{3.56}$$

with some amplitude B and initial phase $-\delta$ to be determined. Let us check if our guess is correct. The easiest way to do so is to consider an equation exactly similar to (3.55) of the complex variable $z(t) = x(t) + i\, y(t)$. The imaginary part $y(t)$ is some function that is irrelevant in our arguments. We then search for a solution of the differential equation, of which (3.55) is the real part

$$\frac{d^2z}{dt^2} + \gamma\frac{dz}{dt} + \omega_0^2 z(t) = \frac{F_0}{m}e^{i\omega t}. \tag{3.57}$$

Considering that the equations are linear, the real parts of the solutions of Eq. (3.57) are solutions of (3.55). The function corresponding to our guessed solution is

$$z(t) = z_0 e^{i\omega t}. \tag{3.58}$$

Let us try it in (3.57)

$$-\omega^2 z_0 e^{i\omega t} + i\gamma\omega z_0 e^{i\omega t} + \omega_0^2 z_0 e^{i\omega t} = \frac{F_0}{m} e^{i\omega t}$$

which must be satisfied in every instant of time. And so it is, because all the terms depend on time by the same factor. Hence, Eq. (3.55) is a solution provided that

$$-\omega^2 z_0 + i\gamma\omega z_0 + \omega_0^2 z_0 = \frac{F_0}{m} \tag{3.59}$$

which is an algebraic equation. The unknown, the parameter we must find to have the solution, is the complex quantity z_0. This is immediately found to be

$$z_0 = \frac{F_0/m}{\omega_0^2 - \omega^2 + i\gamma\omega}. \tag{3.60}$$

We see that the solution is completely determined by the characteristics of the oscillator, ω_0 and γ and of the applied force, F_0 and ω. It does not depend on the initial conditions.

The particular solution of Eq. (3.57) is then

$$z(t) = \frac{F_0/m}{\omega_0^2 - \omega^2 + i\gamma\omega} e^{i\omega t}. \tag{3.61}$$

To have a particular solution of Eq. (3.55) we must now take the real part of this expression. To do that it is convenient to write z_0 in terms of its modulus B and its argument $-\delta$ (we shall soon see the reason for the negative sign)

$$z_0 = Be^{-i\delta}. \tag{3.62}$$

Equation (3.60) gives z_0 as a ratio. The modulus of a ratio is the ratio of the modulus of the nominator and the modulus of the denominator

$$B = \frac{F_0/m}{\sqrt{\left(\omega_0^2 - \omega^2\right)^2 + \gamma^2\omega^2}}. \tag{3.63}$$

The argument of the ratio is the difference between the argument of the nominator, which is null, and the argument of the denominator, and its opposite is

$$\delta = \arctan \frac{\gamma\omega}{\omega_0^2 - \omega^2}. \tag{3.64}$$

The particular solution of Eq. (3.57) is then

$$z(t) = Be^{i(\omega t - \delta)} \tag{3.65}$$

and, taking the real part, the particular solution of Eq. (3.55) is

$$x(t) = B\cos(\omega t - \delta). \tag{3.66}$$

Finally, the general solution of Eq. (3.55) is

$$x(t) = Ae^{-(\gamma/2)t}(\cos\omega_1 t + \phi) + B\cos(\omega t - \delta). \tag{3.67}$$

Let us now discuss the motion we have found. It is the sum of two terms. The first one represents a damped oscillation at the angular frequency ω_1 that is proper for the oscillator. The constants A and ϕ, depending on the conditions from which the motion started, appear in the first term. The second term depends on the applied force. The motion is under these conditions quite complicated. However, the amplitude of the first term decreases in time the faster the greater is γ. It diminishes by a factor of e in every time interval $2/\gamma$. After a few of such intervals, the first term has practically disappeared. Once this transient regime has gone, the regime of the motion is *stationary*. The *stationary oscillation* or forced oscillation is described by our particular solution Eq. (3.66), which is called a *stationary solution*. We write it as

$$x_s(t) = B\cos(\omega t - \delta). \tag{3.68}$$

We repeat that the stationary motion is a harmonic oscillation at the angular frequency of the force, not at the proper frequency of the oscillator. However, both the amplitude B and the quantity δ, which is not the initial phase but the phase delay of the displacement x relative to the instantaneous phase of the force, do depend on the characteristics of both the oscillator and the force as in Eqs. (3.63) and (3.64). An important phenomenon, the *resonance*, happens when the angular frequency of the force is near or equal to the proper angular frequency of the oscillator: the amplitude is very large and the phase delay varies very rapidly.

Figure 3.21 represents the amplitude of the forced oscillation as a function of the angular frequency B of the force. It has a maximum at the *resonance frequency*

$$\omega_R = \sqrt{\omega_0^2 - \gamma^2/2} \tag{3.69}$$

as one obtains with the usual methods finding the derivative of Eq. (3.63). Notice that ω_R is close but not exactly equal both to the angular frequency of the damped oscillations ω_1 and the proper angular frequency of the free oscillator ω_0. However for small damping, namely for $\gamma/\omega_0 \ll 1$, all of them become almost equal.

A simple way to observe the resonance phenomenon, and to understand the reason for the noun, is using two tuning forks. The tuning fork is an acoustic harmonic oscillator that vibrates at a specific frequency when set vibrating by

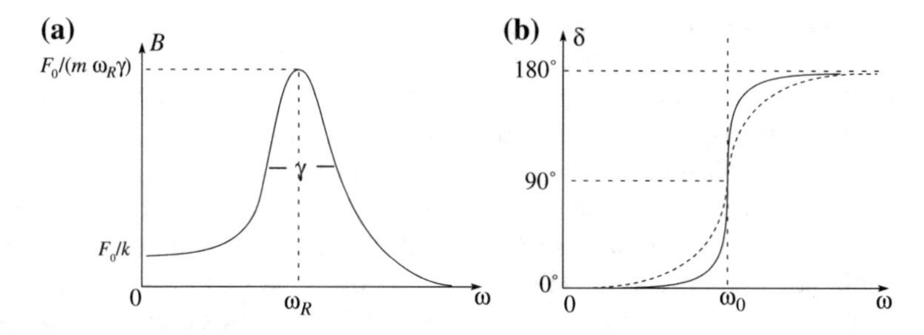

Fig. 3.21 Dependence on the applied force angular frequency of forced oscillations **a** amplitude, **b** phase delay, γ increases from continuous to *dashed curve*

striking it. It is made, like a two-pronged fork, with U-shaped prongs, called tines, and a stem of a metal, usually steel. The instrument is used to have a definite pitch, typically an A at 440 Hz, to tune the music instruments.

We strike a tine of one of the forks to have it vibrating, and we hear the sound, with the other one a few meters far. We then bring the latter nearby and stop the first fork by touching its tines. And we still hear the pitch. The second fork, that has the same proper frequency, resonated. The first fork had excited sound waves in the air, namely pressure oscillations at the frequency of its vibrations (the sound we hear). These pressure oscillations act as a periodic force on the second fork at its resonant frequency. We can double check that this is true as follows. We fix, with a locking screw near the top of one of the tines of the second fork, a small weight and repeat the experiment. This time we do not hear the second fork sound. Its proper frequency is now different and it is no longer in resonance with the first one.

Going back to Fig. 3.21a, we observe that calculations show that the full width of the resonance curve at half maximum (FWHM) is equal to γ and that the maximum is inversely proportional to γ. As a matter of fact, Eq. (3.63) immediately shows that the amplitude is infinite in the ideal case of $\gamma = 0$.

We discuss now the behavior of δ, the phase delay of the displacement relative to the force, given by Eq. (3.64) and shown in Fig. 3.21b. When the frequency of the force is small relative to the proper one, $\omega \ll \omega_0$, then $\delta \approx 0$, namely force and displacement are in phase. On the contrary, if the frequency of the force is much larger than the proper frequency, $\omega \gg \omega_0$, then $\delta \approx \pi$. We can easily understand the physical reasons for that, considering the relative importance of the different terms in Eq. (3.53). At low frequencies the accelerations are quite small and the applied force acts mainly against the elastic force $-kx$ and is consequently in phase with x. At high frequencies, as we have just seen, force and displacement are in phase opposition; when the mass is on the right, the force pushes to the left and vice versa. Accelerations are now very large and the dominant term is $-m(d^2x/dt^2)$, namely the inertia. Force acts mainly against acceleration and is in phase with it, which we know to be in phase opposition with displacement.

We also notice that our calculation shows the transition to be between the two just described regimes and takes place in a angular frequency interval of the order of γ. The less is the damping the more sudden is the transition. In resonance, as immediately seen in Eq. (3.64) $\delta = \pi/2$, namely the displacement is in quadrature with the force, hence it is in phase with the velocity. The power exerted by the force that is the product of the force and the velocity is a maximum.

The resonance phenomenon is very common in nature and in technology, not only in mechanics but also in electromagnetism, optics, atomic physics, nuclear and particle physics. In fact, all the systems oscillate harmonically when displaced close to a stable equilibrium configuration. We shall discuss this in Sect. 3.11. These oscillations take place at definite frequencies characteristic of the system. Engines, for example, have always a rotating part. Irregularities in their structures, even if small, may produce periodic stresses of an axis and of the support structures at the frequency of engine rotation. When this is varied and reaches one of the resonance frequencies of the system (there may be more than one) the amplitude of the vibration may become very large and, if the damping is small, even destroy the engine, if it is not properly designed.

We now consider the energy stored in the oscillator when in its stationary motion, Eq. (3.68). It is the sum of the kinetic and potential energies

$$U_{\text{tot}} = \frac{1}{2}kx_s^2 + \frac{1}{2}m\left(\frac{dx_s}{dt}\right)^2 = \frac{1}{2}mB^2\left[\omega_0^2\cos^2(\omega t - \delta) + \omega^2\cos^2(\omega t - \delta)\right]. \quad (3.70)$$

The expression is similar to what we found for the free oscillator. However, the two terms are now proportional one to ω_0^2 and one to ω^2 while for the free oscillator both were proportional to ω_0^2 and the energy was constant in time. Now the total energy varies periodically. This is because the power delivered by the force is not equal at a single instant to the power dissipated by the viscous force, while their averages on a period are equal. The instantaneous balance exists, however, in resonance, when $\omega^2 = \omega_0^2$ and the total energy

$$U_{\text{tot}} = \frac{1}{2}mB^2\omega_0^2 \quad (3.71)$$

is constant.

3.10 Energy Diagrams in One Dimension

In our previous discussion, the role of the force has been the principal one, while that of the potential energy was somewhat secondary. However, when in a more advanced study of mechanics and in other fields of physics, the potential energy has a central role. We have studied the problem: given a conservative force, find its potential energy. We now consider the inverse problem: knowing the potential energy, find an expression of the force.

For simplicity we consider the motion in one dimension only. The point P moves on a line, which we take as the x-axis. Suppose we have only one force acting on the point and call F_x its x component. Suppose the point moves from x_1 to x_2. In general, the knowledge of x_1 and x_2 is not sufficient to know the work done, but we also need to know the path taken. The point P might have gone directly from x_1 to x_2, or have moved in the opposite direction and after a while have come back, etc. For example, if the force is friction, its work is proportional to the total length of the path. However, if the force is conservative, as we shall assume, its work depends only on x_1 and x_2 by definition. For example the force might be an elastic force, or the weight of the point. In this case we have

$$U_p(x_2) - U_p(x_1) = -\int_{x_1}^{x_2} F_x dx. \tag{3.72}$$

We can fix the arbitrary additive constant by choosing a position x_0 in which the potential energy is null by definition $U_p(x_0) = 0$ and write

$$U_p(x) = -\int_{x_0}^{x} F_x(x) dx. \tag{3.73}$$

We now want to invert Eq. (3.73). To do that we take the derivative of both its members, immediately obtaining

$$F_x(x) = -\frac{dU_p(x)}{dx}. \tag{3.74}$$

In one dimension, the force is the opposite of the derivative of its potential energy with respect to the position. For example, the potential energy of the weight (x is vertical upwards) is $U_p(x) = mgx$ and the corresponding force, by derivation, is the one we know $F_x = -mg$, the elastic potential energy is $U_p(x) = -kx^2/2$ and, by derivation, the force is $F_x = -kx$.

Equation (3.74) can be written as

$$dW = F_x(x)dx = -dU_p(x), \tag{3.75}$$

which shows that the elementary work of a conservative force is the differential of a function, the opposite of the potential energy.

Suppose now that the potential energy $U_p(x)$ of the force $F_x(x)$ acting on our point P (in one dimension) to be the function shown in Fig. 3.22. The study of this type of diagram, called *energy diagrams*, is often useful to understand, even if in a semi-quantitative way, the possible types of motion of the system.

We start from the *equilibrium* positions of point P. A position, more generally a state, is said to be of equilibrium when, if the system was abandoned in that position

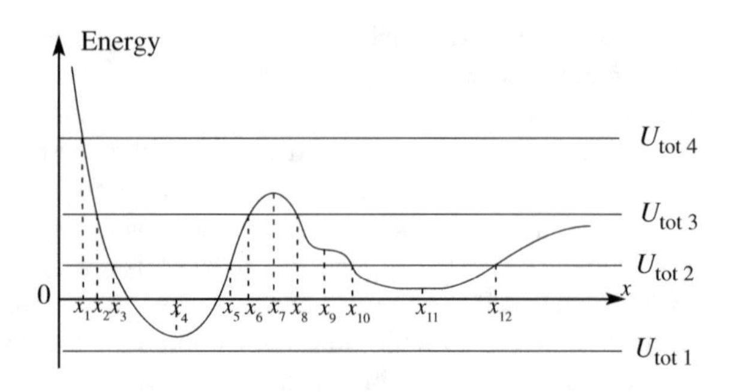

Fig. 3.22 The energy diagram example discussed in the text

with null velocity, it remains there indefinitely. This means that in these positions the force is zero. We can recognize immediately these positions on the diagram as those in which the derivative, i.e. the slope of the curve, is zero, namely where the curve has a maximum, a minimum or a flex, x_4, x_7, x_9, x_{11} in the figure. However, in practice we can never position the body exactly in a position and if we try to do that in a maximum or in a flex, the body will run away. As a matter of fact, there are three types of equilibrium states. To be general (for the material point) we define them in three dimensions.

1. *Stable equilibrium*. A position of a material point is of stable equilibrium if, when it is removed in whatever direction by an infinitesimal distance, the resultant of the forces tends to bring it back towards the equilibrium position (restoring force).
2. *Unstable equilibrium*. A position of a material point is of unstable equilibrium if at least a direction exists such that, when the point is moved in that direction by an infinitesimal distance, the resultant of the forces tends to bring it further away from the equilibrium position.
3. *Indifferent* or *neutral equilibrium*. If the point is removed by an infinitesimal distance in any direction, the point remains there. In other words the equilibrium position is surrounded by other equilibrium positions.

Going back to Fig. 3.10, in one dimension, the position x_4, where the potential energy has a relative minimum, is of stable equilibrium. Indeed, if we move a small distance on the left, the force $-dU_p/dx$ is positive, hence in the direction towards x_4. On the contrary, if we move to the right the force, $-dU_p/dx$, is negative, hence directed to the left.

The position at a maximum, like x_7 in the figure, is of unstable equilibrium. If we move a bit on the left the force is to the left ($-dU_p/dx < 0$), if we move to the right, the force is to the right. In both cases the force tends to pull the point farther from equilibrium.

In fact, one of these conditions is enough to make the equilibrium unstable. This happen in the positions of the flexes, like x_9 in the figure. Moving the point to the left, the force is restoring, but moving it to the right the force is of removal.

Consider finally the position x_{11}. It is on a non-null segment on which $dU_p/dx = 0$, namely it is surrounded by other equilibrium positions and the equilibrium is neutral.

We warn the reader that the curve representing U_p suggests a ball moving on hills and valleys, namely to think of the ordinate axis as the height. This is indeed the case for weight, but not for other forces. Intuition should be controlled.

Three other pieces of information can be extracted from the diagram. For every value of potential energy, the material point may have different values of kinetic energy. The sum of the two

$$U_{\text{tot}} = U_p + U_k = \text{constant} \tag{3.76}$$

is constant during its motion. In Fig. 3.22 we have drawn, as examples, four different values of the total energy. In any case, the kinetic energy is the difference between total energy and potential energy, the distance from the line and the curve.

We now consider that, while total and potential energies may be positive or negative (or zero), the kinetic energy cannot be negative. Consequently, if the total energy is too low, as is U_{tot1} in the figure, for which in every point the kinetic energy would be negative, it is not possible for our system. The total energy cannot be less than the absolute minimum (the deepest in the figure) of the potential energy.

If the total energy is somewhat larger, as is U_{tot2} in the figure, the motion of the point can happen only in two regions, between x_3 and x_5 in and between x_{10} and x_{12}. In this example, the two regions are separated by a non-reachable interval. If a point starts moving with total energy U_{tot2} in one of the two regions, it cannot leave it. One might think that the point might jump from one allowed interval to another, because total energy would remain the same. But this cannot happen because between the two regions there is a forbidden one, in which the total energy would be different or kinetic energy would be negative. However, this type of phenomenon happens in quantum mechanics, in atomic nuclei for example, which is called "tunnelling" even if no tunnel exists, because it looks as though the system would cross under the barrier in a tunnel.

We can learn something more from the diagram in Fig. 3.22. Suppose our material point with total energy U_{tot2} to be at a certain instant in x_3. In this point the line of U_{tot2} intersects the curve of the potential energy. All the energy of the point is potential, its kinetic energy is zero. The point has zero velocity. However, the point does not remain still, because it is not in an equilibrium position. The force $-dU_p/dx$ is positive and accelerates the point in the direction of increasing x, with increasing kinetic energy (the distance on the diagram from the line to the curve increases). The force, and consequently the acceleration, slows down as the slope of the potential energy curve diminishes. They become zero when the point reaches the minimum in x_4. The motion does not stop there, it continues now decelerated (the slope is positive, hence the force is negative, opposite to x). We read from the diagram that the kinetic energy is now diminishing. It does so up to zero when the point is in x_5. This point is reached with zero velocity.

What does happen afterwards? The force $-dU_p/dx$ acting on the point is now negative, namely in the direction opposite to x. The point restarts its motion going back. In conclusion, the motion is an oscillation back and forth between x_3 and x_5. The motion is periodic, but generally not harmonic. We shall see in the next section under what conditions it is so. Clearly also the motion with total energy U_{tot2} between x_{10} and x_{12} is periodic.

Consider now a larger value of the total energy, U_{tot3} in the figure. There are two possible motions. The first one is bounded, a periodic oscillation between x_2 and x_6 similar to that we have just discussed. The second motion is, for example, the motion of a point approaching from infinite distance on the right. Initially it accelerates, then, once x_{11} is passed, decelerated up to stop in x_8. Here it bunches back going through in inverted order all the phases moving farther and never to come back again. The motion is unbounded.

At still higher values of the total energy, as U_{tot4}, no periodic motion is possible, but only unbound motions. A particle coming from far away, once reached x_1 stops and bunches back to infinity.

3.11 Energy Diagrams for Relevant Forces

In this section we shall use the methods described in the previous section to relevant types of motion: the oscillation under the action of a real elastic force, the oscillation of a pendulum and the oscillations of a diatomic molecule.

Let us start with a perfectly elastic spring on the x-axis, which has its origin in the rest position of the spring. We know that the force it exerts on a point in the generic position x is $F_x = -kx$ and the potential energy is $U_p = -kx^2/2$. In practice, as we saw in Sect. 3.1, no spring is perfectly elastic. For large deformations the dependence of the force on the deformation is no longer linear (see Fig. 3.2) or, in other words, the curve of the potential energy is not a parabola, but as shown in Fig. 3.23.

The equilibrium position $x = 0$ corresponds to the minimum of the potential energy. If the material point is abandoned outside the equilibrium position, its oscillations are periodic. They are also harmonic if the displacement is not too big,

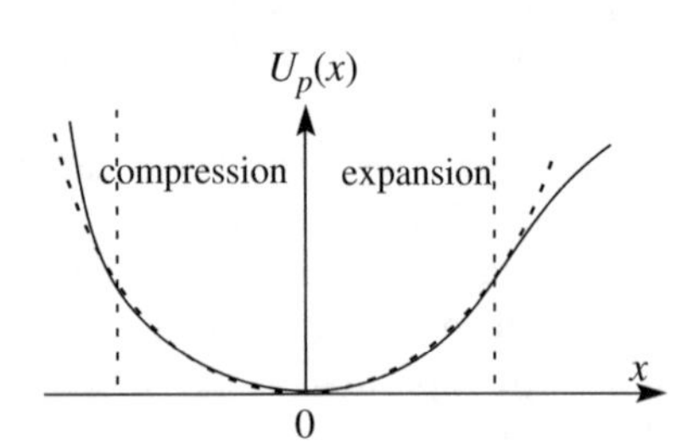

Fig. 3.23 The potential energy versus deformation for an ideal (*dashed curve*) and real (*continuous curve*) spring

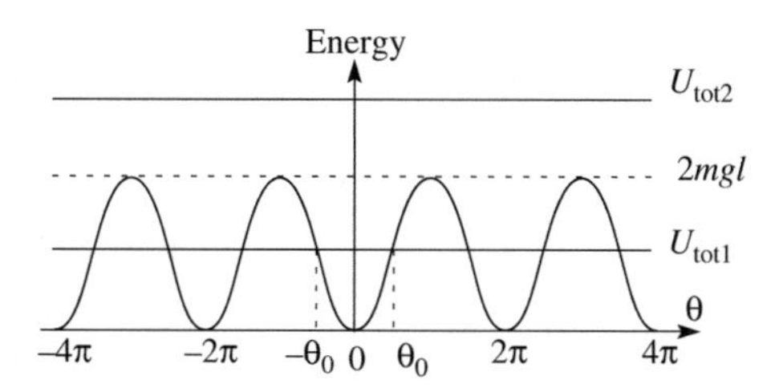

Fig. 3.24 The energy diagram for a pendulum

say within the vertical dashed lines that mark the region in which the potential energy curve is at a good approximation a parabola.

Consider now a simple pendulum. We attach a small sphere of mass m to a wire of length l and negligible mass with the other extreme fixed. If the sphere is abandoned without velocity from out of equilibrium it will move on the arc of a circle of radius l. The position can be measured with one variable, the angle θ between the wire and the vertical. The potential energy is

$$U_p(\theta) = mgl(1 - \cos\theta) = 2mgl\sin^2(\theta/2). \tag{3.77}$$

Figure 3.24 shows this function. The variable θ can take any value from $-\infty$ to $+\infty$. However, the function is periodic and all the possible physical positions are already described by the values of θ between $-\pi$ and π.

In the figure we have taken the minimum potential energy as the zero of the energy scale. We can see that the motion can be unbounded (in angle), if the total energy is larger than $2mgl$, which is the maximum potential energy, as U_{tot2} in the figure. The angle θ grows indefinitely in time, the pendulum rotates on the circle of radius l (in practice the wire would tangle around the nail). The velocity varies from a minimum when the ball is in its highest position ($\theta = \pi$, 3π, 5π,..), to a maximum when it passes through the equilibrium position ($\theta = 0$, 2π, 3π,..).

If $U_{tot} < 2mgl$, as for example U_{tot1} in the figure, the motion is limited. The ball oscillates between the angles $-\theta_0$ and $+\theta_0$. In general however, the motion is not harmonic, because the potential energy curve is not a parabola. If the oscillations are small, however, the curve is approximately parabolic, as shown in Fig. 3.25, and

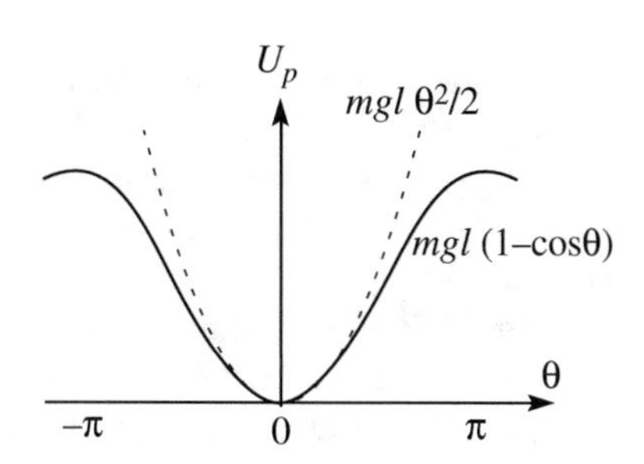

Fig. 3.25 The potential energy of a pendulum and its parabolic approximation

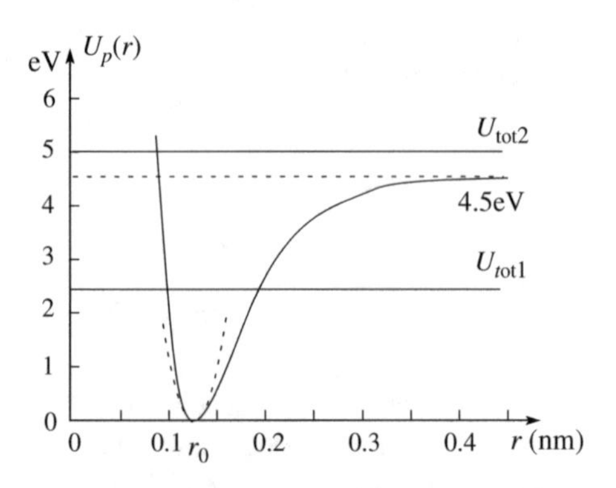

Fig. 3.26 The potential energy of the HCl molecule

the motion is harmonic. The same thing can be seen analytically. If we develop Eq. (3.77) in series and stop at the first term we obtain

$$U_p(\theta) = mgl\theta^2/2.$$

Notice that the approximation is quite good because the next term in the development, the term in θ^3 is null, hence the first neglected term is the one in θ^4.

The last example is the diatomic molecule. To be concrete we consider HCl. With a good approximation we can consider the two components as point-like. The atomic clouds of the two atoms keep the two nuclei at the stable equilibrium distance r_0. If the distance is different, a force appears, which tends to bring back the equilibrium. These forces, which are responsible for chemical bonds, are electromagnetic and of quantum nature. They are different from the van der Waals forces we considered in Sect. 3.3. Figure 3.26 shows the potential energy as a function of the distance between the nuclei of H and Cl. It is known as *Morse potential*. The curve has a minimum, corresponding to the equilibrium distance between the nuclei. The distances are of the order of the nanometers. The energy is given in *electronvolt* (eV), which is a practical unit for atomic energies. An electronvolt is the kinetic energy gained by an electron falling under the electric potential difference of one volt. Its value is in round figures

$$1 \text{ eV} = 1.6 \times 10^{-19}\,\text{J}. \tag{3.78}$$

Suppose now we communicate a certain energy to the system, for example by striking with another molecule. Also in this case there are two types of motion. If the energy given to the molecule is large enough, as U_{tot2} in the figure, the motion is unbounded. The two ions separate and the molecule dissociates. If the energy is smaller, like U_{tot1}, the molecule remains bound and performs a periodic oscillation. As seen in Fig. 3.26, the potential energy curve is not symmetric about its minimum. However, if the total energy is small enough and the curve can be approximated with a parabola, the oscillation is almost harmonic.

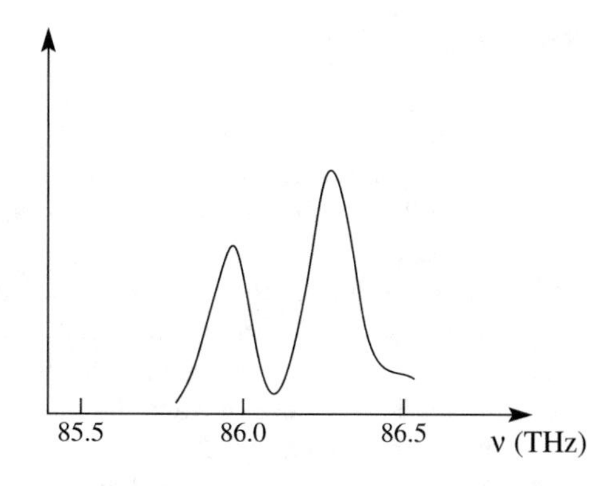

Fig. 3.27 Absorption probability for HCl molecules versus frequency

We also observe that the potential energy curve grows more rapidly at energies smaller than the minimum than at higher ones. In other words, the restoring force is larger than it would be if elastic for compression, smaller for expansion. Macroscopically this translates in the asymmetry of the deviations from the behavior described by Fig. 3.2.

The resonance phenomenon is present also in the molecular oscillators, at quite high frequencies, of the order of 10^{13} Hz (10 THz). These are the frequencies of the electromagnetic waves in the infrared. Imagine doing the following experiment. We radiate a container with transparent walls containing a HCl gas with an infrared radiation, of which we can vary the frequency and we measure the intensity of the radiation transmitted by the gas in correspondence. Taking the ratio between the transmitted and the incident intensities we have the quantity of radiation absorbed by the gas as a function of frequency. We obtain Fig. 3.27. It is a resonant curve, because in resonance much more energy is transferred from the radiation to the molecular oscillators than for other frequencies. However, two peaks, not one, are observed. The reason is that Chlorine has two isotopes, ^{35}Cl and ^{37}Cl of atomic masses 35 and 37 respectively. The two proper frequencies squared ω_0^2 are different, as the forces are equal, the masses different, in the two cases. To be complete, in the spectrum several doublets like the one in Fig. 3.27 are present. This is because quantum oscillators have several, rather than a single one, proper oscillation frequencies.

From the examples in this section we can draw an important conclusion. The physical systems are found naturally in their (or one of their) stable equilibrium state(s) corresponding to the minimum (or one of them) of the potential energy. A small perturbation can take them out of equilibrium. The potential energy curve is not in general a parabola. However, if the displacement from equilibrium is small enough it can be well approximated by a parabola. In these conditions the system oscillates harmonically. Consequently, the largest fraction of natural oscillations are indeed harmonic.

Problems

3.1. Consider the oscillator of Fig. 3.5 with $m = 0.3$ kg and $k = 30$ N/m. Calculate the proper angular frequency, the period and the frequency of its oscillation. Write the equation of motion for an initial displacement, with zero velocity, of 4 cm.

3.2. Show that the amplitude of a damped oscillator is halved in a time of $1.39/\gamma$. How much is the energy variation in this time?

3.3. A damped oscillator has the proper angular frequency $\omega_0 = 300$ rad/s and $\omega_0/\gamma = 50$. Calculate the angular frequency of the free oscillations ω_1 and the resonance frequency ω_R. Compare the values.

3.4. We build a mechanical oscillator as in Fig. 3.5. We can use a body with a certain mass and two identical springs. We separately attach to the mass: (a) one spring, (b) two springs in series, (c) two springs in parallel. What are the ratios of the proper angular frequencies in cases (b) and (c) to case (a)?

3.5. A perfectly elastic spring stretches 10 cm when it hangs a mass of 10 kg. (a) what is the value of the spring constant? (b) Lay the spring and the mass on a horizontal plane without friction. Move the mass so as to stretch the spring 5 cm and let it go at $t = 0$. Write the equation of motion if (a) the initial velocity is zero, (b) the initial velocity is 1 m/s in the direction of increasing x.

3.6. Consider a forced oscillator vibrating at the angular frequency ω in its stationary regime. Show that its energy is mainly potential when $\omega \gg \omega_0$, mainly when $\omega \ll \omega_0$, exactly half and half when $\omega = \omega_0$.

3.7. We know the oscillation amplitudes of the displacement and the velocity of a harmonic oscillator. How can we know the angular frequency?

3.8. A force with sinusoidal dependence on time acting on an oscillator makes it oscillate, in a stationary regime, with amplitude $A_1 = 20$ mm. A second force, acting alone on the same oscillator, makes it oscillate in the stationary motion with amplitude $A_2 = 40$ mm. If both forces act together, the amplitude in the stationary motion is $A = 30$ mm. What is the phase difference between the forces?

3.9. A car of mass $m = 1000$ kg travels horizontally at 100 km/h. Suddenly an obstacle appears at 100 m. The driver brakes immediately (neglecting the reaction time, which is 1–2 s) and stops 10 m before the obstacle. Assuming the force to have been constant how much was the magnitude of the force? If the road were downhill with a slope of 15 % at which speed the car would have hit the obstacle?

3.10. A block of mass $m = 1$ kg lies on a horizontal plane attached to a rope, the other extreme of which is fixed to the point O of the plane. The block under these constraints is moving on a circle of center O and radius $l = 1$ m and velocity at the considered instant $\upsilon = 2$ m/s. The coefficient of kinetic friction between block and plane is $\mu_d = 0.4$. What is the magnitude of the resultant of the forces in that instant? What is the direction relative to velocity?

Fig. 3.28 Body sliding on a wheel, problem 3.13

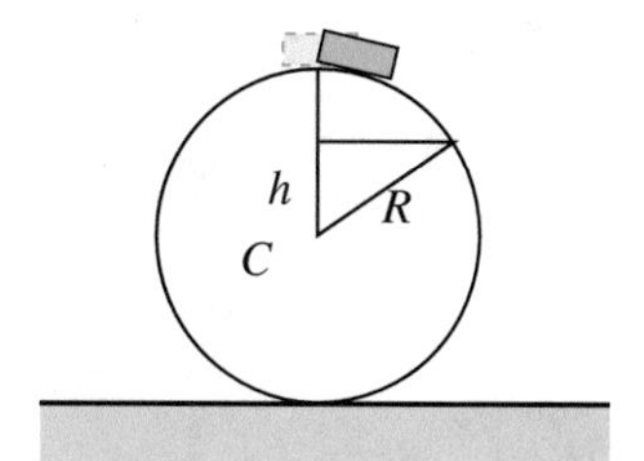

3.11. A sphere of radius a moving with velocity υ acts in air with a drag force R. The latter depends on the radius as $R(a, \upsilon) = C_1 a\upsilon + C_2 a^2 \upsilon^2$ with $C_1 = 3.1 \times 10^{-4}$ kg m^{-1}s^{-1} and $C_2 = 0.87$ kg m^{-3}. Consider a raindrop falling starting from null velocity. The drop moves under the action of its weight and the resistance. When the velocity is small, the weight is larger than the resistance and the drop accelerates. However, at a certain velocity the two forces become equal and opposite and the velocity becomes constant. It is called limit velocity. Calculate the limit velocities for a drop of radius $a = 0.1$ mm and for one of radius $a = 1$ mm. In both cases assume the second term in the above expression can be neglected. Verify a posteriori if the assumption is reasonable. For a drop of radius $a = 1$ mm, now assume that the first term is negligible and again verify a posteriori if the hypothesis was reasonable.

3.12. A body of mass m is attached to an extreme of a rope of length R. The other extreme is fixed. The body rotates in a vertical plane. a) Find the expression of the tension T of the rope when the body passes, with velocity υ, in the highest point of the trajectory. What is the agent of the centripetal force in this point? Study the meaning of the found expression for decreasing values of υ. What does $T > 0$, $T = 0$ and $T < 0$ mean? What does happen when the velocity is such that $T = 0$? Repeat for the lowest point.

3.13. A small body starts sliding, with negligible initial velocity, on a frictionless wheel starting from its highest point, as in Fig. 3.28. The radius of the wheel is R. (a) at what height h, measured from the center of the wheel does the body detach and fall freely? (b) how would the result change on the moon?

Chapter 4
Gravitation

The first two books of Newton's Principia establish the mechanics laws for phenomena on the surface of earth. The third book, titled "The system of the word", applies the same laws to interpret the motions of extra-terrestrial bodies. The grand unification of terrestrial and heavenly physics, started by G. Galilei and J. Kepler, was completed. In the introduction to the volume, I. Newton wrote

> It was the ancient opinion of not a few, in the earliest ages of philosophy, that the fixed stars stood immovable in the highest parts of the world; that under the fixed stars the planets were carried about the sun; that the earth, as one of the planets, described an annual course about the sun, while by a diurnal motion it was in the meantime revolved about its own axis; and the sun, as the common fire which served to warm the whole, was fixed at the centre of the universe.
>
> This was the philosophy taught of old by Phylolaus, Aristarchus of Samos, Plato in his riper years, and the whole sect of the Pythagoreans; and this was the judgment of Anaximander, more ancient still …

A few lines below, after having mentioned the contributions of the Romans and of the Egyptians, he added

> It is not to be denied that Anaxagoras, Democritus, and others, did now and then start up, who would have it that the earth possessed the centre of the world, and that the stars of all sorts were revolved towards the west about the earth quiescent in the centre, some at a swifter, others at a slower rate.
>
> However, it was agreed on both sides that motions of the celestial bodies were performed in spaces altogether free and void of resistance. The whim of solid orbs was of a later date, introduced by Eudoxus, Calippus and Aristotle; when the ancient philosophy began to decline, and to give the place to the new prevailing fictions of the Greeks

Observation of the night sky, with its moon and countless stars has, since ancient times, never failed to astonish humanity throughout the world. Along with astonishment, a deep curiosity aroused about the nature of these heavenly bodies and the

---

The original version of this chapter was revised: Belated corrections have been incorporated. The erratum to this chapter is available at https://doi.org/10.1007/978-3-319-29257-1_9

A. Bettini, *A Course in Classical Physics 1—Mechanics*,
Undergraduate Lecture Notes in Physics, DOI 10.1007/978-3-319-29257-1_4

reasons of their existence. Along with the myth, truly scientific activities developed in time in several cultures. Since the second millennium B.C. mankind accurately and systematically registered the positions of the stars in the sky. However, the mystical charm of the starry sky contributed to the suggestion, in several periods, that the motion of the heavenly bodies should have obeyed symmetry rules of a higher, often divine, order. This is the case of the solid orbits of Aristotle, mentioned by Newton, and of the uniform circular motions of Ptolemy and Copernicus. Gradually, beginning in the Renaissance, there developed an inquiry leading to establishment of the physics laws that rules the motions in the cosmos.

In this chapter we shall study universal gravitation, the physical law that describes motions of the planets and their satellites, of the solar system and of the galaxies and their clusters as well as the motions of all bodies up to the boundaries of the Universe. We might start from the Newton law of gravitation and analyze its consequences. We prefer to reach it following, albeit briefly, the historical process that led to discovery of the law. Indeed, the path leading to these discoveries has never been straight, but rather tortuous, through lateral, sometimes wrong, paths, with successes and failures, laborious in any case. Universal gravitation is one of the grand theories built by several scientists. Knowledge, even if in a summary, of the historical roots of the process adds to the depth of the physics laws. As a matter of fact, physics can be understood even without knowing its history. The historical part of the chapter should be considered as a, hopefully interesting, reading adventure. The parts to remember are the laws and their experimental proofs.

Figure 4.1 shows the lifetime spans of the great authors of the development of mechanics and astrophysics from the XVI to the XVIII century, the period of the construction of a vast theoretical edifice.

In Sect. 4.1 we shall briefly describe the geocentric and heliocentric models. In Sect. 4.2 we shall see how the periods and diameters of the orbits of the planets were measured from Greek civilization to the Renaissance. We shall then see the fundamental contribution of Tycho Brahe with his systematic measurements, with precision increased by an order of magnitude, of the positions of the planets and how Johannes Kepler, based on those measurements, discovered that the orbits of the planets are ellipses, rather than circles, and established his three laws. The Kepler laws are very important but still phenomenological. The dynamical theory was later established by Newton, as discussed from Sects. 4.4 to 4.6.

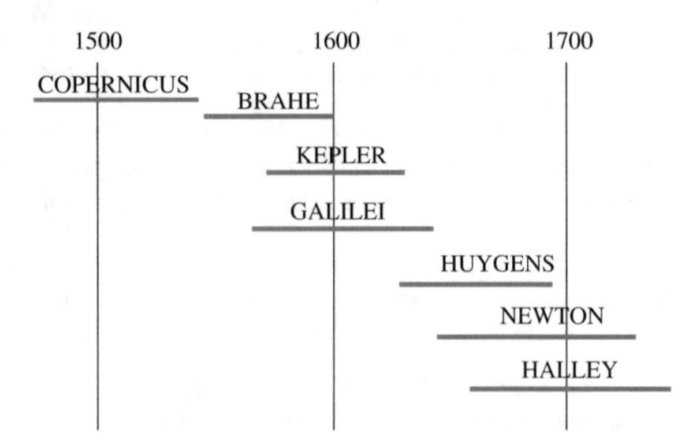

Fig. 4.1 Life spans of the principal contributors

The Newton law is a simple and symmetric mathematical expression. In the fundamental physical laws the harmony of the world takes on an abstract character, appearing as the simplicity of the mathematical expression that is able to describe an enormous quantity of different phenomena, which, when that law was not known, appeared to be uncorrelated.

The Newton law contains a universal constant, which is the same on earth and in the Cosmos. In Sect. 4.7 we see how it was measured in the laboratory.

The gravitational force acts between bodies that are not in contact, rather they may be very far from each other. The force acts through a vacuum. This is also the nature of all the other fundamental forces, in particular of the electromagnetic one. For all of them the concept of field of force is important. The source of the force, for example the sun, creates a field of force in all the space around it. The field then acts on every massive object as a force. We shall see that in Sect. 4.8.

In Sect. 4.9 we shall go back to history and show how G. Galilei discovered the satellites of Jupiter, discussing some of his data.

In Sect. 4.10 we shall see how the Newton law describes the motions of cosmic objects of the most different sizes and distances and how it shows that the nature of the largest fraction of matter is still unknown. It is called dark matter.

In the first part of the chapter we assumed for simplicity the orbits of the planets to be circular. In the final three sections, we relax this assumption and discuss fully the problem of elliptic orbits. This is known as the "direct Kepler problem": knowing that the orbit is an ellipse with the centre of force in one of the foci, find the force. We shall do that first using modern calculus formalism (Sect. 4.10), then, in (Sect. 4.11) we shall read and explain, line by line, the original demonstration of Newton, as a beautiful example of his thought. In the last section, we shall consider the energy of a body in the gravitational field of a central body.

4.1 The Orbits of the Planets

Observational astronomy is a very ancient science, dating back to the most ancient civilizations to the third millennium B.C. The varying celestial co-ordinates of the stars, of the moon and of the planets were accurately and systematically measured and registered. The problem has always been to understand what the data meant. Particularly complicated are the motions of the planets, which owe their name to the Greek word for tramp.

The Heavenly bodies, including the planets, are so far away that their distances could not be measured in ancient times, with the exception of the moon. What we measure, for each body, is the direction at which it appears as a function of the time of the observation. The directions are given by two angular co-ordinates. However, it was natural to think of the stars as points on a sphere of very large, but arbitrary, radius, which was called a *celestial sphere*. Its center is on the earth and its axis coincides with the rotation axis of the earth. The circle cut on the celestial sphere by the plane through the earth equator is called the *celestial equator*.

The annual motion of the Heavenly bodies appears to an observer on earth as a rotation around a common center, the earth. The stars, as different from the planets, do move on the celestial sphere, but keeping all the distances between them invariable. For this reason they have been called fixed. We know now that the stars are not fixed at all and that they are at very different distances. They appear to be fixed because the distances are enormous. The most striking (apparent) motion is the diurnal one due to the rotation of the earth on its axis. A further apparent motion of the fixed stars, due to the revolution of the earth around the sun, is a rotation with a period called the *sidereal year*. The sidereal year is also the time taken by the sun, in its apparent motion, to return to the same position relative to the fixed stars. As such it is almost, but not exactly, equal to our common year. As we shall see, the moon and the planets have more complicated apparent movements (which are combinations of their own and of earth).

As for distances, Aristarchus of Samos (310–230 BC) developed a brilliant method to extract the distances from earth to the moon and to the sun by angular measurement. Due to the insufficient resolution in the measurement of the angles, he evaluated that the distance of the moon is about 20 times the radius of the earth, instead of 60 as it is, and that the sun is 20 times farther than the moon, rather than about 400 as it is. This was enough to conclude that, considering the moon and the sun to have the same apparent size, the real size of the sun had to be enormous. Aristarchus concluded that his finding confirmed that the sun must be the center of the system. He then found the correct order of the distances of the five planets around the sun that are visible with the naked eye around the sun, which was standing still at the center of the system. However, at least to our knowledge, he did not fully develop a quantitative model of the planetary system.

A powerful quantitative model was developed three centuries later by Claudius Ptolemy (90–168 AD), who lived at Alexandria in Egypt. By that time the idea that the heavenly bodies had to move with constant (in magnitude) velocity on circles, or combinations of circles, being brought around on a system of solid spheres, had become dominant, as Newton recalls with the words quoted in the introduction to this chapter. His book, originally written in Greek and titled "Mathematical syntaxes" came to us through its Arabic translation and is universally known as *Almagest*.

The "planets" were seven: sun, moon, Mercury, Venus, Mars, Jupiter and Saturn. Figure 4.2a shows the basis of the model. Earth is at rest at the center of the system. The sun describes a circle around the earth. The path of the sun on the celestial sphere, through the fixed stars, is the *ecliptic*. The motion of each planet, like P in the figure, is more complicated. In a first approximation it is described by a circular uniform motion around earth performed by the point C and by a second circular uniform motion of the planet itself around C. The former circle is the *deferent* the latter the *epicycle*. The two motions are (approximately) in the same plane and their combination is a curve, called an *epicycloid*, shown in Fig. 4.2a. Clearly, both the deferent and the epicycle are different for different planets. The observed trajectory of the planet is the projection of its epicycloid on the celestial sphere, taking into account the angle between its orbit and the plane of the celestial equator, which is also somewhat different for different planets. Notice that for the

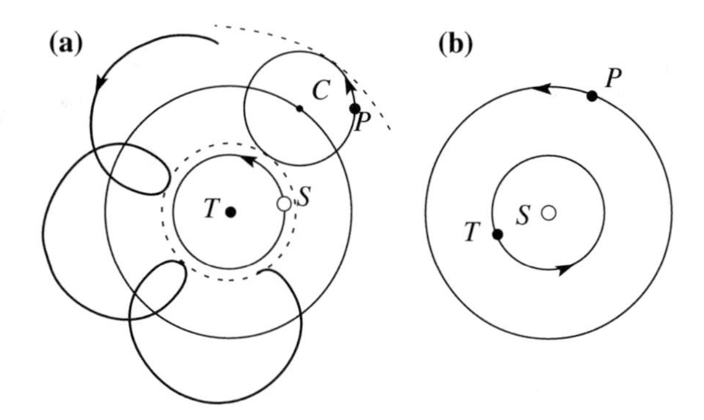

Fig. 4.2 Motion of an external planet relative to **a** the earth, **b** the sun. Figures are approximate

largest fraction of its period the planet moves forward, from West to East relative to the fixed stars. However, in correspondence with the smaller loops of the epicycloid it moves, for some time, backwards. This is in accord with observations.

Ptolemy calculated, on the basis of the available measurements, results of centuries of observations, the radii of the primary and secondary circles (with the solar orbit radius as unit) and the corresponding periods. He found however, that this relatively simple model did not work, namely did not explain all the data. To make it work he added two features.

1. The primary circle (deferent) of each planet is not centered exactly on earth but in a point not very far from her and different from planet to planet. It is called *equant*, because it makes the motion on the deferent uniform. We now know that the equant is the empty focus of the elliptical orbit of the planet. We shall understand in Sect. 4.3 how it works.

2. A number of tertiary and quaternary circles, all called epicycles.

The model of Ptolemy, though even not particularly simple, was able to reasonably explain all the observational facts and would remain such till the accuracy in the measurements of the planets positions will be improved by an order of magnitude by Tycho Brahe (1546–1601).

We can notice that the period of the deferent in two cases (Mercury and Venus) and of the (first) epicycle in the other three cases (Mars, Jupiter and Saturn) are all equal to a sidereal year. We know that the orbit of the first two planets is smaller, the orbit of the other three is larger than the orbit of the earth (Fig. 4.2). Ptolemy did not notice this feature. In his model all the circles are independent. This important discovery is due to Nicolaus Copernicus (1473–1543).

Another feature that is not explained by the model is why both Mercury and Venus never depart much from the sun. The maximum angle between Mercury and the sun is $\theta_m = 22.5°$ and Venus and the sun $\theta_m = 46°$.

Let us now go back for simplicity to the model with only one primary and one secondary circle. Let us change our reference frame by choosing the sun at rest at its center. We assume that the earth moves uniformly on a circle around the sun with the radius of the epicycle of the planet and that the planet *P* moves uniformly on the circle centered on the sun and radius equal to the deferent radius, as in Fig. 4.2b. The relative positions of earth and planet are exactly the same as before, but the

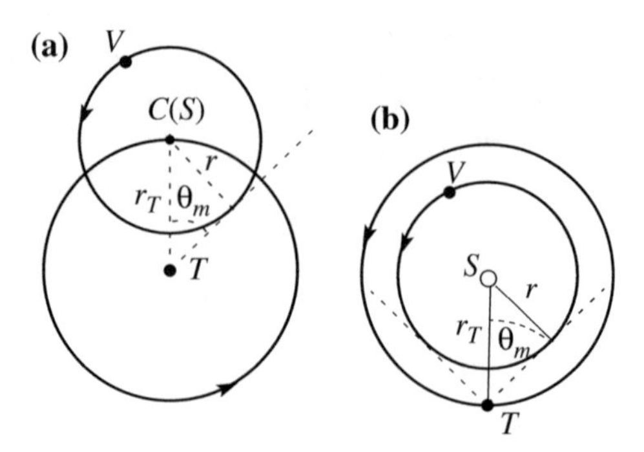

Fig. 4.3 The orbit of an internal planet, **a** viewed from earth, **b** viewed by the sun

description is now logically simpler. In addition, the reason for which the period on the epicycle is the sidereal year becomes obvious. Figure 4.2 represents an external planet. The reader can easily verify that an analogue explanation works for the internal planets by simply exchanging the roles of primary and secondary circles.

In the heliocentric frame the reason why Mercury and Venus, which have orbits smaller than earth around the sun, cannot be very far from the sun when viewed by earth is also clear, as shown in Fig. 4.3. This argument was already known to the Greeks, in particular to Aristarchus.

The heliocentric description we have sketched, in its modern form, is due to Nicolaus Copernicus. He gave a preliminary version of his model in the "*Commnetariolus*" distributed privately to his friends in 1514, and the final one in "*De Revolutionibus Orbium Caelestium*" published in 1543, the year of his death. Differently from Aristarchus, Copernicus developed a full mathematical model able to explain the observational facts.

The Copernicus model, as we have presented it so far, looks much simpler than the Ptolemy model. One can then ask why it took so long to be accepted. The reason is that, as for Ptolemy before him, such a simple model does not work. The main reason was that Copernicus still believed that the orbits had to be circles or combinations of circles and the motions on them uniform. The reason of the belief was dogmatic, rather than scientific: the heavenly bodies being the creation of God, their motion must be perfect. The bodies must be on a rotating sphere, because, in his words, the sphere in its rotation moves

> on itself through the same points, it expresses its form in the simplest body, in which it is impossible to find either a beginning or an end or distinguish the points from each other.

The consequence was that, to agree with the data, Copernicus, as long before him Ptolemy, had to introduce a rather large number of epicycles. Indeed, the Copernicus model, in the form he presented it, was not less arbitrary than the Ptolemy model.

4.2 The Periods of the Planets and the Radii of Their Orbits

As we have already mentioned, two planets, Mercury and Venus, in their motion as seen from earth never go far from the sun. The maximum angle between Mercury and the sun is $\theta_{\mathrm{m}} = 22.5°$ and Venus and the sun $\theta_{\mathrm{m}} = 46°$. The model of Copernicus allows us to calculate the radii of the orbits of these planets. Here the model shows its superiority to Ptolemy.

From Fig. 4.4, which is drawn for Venus, we have

$$r/r_E = \sin\theta_m. \tag{4.1}$$

Notice that the condition is for the ratio of the radius of the planet with the radius of the earth orbit. Indeed the latter is the natural unit in astronomical measurements and is called *astronomical unit* (*au*). To be precise the astronomical unit is the mean distance of the earth from the sun. We shall not discuss the different methods to measure r_E. We simply mention that the problem of the scales of the distances is a central one in astronomy.

The value of the astronomical unit was not known even to Kepler. He was able to determine a lower limit (on the basis of the parallax of Mars) as 1 au > 15 Gm. The first measurements were made at the beginning of the XVII century by Giovanni Domenico Cassini (1625–1712) and by Edmund Halley (1656–1742), who found values between 140 and 150 Gm.

The value known today is

$$1\ \mathrm{au} = 1.49597871 \times 10^{11}\ \mathrm{m} \simeq 149.6\ \mathrm{Gm}. \tag{4.2}$$

From the above values of θ_{m} we have for Mercury $r_M \approx 0.34$ au and for Venus $r_V \approx 0.72$ au.

For the external planets, the three known to Copernicus, the argument is similar, but now the radius of the orbit of the planet is larger than that of the earth. Figure 4.4 gives the geometry. The Copernicus interpretation is that the larger circle, the deferent, is the orbit of the planet, and the smaller one, the epicycle, is the

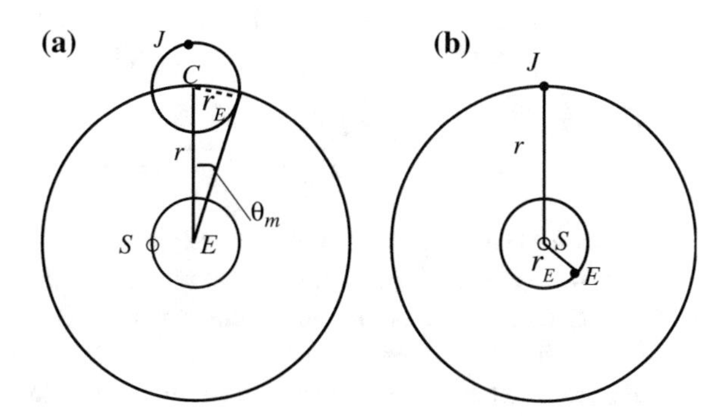

Fig. 4.4 The earth and the Jupiter orbits

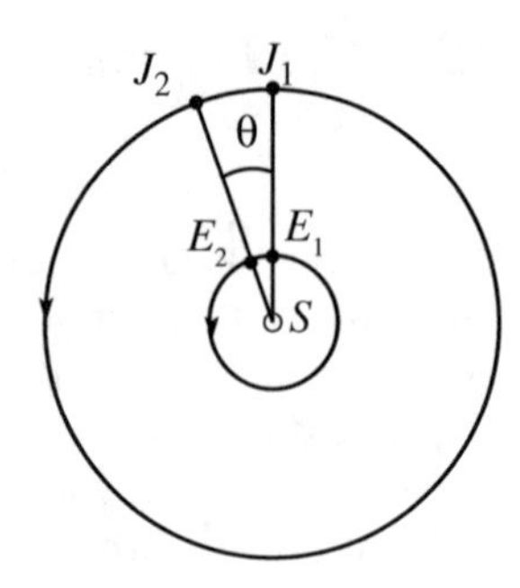

Fig. 4.5 Two consecutive transitions of Jupiter on the celestial meridian

orbit of earth. Consequently the angular diameter under which the latter is seen from earth is $2\theta_{\mathrm{m}}$. From the figure we see that

$$r_E/r = \sin\theta_m. \tag{4.3}$$

Already Ptolemy knew the angles for the three planets, $\theta_{\mathrm{m}} = 41°$ for Mars, $\theta_{\mathrm{m}} = 11°$ for Jupiter and $\theta_{\mathrm{m}} = 6°$ for Saturn. Equation (4.3) gives for the radii of their orbits $r_{Ma} \approx 1.5$ au for Mars, $r_J \approx 5.2$ au for Jupiter and $r_J \approx 9.5$ au for Saturn.

Let us see how to extract the periods from the observational data. For that we must take into account that the observations are done from a frame moving in the solar system. This problem is solved a little differently for the internal and for the external planet, as in the case of the radii of the orbits. For the sake of brevity we shall consider only one external planet, for example Jupiter.

Consider the two situations represented in Fig. 4.5. In both of them the relative positions of earth, sun and Jupiter is the same. It is also such, being the three bodies on the same line, to be easily and precisely recognized. This is done, for a given observer, by taking the date at which Jupiter crosses the celestial meridian at midnight. The celestial meridian is the projection of the local meridian on the *celestial sphere*.

The intervals between two consecutive recurrences of the phenomenon are all equal and called the *synodic period*. Consequently, we can average on several measurements and increase the precision. The synodic period of Jupiter is $\tau = 399$ d. In this period Jupiter travels through the angle θ (Fig. 4.5), the earth travels that plus a revolution, namely $360° + \theta$. The number of revolutions of Jupiter per unit time $n_J = 1/T_J$, where T_J is its period. Similarly for earth, $n_E = 1/T_E$. We can then write $n_E\tau = n_J\tau + 1$, that is $\tau/T_J = \tau/T_E + 1$, and also

$$T_J = \frac{T_E}{1 - T_E/\tau} = \frac{365}{1 - 365/399} = 11.8 \text{ year}.$$

Question Q4.1. Find the equivalent expression for an internal planet.

Table 4.1 gives the values of the orbit radii in astronomic units and of the periods of the first six planets as known to Copernicus and as it is today.

We see that the values known to Copernicus, in particular for the periods, were already close to the modern ones. We add that the values that can be extracted from

Table 4.1 Orbit radii and periods of the first six planets

Planet	Orbit radius (au)	Orbit radius (au)	Period	Period
	Copernicus	Modern	Copernicus	Modern
Mercury	0.376	0.387	87.97 day	87.97 day
Venus	0.719	0.723	224.70 day	224.70 day
Earth	1.000	1.000	365.26 day	365.26 day
Mars	1.520	1.524	1.882 year	1.881 year
Jupiter	5.219	5.203	11.87 year	11.862 year
Saturn	9.174	9.539	29.44 year	29.457 year

the data of Ptolemy are quite similar too. A millennium of observations before 150 AD did allow great precision.

4.3 The Kepler Laws

As we have seen the (almost) heliocentric Copernicus system was not much simpler than the Ptolemy (almost) geocentric one. To be precise, the centre of the Copernicus system is not the sun, but the centre of the elliptical orbit of the earth. In both cases, beyond a primary circle, several secondary and tertiary ones were necessary to fit the data. Since his youth, Tycho Brahe (1546–1601) started his study of the astronomical texts and his observations of the night sky. He soon found out that neither the tables of Ptolemy nor those of Copernicus were very accurate. Both of them were in contradiction with the facts. When he was 17 year old he had the opportunity to observe a not very frequent phenomenon, the conjunction of Jupiter and Saturn (the two planets appear very close to each other). Brahe calculated the conjunction time predicted by the Ptolemy tables finding it to be off by about one month (which is not really so much considering it is based on observations 1400 old) and that predicted by the Copernicus tables finding it off by several days (being an extrapolation over a few decennia, the relative error of Copernicus is much larger than that of Ptolemy). Brahe was now sure that a correct model of the Cosmos (then the solar system) could be found by planning and performing a systematic series of measurements as accurate as possible, rather than interpreting the classic texts.

The observations still had to be done with the naked eye because the telescope, as a scientific instrument, will not be invented by Galilei until 1609. One of his first instruments is shown in Fig. 4.6. The star under consideration must be seen through two small holes (*D* and *E* in the figure) fixed at the extremes of a bar that can rotate over the arc of a circle. The angle of the bar relative to the vertical, defined by the plumb line *AH*, is measured with a goniometer on a scale giving the arc minute. To increase the sensitivity the instrument had to be large. The graduated circle was almost seven meters in diameter. The instrument had to be robust and accurately

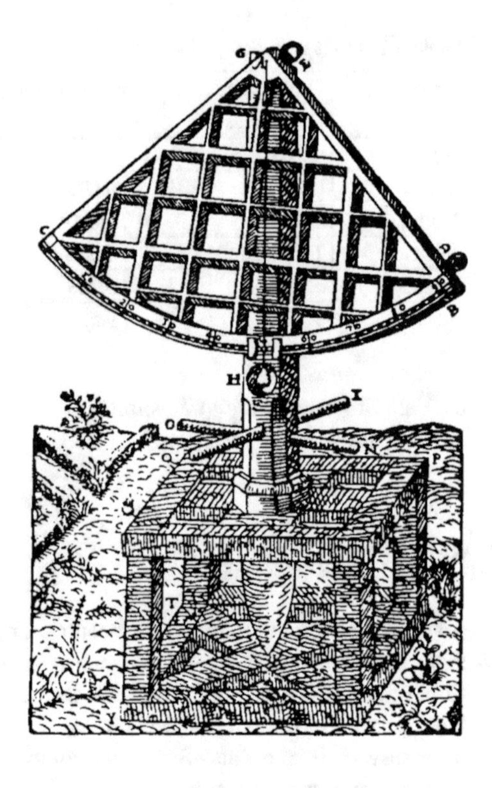

Fig. 4.6 Instrument of Brahe to measure the position of the stars

built to reduce systematic errors. The instrument was built of timber and was so heavy that twenty men were needed to install it in a garden.

Somewhat later Brahe succeeded to be funded by king Frederic II of Denmark and Norway for the construction of a big astronomic observatory on the island of Hveen near Copenhagen, the *Uraniburg observatory*. The castle in which the observatory was built had a rich library, bedrooms, kitchens and dining rooms. Brahe designed, built and installed a dozen different instruments, apt at various types of observation. For the next 20 years, at Uraniburg and later in Prague, Brahe continued his systematic observations. Before Brahe the angular resolution had not improved from Grecian times, being about 10′. He was determined to improve down to 1′ or better. He gathered the data in a series of tables, which became the database that allowed Kepler and Newton to solve the problem of the heavenly bodies' motions.

Johannes Kepler (1571–1630) started his studies in the school of Tycho Brahe in 1600. He began by searching through a large amount of available data to determine if he could find any simple relation. Tables such as Table 4.1 pointed to existence of a relation between orbit radii and periods. The larger the radius the larger is the period. But, is there really a mathematically simple relation? Kepler finally found it and published it in the book "Harmonice mundi" in 1618. He writes with confidence:

Table 4.2 Ratios of the cubes of the orbit radii and the squares of the period for the first six planets

Planet	r^3/T^2 (au^3 d^{-2})
Mercury	7.64×10^{-6}
Venus	7.52×10^{-6}
Earth	7.50×10^{-6}
Mars	7.50×10^{-6}
Jupiter	7.49×10^{-6}
Saturn	7.43×10^{-6}

> initially I thought I was dreaming…but it is absolutely certain and exact that the ratio existing between the periodic times of any pair of planets is exactly the ratio of the mean distances [*from sun*] to the power 2/3.

We can do the calculations ourselves. Starting from Table 4.1 we obtain the data in Table 4.2. We can easily understand Kepler's pride and satisfaction when he found such a simple relation. We know it as the 3rd Kepler law, because it came 10 years later than the discovery of the first two. The first two laws regard the orbits of a single planet, the third gives a relation between different planets.

Let us now briefly see how Kepler established that the orbits of the planets are not complicated combinations of circles, but, simply, ellipses. Its great discovery was based on the study of a single planet, Mars. The choice fell on Mars because its deviations from the predictions of both models based on circles where larger than for the other planets. Its strange behavior was the object of study of several astronomers, but its anomalies remained unexplained. Brahe had taken Kepler as his assistant in 1600 and charged him with a solution to this problem. Kepler worked on the problem for 6 years, in which partial successes alternated to partial failures, wrong paths were followed and retraced back, before reaching the solution that we know.

Kepler fully accepted from the start a heliocentric view with the guiding idea that the orbits should be a simple curve around the sun, but not necessarily a circle. The problem to find the curve was made difficult by the fact that the positions of the planet, Mars in his analysis, were measured in a frame fixed to the earth, which moves in a non-uniform and unknown motion around the sun. It took several years to solve this first problem, to find accurately enough, the motion of earth. We shall not describe here the various mathematical methods he employed, some of which are really elegant. We simply state that he found that the earth orbit is indistinguishable from a circle. However, its center is not the sun and its angular velocity about the sun is not uniform. The dogma that had resisted from Aristotle to Copernicus included was broken.

With reference to Fig. 4.7, d is the distance from the center of the sun to the center of the circle and R its radius. From the data of Brahe, Kepler found that $d/R = 0.018$. The angular diameter of the sun, as seen from earth, varies periodically during the year between a minimum and a maximum. Kepler had Brahe's measurements for that. With the above value of d/R, Kepler calculated the variations of earth sun distance during the year and the consequent variations of the apparent sun

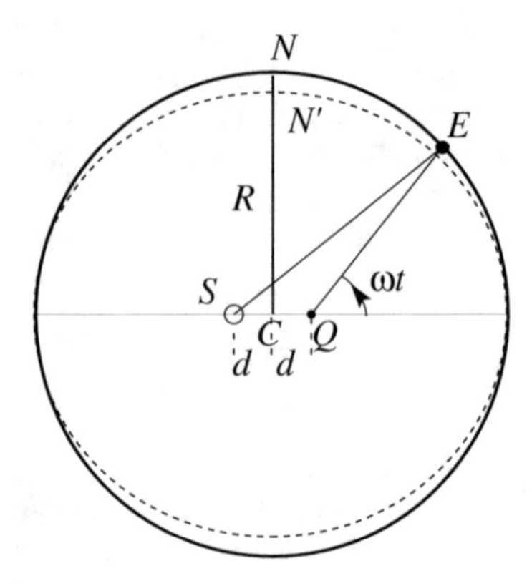

Fig. 4.7 Scheme of the earth's orbit. First approximation by Kepler. *Continuous line* is a circle, *dotted line* an ellipse; the difference between them is exaggerated

diameter. He found his results in agreement with the data. He gained confidence that he was on the correct path.

In retrospect we know now, and Kepler himself was to learn that in a while, that this model of the earth orbit is not correct, because the orbit is an ellipse. However, the eccentricity of the earth orbit is so small that the maximum difference between the preliminary Kepler model and the true orbit was smaller than the experimental uncertainty. To fix the orders of magnitude, the distance NN' is about one half of a per cent of R. In conclusion the error introduced in the analysis by the preliminary model is irrelevant.

Having defined the geometry of the orbit, Kepler had to find the motion. He did that using a trick invented by Ptolemy, and that we have already quoted, the equant. This is the point Q in the figure, lying on the line joining the center of the sun and the center C of the circle, at the same distance d as the sun but on the other side. Then the angular velocity of the position vector from Q to the earth is constant. It is called equant for this reason. We shall see soon why it works.

Kepler now knew the motion of earth in a reference frame in which the sun stood still. He could then calculate the positions of Mars at all times. It was an enormous amount of calculations (by hand obviously). Once more, he assumed the orbit of the planet to be an eccentric circle and a uniform angular velocity around an equant (different from that of earth). He calculated 40 points on the Mars orbit and compared it with the Brahe data. The maximum disagreement was only 8′, a very small one, but larger than the uncertainties in the Brahe measurements. Kepler knew he could trust Brahe. The model had to be wrong.

Kepler had to find another curve. Finally, his enormous computing effort showed the light. Suddenly, everything became clear: the curve is the ellipse. The first two Kepler laws were found. Kepler continued his work finding the parameters of the ellipse of the orbits of the other planets, including earth, calculating their positions and finding them in agreement with the rich and precise Brahe data.

We notice now that the reason why an eccentric circle had worked for the earth and not for Mars is the relatively large eccentricity of its orbit, which is 0.09, which is five time larger than that of the earth.

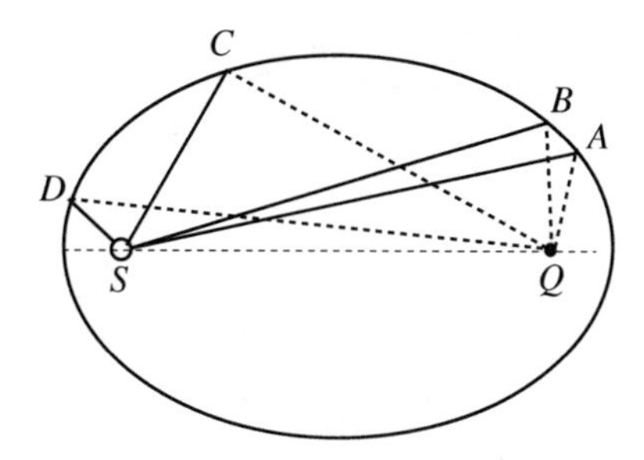

Fig. 4.8 Geometry plus area law explain the equant

He published his results in 1609 in his book *Astronomia nova*.

The three Kepler laws are:

1. The orbits of the planets are ellipses, the sun occupying one of their foci.
2. The position vector from the sun to the planet sweeps out equal areas in equal times
3. The ratio of the squares of the periods of any two planets is equal to the ratio of the cubes of their average distances from the sun.

We can now show the reasons that make the equant work in a first approximation. Indeed, the reason is in the second Kepler law. Consider Fig. 4.8 where an ellipse, in fact much more different from a circle than the real cases, is shown. The equant, which is the center of a circle that tries to represent the ellipse, is just the empty focus of the ellipse. In Fig. 4.8 the areas *SCD* and *SAB* are travelled in the same time by the planet and are equal for the second Kepler law. Consequently the arc *CD* is longer then *AB* proportionally at its distance from the sun. However, there is a second effect. A given path length on the orbit appears from the sun to be smaller, in its angular span, when it is closer than when it is farther, once more proportionally to the distance. The two effects, one due to the law of the areas and the geometrical one are identical. Consequently, if we look to the planet from the other focus, the former effect remains while the second inverts and the two cancel each other.

The contribution of Brahe had been a systematic and accurate experimental work, the work of Kepler an ingenious and superb analysis of the data. Both were needed to discover three simple laws, which were able to interpret all the available data. The work was not yet complete however. The marvelous Kepler laws were still purely phenomenological. A fundamental step was missing: their dynamical interpretation, which was going to lead to universal gravitation, one of the highest creations of human genius, the genius of Isaac Newton (1642–1727).

4.4 The Newton Law

We begin by showing that a consequence of the Kepler laws is that the angular momentum, **L**, of any planet *P* about the position of the sun is constant. With reference to Fig. 4.9, let **r** be the position vector, **v** the velocity and *m* the mass of the planet. Its angular momentum is then

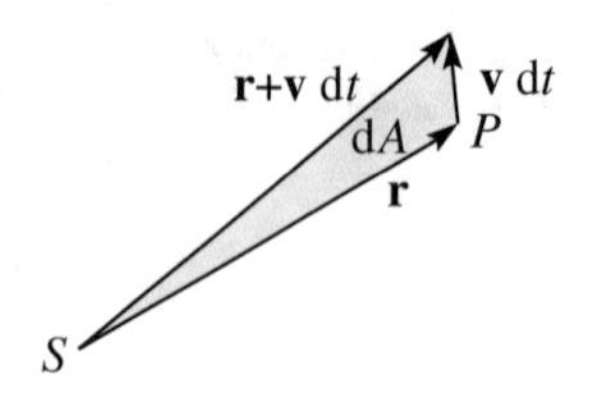

Fig. 4.9 The elementary area swept by the radius vector in *dt*

$$\mathbf{L} = \mathbf{r} \times m\mathbf{v}. \tag{4.4}$$

L is always perpendicular to both **r** and **v**, hence to the plane of the orbit that is constant for the first Kepler law. Hence the direction of **L** is constant.

In addition **L** is constant also in magnitude for the second law. Indeed, consider the area *dA* swept by the position vector in the time *dt*, which is the area of the triangle in Fig. 4.9. Two of its sides are **v** *dt* and **r**. Remembering the geometric meaning of the vector product we have

$$dA = \frac{1}{2}|\mathbf{r} \times \mathbf{v}\,dt| \tag{4.5}$$

or

$$\frac{dA}{dt} = \frac{1}{2}|\mathbf{r} \times \mathbf{v}|. \tag{4.6}$$

The quantity *dA*/*dt* is the area swept by the position vector in the unit of time and is called *areal velocity*. It is constant for the second Kepler law. We immediately recognize that the second member is proportional to the magnitude of the angular momentum, namely

$$L = |\mathbf{r} \times m\mathbf{v}| = 2m\frac{dA}{dt}. \tag{4.7}$$

The areal velocity being constant, the magnitude of the angular momentum is constant too. In conclusion the angular momentum vector about the sun is constant. On the other hand, the planet is certainly subject to a force, because it accelerates, but this force does not vary the angular momentum about a point fixed in an inertial frame. Consequently, its moment about that pole must be zero, namely its direction must be parallel to the position vector from the sun to the planet. It must be towards the sun because in a curved motion the force is always directed on the side of the curvature center.

In conclusion, the force on every planet must be directed towards the sun. The conclusion suggests, better forces, us to think the sun to be the source of the forces acting on all the planets.

We now consider the magnitude of the force. The symmetry of the problem suggests choosing a reference frame with origin in the sun and polar co-ordinates

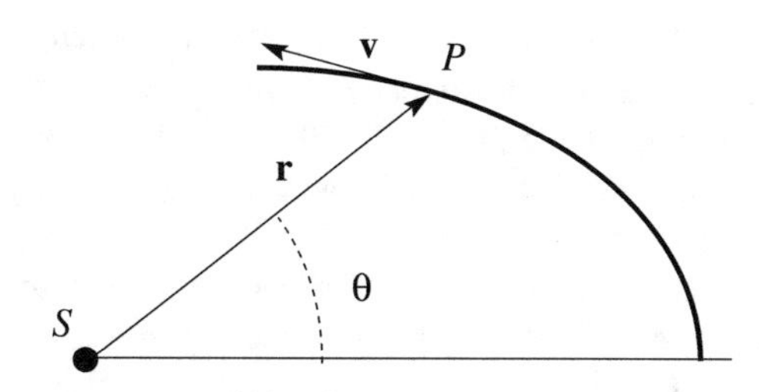

Fig. 4.10 The reference frame to study the motion of the planet

with an arbitrary polar axis. Let r be the magnitude and θ the azimuth of the position vector of the planet **r**. Data show that the motion of the planets does not slow down through the centuries, hence the force should be conservative. Having just shown that it is also central, for the theorem we demonstrated in Sect. 2.15, its magnitude cannot depend on θ, but depends only on the distance from the center of the sun r (Fig. 4.10).

To make the demonstration as simple as possible we shall assume the orbits to be circumferences rather than ellipses. In Sects. 4.11 and 4.13 the problem of the ellipse will be treated exactly.

If the motion is circular, the area law implies that the angular velocity ω is constant. The force should be the centripetal force of such a motion

$$F(r) = m\omega^2 r = 4\pi^2 \frac{r}{T^2} m \tag{4.8}$$

where m is the mass of the planet and T is its period. The third Kepler law states that

$$T^2 = K_S r^3 \tag{4.9}$$

where K_S is the proportionality constant, the same for all the planets of the solar system (but not necessarily for other systems) and that, substituted in Eq. (4.8), gives

$$F(r) = \frac{4\pi^2}{K_S} \frac{m}{r^2}. \tag{4.10}$$

We have found two fundamental properties of the force: 1. It is inversely proportional to the distance from the sun, which is its source, 2. is proportional to the mass of the planet. We now show the third property: the force is proportional to the mass of its source. To find it, observational data on systems similar to the solar one, but with a different central body, are needed. Newton, had already compared the force exerted by earth on bodies on its surface, namely the weight, and on the moon, as we shall see in Sect. 4.5. He had established that, taking into account the difference in the distances from the center, the force is the same. The characteristics of the gravitational force are universal.

Two small "solar systems" were known, Jupiter with its four principal satellites (Io, Europa, Ganymede, Callisto), which had been discovered by Galileo Galilei (1564–1642) (we shall tell of the discovery in Sect. 4.9), and Saturn with its two larger satellites, which had been observed by Christiaan Huygens and by Giovanni Domenico Cassini. These observations had established the validity of the 3rd Kepler law for the systems (in both cases more, smaller, satellites were discovered in recent times with the space missions).

Gravity, Newton concluded, is of all the planets and satellites, and continued:

> And since all attraction (by Law III) is mutual, Jupiter will therefore gravitate towards all his own satellites, Saturn towards his, the earth towards the moon, and the sun towards the primary planets.

To be concrete, consider one of the Jupiter satellites, Callisto. Jupiter is attracted by the sun with a force proportional to its mass and attracts Callisto with a force proportional to the mass of Callisto. For the 3rd Newton law Callisto attracts Jupiter with a force equal and opposite. But, this latter force has the same characteristics as the force that Jupiter receives from the sun, including being proportional to the mass of Jupiter. We can conclude that the force that Jupiter exerts on Callisto is proportional to the mass of Jupiter (beyond that of Callisto). The property is general, namely the gravitational force between any two (point-like) objects of masses m and M is proportional to the product of the masses. We write

$$F(r) = G_N \frac{mM}{r^2} \tag{4.11}$$

where G_N is a universal constant, the *Newton constant*, that we shall soon determine. This equation gives the magnitude of both the forces of mass M on m and of m on M. Their directions are equal and opposite. If $\mathbf{r}$ is the position vector from M to of m and $\mathbf{u}_r$ is its unitary vector, the force exerted by M on m is

$$\mathbf{F}(r) = -G_N \frac{mM}{r^2} \mathbf{u}_r. \tag{4.12}$$

This is the Newton law of *universal gravitation*.

We first observe that, as written, the law is valid for point-like objects. In the cases of the solar system and in the systems of Jupiter and Saturn, all the bodies, sun included, can be considered as points because their distances are always very much larger than their diameters. However, also two extended objects, for example two bricks one close to the other, attract gravitationally one another. To find the force we must ideally divide each body in infinitesimal parts. Every pair of infinitesimal elements attracts each other with the force of Eq. (4.12) where $\mathbf{r}$ is the position vector of one element relative to the other and the masses are those of the two elements. The total force is obtained by taking the vector sum (integrating) of all the pairs. There is certainly a case in which such an integration is needed, namely the weight. Indeed, we state that the weight of an object on the surface of earth is the gravitational force of the earth considered as a point in its center. Why is this possible? The answer is in Sect. 4.6.

A second observation is on the masses in the Newton law Eq. (4.11). They are clearly gravitational masses. However, in our demonstration we have started from Eq. (4.8) where the mass is the inertial one. As we have seen in Sect. 2.9, the equality of inertial and gravitational masses had been established by the experiments of Galilei, which Newton had repeated. However, the experiments had been done on terrestrial bodies and the question arises: does the same relation hold for celestial bodies? Newton showed this to be true considering the system of Jupiter and its four Galileian satellites. The system is a small replica of the solar system, but is part of the solar system too. Observations had shown that the satellites perform "exceedingly regular motions". The radiuses of the orbits about Jupiter and the periods had been measured. The periods turned out to be proportional to the 3/2rd power of the orbits radiuses. Consequently, the force exerted by Jupiter is inversely proportional to the distance. Suppose now the ratio between gravitational and inertial mass of Jupiter and any of its satellites, Callisto for example, to be different, say as

$$\left(m_{Cg}/m_{Ci}\right) - \left(m_{Jg}/m_{Ji}\right) = \pm\varepsilon$$

where ε is a positive small number. Then, Newton argues, the forces of the sun on Jupiter and on Callisto, at equal distances from the sun, will differ by $\pm\,\varepsilon$ also, and this would have an effect on the orbit of Callisto about Jupiter. The calculation of the effect needs to solve a three-body problem, Jupiter, Callisto and the sun, which cannot be done analytically. But Newton was able to find that, if the forces of the sun on Jupiter and Callisto would differ in a certain proportion, then the distances of the center of the orbit of Callisto (call it r_{CS}) about the sun and the center of Jupiter (r_J) from the sun would differ "nearly" as the square root of the same proportion "as by some computations I have found", namely,

$$\frac{r_{CS} - r_J}{r_J} = \sqrt{1 \pm \varepsilon} \cong 1 \pm \varepsilon/2.$$

He writes

> Therefore if, at equal distances from the sun, the accelerative gravity (*he means the gravitational force*) of any satellite towards the sun were greater or less than the accelerative gravity Jupiter towards the sun but by one 1/1000 part of the whole gravity, the distance of the centre of the satellite's orbit from the sun would be greater or less than the distance of Jupiter from the sun by one 1/2000 part of the whole distance; that is the fifth part of the utmost satellite (*Callisto*) from the centre of Jupiter; an eccentricity of the orbit which would be very sensible. But the orbits of the satellites are concentric to Jupiter, and therefore the accelerative gravities of Jupiter, and of all its satellites towards the sun, are equal among themselves.

Newton adds that if the ratios of gravitational to inertial mass of the earth, m_{Eg}/m_{Ei}, and of the moon, m_{Mg}/m_{Mi}, would be different, the above-described effect should be present and a deformation of the moon orbit should be observable. Today, the moon-earth distance is measured with extreme precision with LASER ranging techniques. In 1969 the Apollo 11 astronauts and later other lunar missions deployed on the surface of the moon systems of mirrors able to reflect back a LASER pulse sent from earth. The measurement of the round-trip time of the pulse gives the moon distance with a few millimeter precision as a function of time. The extremely sensitive technique did not detect any effect, providing the very low upper limit

$$\left|\left(m_{Eg}/m_{Ei}\right) - \left(m_{Mg}/m_{Mi}\right)\right| \le 5 \times 10^{-13}.$$

We now come back to the universality of the Newton law. If it is so, the constant G_N must be the same in any circumstance and is one of the fundamental constants of physics, called the gravitational Newton constant. At laboratory scale, between everyday life size objects, the Newton law is very small and difficult to measure. This was first done by Henry Cavendish (1731–1810) (see Sect. 4.7) leading him to a laboratory measurement of G_N (which is also called a Cavendish constant).

The universality of the Newton law needs to be verified experimentally. This has been done at all the length scales in many different conditions, finding it valid. We shall discuss a few examples further in the chapter. However, a limit of validity exists, as we shall see.

Equation (4.12) is mathematically very simple and symmetric in its elements. It interprets a huge amount of phenomena, from the motion of planets to the free fall of objects on earth, from the motion of the satellites, to that of the stars and the galaxies. The expression shows us how Nature can be described in its most fundamental aspects in simple and elegant mathematical form. The harmony of the world that up to the Middle Age, and to Copernicus, was believed to be substantiated in the existence of a mechanism of solid spheres, symmetric objects, that rotate uniformly (simple motion), comes back, in an abstract form, in the harmony, so to speak, of the physical law.

We finally come back on the constant K_S in Eq. (4.9). From Eq. (4.12) we can write, for the solar system

$$K_S = \frac{T^2}{r^3} = \frac{4\pi^2}{G_N M}. \tag{4.13}$$

We see that the constant depends on the mass of the sun, namely the mass of the central body. It is not universal. For example for the Jupiter system it is the mass of Jupiter, for the earth-moon systems it is the mass of the earth, etc.

4.5 The Moon and the Apple

If Eq. (4.12) is universal, the force that earth exerts on the moon, the centripetal force corresponding to her motion, must be the same as the force she exerts on a body on her surface, for example an apple, which is its weight. In particular the constant G_N should be the same. As Newton himself recalls, in 1665 he started to ask himself this question. He developed the following argument. Indeed, in her circular motion the moon continuously falls accelerating towards earth. This is simply another way to look at centripetal acceleration.

Suppose that the moon is at the point A of her orbit, as in Fig. 4.11, at a certain instant. In the figure we have taken a reference frame with the origin O in the center of earth and y-axis directed towards the moon in the considered instant. After a certain short time, say after one second, if no force were present, the moon would have moved to point B. On the other hand, if the moon would be abandoned still in B, she would fall in a second, under the action of gravity, from B to P. Point P is at the same distance r from the center of earth as A. Let us calculate the drop h, taking into account that the angle θ is very small. The Pythagorean theorem for the triangle ONP gives

$$r^2 = (r-h)^2 + x^2 = r^2 + h^2 - 2rh + x^2.$$

If θ is infinitesimal, h^2 is an infinitesimal of second order and can be neglected. We can also consider x equal to s and write

$$h = \frac{x^2}{2r} = \frac{s^2}{2r}. \tag{4.14}$$

To evaluate the displacement of the moon in one second we can use the proportion $s{:}2\pi r = 1{:}T$, where T is the period of the moon revolution, $T = 27.3$ d $= 2.4 \times 10^6$ s and $r = 3.8 \times 10^8$ m. We have $s = 2\pi r/T \approx 1000$ m and $h \simeq s^2/(2r) = 1.34$ mm. In a second the moon falls a little more than a millimeter. We now

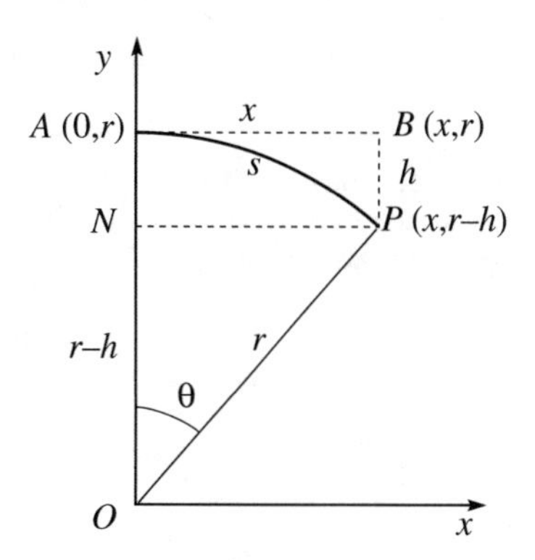

Fig. 4.11 How the moon falls

compare this with the drop length of an object on earth, the famous apple for example, which is

$$h_a = \frac{1}{2} g t^2 = 4.9 \text{ m}. \tag{4.15}$$

The ratio of the two drops in one second is equal to the ratio of their accelerations. The latter, if the Newton law is valid, should be in the inverse ratio of the squares of their distances. The ratio of the drops is $4.9/(1.34 \times 10^{-3}) = 3.65 \times 10^3$. Newton knew that the ratio of the distance of the moon is about 60 times the radius of the earth and what we have just found is about 60^2.

However, Newton had still the problem that we already mentioned. While moon and earth can be considered as points, considering their large distance, for what reason we should consider the apple, on a visually flat ground, should be attracted towards a point 6380 km under the ground as if all the mass of earth would be concentrated there?

This is a "miracle" true only for forces inversely proportional to the distance square. In the next section we shall prove the following theorem: the force exerted by a homogeneous spherical mass in any point outside its surface is equal to the force that would be exerted if all the mass were in a point at its center.

Newton did not publish any result until he had made everything clear, complete and perfect, in the *Principia* published in 1687.

4.6 The Gravitational Force of the Homogeneous Sphere

We shall calculate the force of a sphere of mass M on a point-like particle of mass m outside the sphere at a distance r from the center. We assume that the density of the sphere, if variable, depends only on the distance from the center (spherical symmetry). We shall prove that the force is equal to that which would exert all the mass M concentrated in the center.

We start by observing that is enough to prove the thesis for a spherical shell of infinitesimal thickness. Indeed if it is true for one shell it is also true for the sphere, which can be considered as composed of shells with the same center.

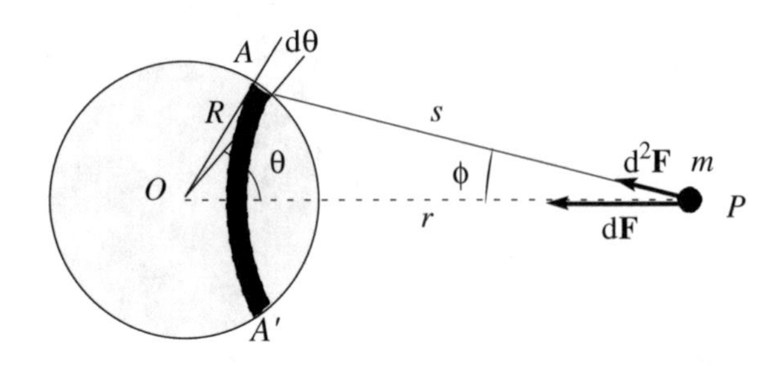

Fig. 4.12 Elements for calculation of the force of a *spherical* shell on an external point

Consider the spherical shell, of radius R and center O, shown in Fig. 4.12, having radius R and on it the ring AA' limited by cones with their vertex in O and semi-vertex angles θ and $\theta + d\theta$. Let ϕ be the semi-vertex angle of the cone with vertex in P and the ring AA' as base.

All the elements of the ring AA' are at the same distance from P and consequently they exert on P forces, call them $d^2\mathbf{F}$, equal in magnitude, but not in direction. The symmetry of the problem tells us that the resultant of these forces, $d\mathbf{F}$, is directed as OP. The contributions normal to it cancel each other. The component in the direction OP of the force is proportional to the mass of the element, to $\cos\phi$ and inversely to the square of the distance s^2. The resultant of the forces on m in P due to the ring being

$$dF = -G_N \frac{mdM}{s^2} \cos \phi$$

where dM is the mass of the ring. Now the mass of the ring is to the mass of the shell as the area of the ring is to the area of the shell:

$$dM : M = 2\pi R \sin \theta \times R \, d\theta : 4\pi R^2,$$

which gives us $dM = (M/2)\sin\theta \, d\theta$. The force of the ring on the mass m is then

$$dF = -G_N \frac{mM}{2s^2} \cos \phi \sin \theta \, d\theta. \tag{4.16}$$

The force of the shell is the integral of this expression for θ varying from 0 to π, namely

$$F = -G_N \frac{mM}{2} \int_0^{\pi} \frac{\cos \phi \sin \theta}{s^2} d\theta. \tag{4.17}$$

Both s and ϕ are functions of the integration variable θ. It is convenient however to express everything as functions of s. The Carnot theorem applied to the triangle OAP gives

$$\cos \theta = \frac{r^2 + R^2 - s^2}{2rR}, \quad \cos \phi = \frac{r^2 + s^2 - R^2}{2rs}. \tag{4.18}$$

We differentiate the first equation, remembering that r and R are constant, obtaining

$$\sin \theta \, d\theta = \frac{sds}{rR}.$$

We substitute this expression and the second Eq. (4.18) in the integral of Eq. (4.17) and take into account that now the variable is s and the limits must be changed in accord, obtaining

$$F = -G_N \frac{Mm}{4r^2R} \int\limits_{r-R}^{r+R} \frac{r^2 + s^2 - R^2}{s^2} ds.$$

The integral does not present difficulties. The indefinite integral gives

$$\int \frac{r^2 + s^2 - R^2}{s^2} ds = \int ds + \left(r^2 - R^2\right) \int \frac{ds}{s^2} = s - \frac{r^2 - R}{s},$$

which, evaluated in its limits, gives $4R$. In conclusion the force of the shell on a point P of mass m is

$$F = -G_N \frac{mM}{r^2}, \tag{4.19}$$

which is, in particular, independent of the radius R of the shell. This proves the theorem.

Consider now a point P of mass m inside the shell. What is the force on P exerted by the shell? The reasoning remains exactly the same, but for the limits on the integration on s. Now the angle θ varies between 0 and 2π and correspondently s between $R + r$ and $R - r$. The definite integral is zero. The gravitational force exerted by a spherical shell on a point inside it is zero. This is another property of the inverse square law forces.

Newton gave another proof of the last property using a simple geometric argument. Consider point P inside the shell as shown in Fig. 4.13 and the cone with vertex in P of very small vertex angle. The two napes intercept on the shell's two surfaces ΔS_1 and ΔS_2. As the density is constant, the masses of the two surfaces are proportional to their areas. The latter are proportional to the squares of their distances from P, say r_1^2 and r_2^2 . But the forces they exert in P are proportional directly to the masses and inversely to the square distances. The two forces are equal in magnitude. As their directions are opposite, their resultant is null. As the shell can be divided in pairs giving null contribution, the resultant is zero.

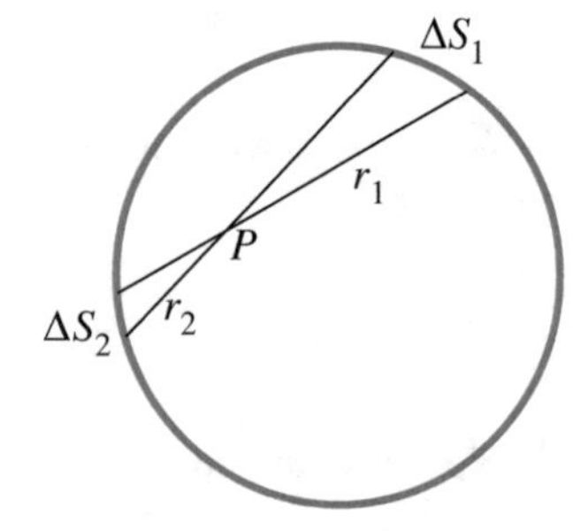

Fig. 4.13 The geometry to calculate the gravitational force of a spherical shell on an internal point

4.7 Measuring the Newton Constant

The Newton gravitational force Eq. (4.12) exerts between every pair of point- like, or spherical, masses. It is important to control experimentally its validity not only at the astronomical scales, but also at the laboratory scale. The laboratory experiments are difficult because the force is, at these scales, very small. Any disturbance such as small air currents, spurious electric forces, the movement of the experimenter itself, is a possible cause of errors and must be eliminated.

However, if we want to know the Newton constant, we must measure the force between two known masses at a known distance. In the case of the heavenly bodies in fact we do not know a priory the masses, but we infer them from the Newton law.

The gravitational force was first measured by Henri Cavendish (1731–1810) in 1798. His experiment is shown schematically in Fig. 4.14. A rigid metal bar suspended on a very thin metal wire, carries to equal lead spheres at equal distance from the wire. The system is in equilibrium and free to rotate about the wire. This type of arrangement is called *torsion balance* and will be further discussed in Sect. 8.9.

Two more larger and heavier equal spheres, of mass M, are arranged symmetrically, each at the same distance from one of the small ones. Consequently each of the large spheres attracts the small one nearby with an (equal) gravitational force. The arm of the couple is the distance between the centers of the small spheres and can be accurately measured. The moment of the couple induces a rotation to the bar. The wire reacts with an elastic torsion moment, which is proportional to its rotation angle. The equilibrium is at an angle at which the torsion moment and the moment of the gravitational couple are equal. Hence, the measurement of this angle gives the moment of the couple and, the arm being known, the forces.

The rotation angle is measured with the technique of the *optical lever*. A narrow light beam is sent to a very light mirror, fixed to the wire. The mirror reflects the beam on a scale located at a certain distance. The device is very sensitive. Even a very small change in the orientation of the mirror causes a sizeable movement of the light spot on the scale. Indeed, the moments are very small. The wire must have a

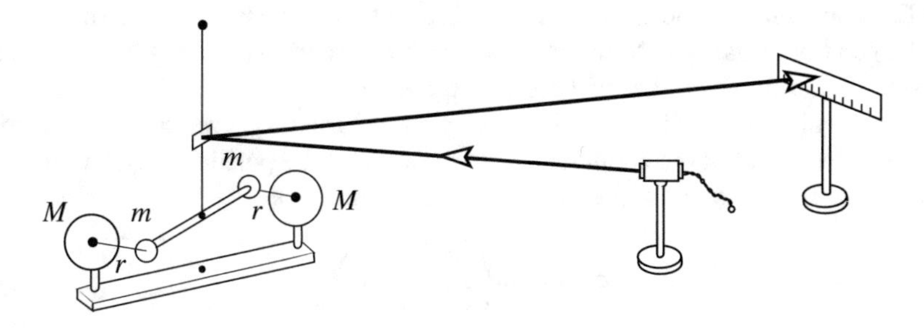

Fig. 4.14 The Cavendish experiment

very small elastic constant and consequently be very thin, but still capable of holding the weight of the small spheres and bar. All the apparatus must be closed in a container to avoid air currents. The presence of electrostatic charges must be avoided, etc.

The value of the gravitational constant obtained by Cavendish was

$$G_N = 6.67 \times 10^{-11}\ \mathrm{m^3\,kg^{-1}\,s^{-2}}. \tag{4.20}$$

The present value is

$$G_N = (6.67384 \pm 80) \times 10^{-11}\ \mathrm{m^3\,kg^{-1}\,s^{-2}}. \tag{4.21}$$

To have a quantitative idea, consider that the large spheres of Cavendish had a mass M = 158 kg, the small ones m = 0.73 kg and that the distance between one small and one large was r = 0.225 m. The two forces to be measured are about 10^{-7} N. This is about the weight of a hair.

4.8 The Gravitational Field

We interrupt in this section our discussion of experimental proofs of the Newton gravitational law, to discuss an important property of the gravitational, and of the other fundamental forces. Namely they are action at a distance. Other examples are the electric force, which operates between electrically charged bodies, and the magnetic force, for example between a magnet and a piece of iron.

In all these cases an extremely useful concept is the *field of force* or simply *field* (gravitational, electric, magnetic field).

Consider the gravitational force exerted by the earth, on different objects. It depends on the mass of the object (is proportional to it) and on the position of the object. If we consider two particles of different masses in the same position and divide the forces acting on each of them by its mass, we find the same result. This vector function of the position is the *gravitational field*.

The gravitational field generated by a distribution of masses is a vector function of the position. It is equal to the force acting on the unit mass in that position. The masses giving origin to the field are called the *sources of the field*.

In particular to describe the field of the earth we can take a reference frame with origin in the center of the earth. Consider a point P at the position vector $\mathbf{r}$ with the unit vector $\mathbf{u}_r$. If we put a mass m in P, it feels the force

$$\mathbf{F}(\mathbf{r}) = m\left(-G_N \frac{M}{r^2}\mathbf{u}_r\right). \tag{4.22}$$

where M is the mass of the earth. The gravitational field is the vector function of the position

$$\mathbf{G}(\mathbf{r}) = -G_N \frac{M}{r^2} \mathbf{u}_r. \tag{4.23}$$

This expression is valid for points outside the earth in the approximation of earth being spherical and with a spherically symmetrical distribution of masses. The physical dimensions of the gravitational field are a force divided by a mass, hence the dimensions of the acceleration. As a matter of fact, it is just the gravity acceleration **g**.

The concept of field eliminates from our reasoning the idea of action at a distance. We can think as follows. The earth, or any distribution of masses, creates in all the space around it a physical entity, the gravitational field, which extends, even if with decreasing intensity, to infinity. The field exists independently of being perceived as a force. But if we place in a point of the field a test body of mass m, it will feel like a force equal to the product of m times the gravitational field in that point. By means of the field the gravitational action becomes local.

We can now consider the potential energy of our test mass in the field of the earth. Defining the potential energy to be zero at infinite distance, we have

$$U_p(\mathbf{r}) = -G_N \frac{M}{r} m. \tag{4.24}$$

The physical meaning is: the potential energy of the mass m in the point P is the work to be done against the forces of the field to move the mass m from infinity to P.

Obviously the potential energy, as the force, is proportional to m. If we divide it by m we find a function of the position, independent of the body

$$\phi(\mathbf{r}) = -G_N \frac{M}{r}. \tag{4.25}$$

This function is the *gravitational potential.* The relationship between potential and field is the same as between potential energy and force. The gravitational potential in a point is the work to be done against the forces of the field to carry from infinity to that point a unitary mass. The physical dimensions of the gravitational potential are a velocity squared. It is measured in m^2/s^2.

Consider now our mass m moving on a circular orbit of radius r with velocity υ. It might be for example our moon. There is a simple relation between kinetic and potential energy. Recalling that $\upsilon = 2\pi r/T$, where T is the period, the kinetic energy is

$$U_k = \frac{1}{2} m\upsilon^2 = \frac{1}{2} m 4\pi^2 \frac{r^2}{T^2}$$

and for the 3rd Kepler law Eq. (4.13)

$$U_k = \frac{1}{2} G_N \frac{Mm}{r} = \frac{1}{2} |U_p|. \tag{4.26}$$

This result, valid for circular orbits, is that the kinetic energy is one half of the potential energy in absolute value. Consequently, the magnitude of the gravitational potential in the points of the orbit is equal to the square of the velocity of the body on that orbit

$$|\phi(\mathbf{r})| = \upsilon^2. \tag{4.27}$$

This expression will be useful in Sect. 7.13.

To appreciate the orders of magnitude, consider the motion of the earth around the sun. The velocity is $\upsilon = 3.3 \times 10^4$ m/s. For Eq. (4.27), the potential of the field of the sun in the points of the earth orbit is $\phi \approx 10^9$ m^2 s^{-2}.

We have already mentioned that limits of the validity of the Newton law exist, when it must give place to general relativity. More precisely, the effects that are in contradiction with the Newton law, and that are explained by general relativity, are of the order of the gravitational potential compared to the square of the speed of light, namely ϕ/c^2. Considering that $c^2 \approx 9 \times 10^{16}$ m^2 s^{-2}, these effects are usually very small (of the order of 10^{-8} on the earth orbit), but can be detected with high precision observations, as in the case of the anomalous precession of the Mercury perihelion (see Sect. 4.11). The effects become large at very high gravitational potentials, near massive and compact objects, like black holes.

The graphic representation of the gravitational field is very useful to have a visual idea of its main features. It is done with the *lines of force* and with the *equipotential surfaces.*

A line of force is drawn as shown in Fig. 4.15. We start from a point, 1 in the figure, where we evaluate the vector of the field. Then we make a small step $\delta\mathbf{s}$ in the direction of the field, reaching point 2. We calculate the field in this point and proceed another step as above, etc. In this way we obtain a broken line. It becomes a field line for δs tending to zero. It is a continuous curve, in all the points of which the field is tangent. Obviously the field lines are infinite in number. However, there is only one line through any given point. If they were, say, two, the field should have had two directions contemporarily. Graphically, we draw a number of lines, which is enough to see the features of the field.

The equipotential surfaces are the loci of the points that satisfy the equations $\phi(x, y,z)$ = constant, one for each value of the constant. These are infinite in number too. It is convenient to draw a set of surfaces at constant steps of the potential. An analogy are the geographic maps in which the level curves are drawn every, say,

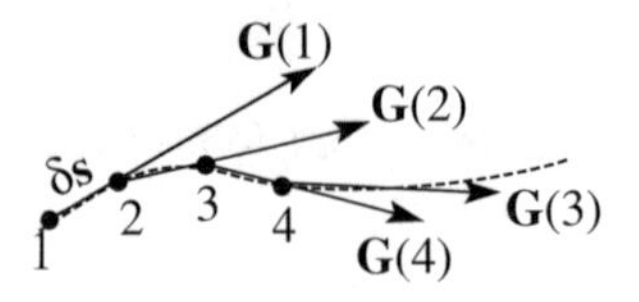

Fig. 4.15 Construction of a line of force

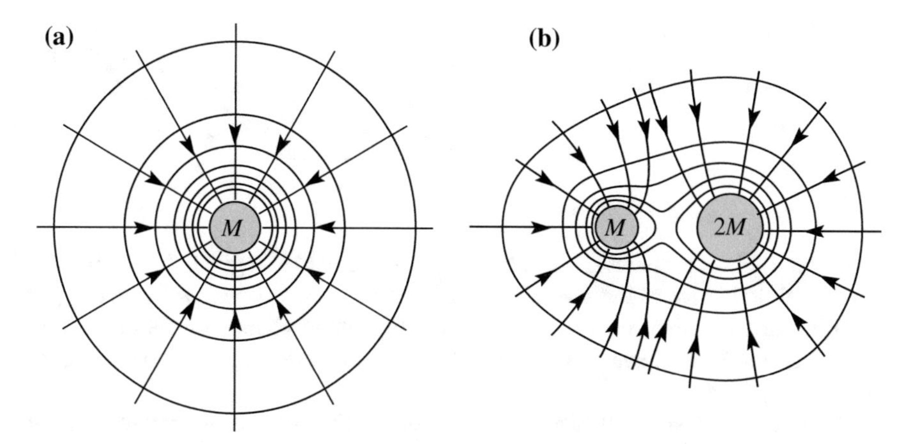

Fig. 4.16 Equipotentials and field lines for **a** a spherical mass M, **b** two masses one twice the other

one hundred meters of elevation. In the regions where the level curves are denser, the elevation varies more rapidly and the slope of the surface is steeper. The situation is analogous for equipotential surfaces.

Figure 4.16a shows some lines of force and equipotential surfaces for a spherical mass M. The lines of force are radial and point to the mass, because the force is attractive. The equipotentials are spherical and become denser getting closer to the mass, which is the source of the field.

Figure 4.16b represents the field originated by two spherical masses, one double the mass of the other. In every point the field is the vector sum of the fields of the two masses taken separately, the potential is simply the sum of the potentials. Notice the "saddle" point on the line joining the two centers. Here there is a minimum moving in that direction, a maximum moving perpendicularly to it.

One sees that the lines of force are always perpendicular to the equipotentials. This is a general property. Indeed, suppose we are moving with the infinitesimal displacement $d\mathbf{s}$. The potential difference between the two points is $d\phi = -\mathbf{G} \cdot d\mathbf{s}$. If the displacement is on the equipotential, $d\phi = 0$ by definition, hence $\mathbf{G}$ must be perpendicular to $d\mathbf{s}$. The lines of force that have the direction of $\mathbf{G}$ are perpendicular to the equipotential.

If we call G_s the projection of $\mathbf{G}$ on the direction of the displacement we can write

$$d\phi = -\mathbf{G} \cdot d\mathbf{s} = -G_s\, ds, \tag{4.28}$$

which can be also written as

$$G_s = -\frac{\partial \phi}{\partial s}. \tag{4.29}$$

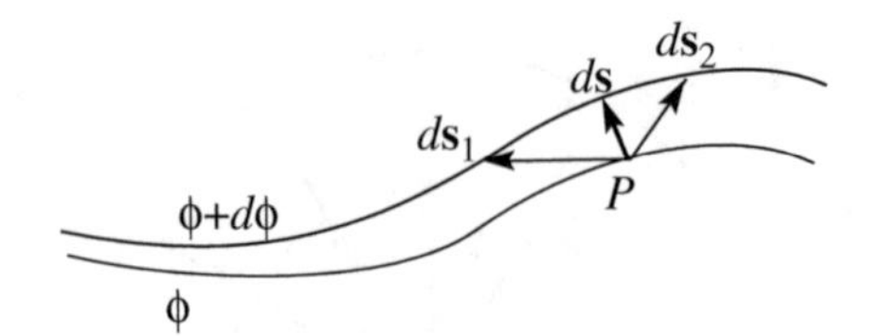

Fig. 4.17 Different steps between the same equipotentials

We read this expression as: the component of the field in a given direction is the directional derivative of the potential in that direction. *Directional derivative* is just the name of the derivative in Eq. (4.29), it is the rate of change of the function in that direction. As we have just seen the directional derivative is null for directions on the equipotentials.

Consider infinitesimal displacements as those in Fig. 4.17, which are in different directions but all leading from the equipotential ϕ to $\phi + d\phi$. The directional derivative is different for each of them because $d\phi$ is the same and ds is different. The derivative is a maximum when the direction is normal to the surfaces because ds is there a minimum. The vector having the magnitude of the maximum directional derivative and the direction of the normal to the equipotential towards increasing potential is called the *gradient* of the potential. Its symbol is **grad** ϕ. In conclusion we have

$$\mathbf{G} = -\mathbf{grad}\,\phi. \tag{4.30}$$

If we think of the level curves of a geographic map, the gradient is directed as the line of maximal slope of the ground; its magnitude is greater the greater is the slope.

On earth, the equipotential surfaces are materialized by the surfaces of the lakes and of the seas (neglecting the waves).

We now see how to calculate the gradient starting from the potential. We start from Eq. (4.28) and use the total differential theorem

$$d\phi = \frac{\partial \phi}{\partial x}dx + \frac{\partial \phi}{\partial y}dy + \frac{\partial \phi}{\partial z}dz = -\mathbf{G}\cdot \mathbf{d}s = -\left(G_x dx + G_y dy + G_z dz\right), \tag{4.31}$$

where dx, dy and dz are the Cartesian components of $\delta\mathbf{s}$. It immediately follows that the Cartesian components of the gradient are the partial derivatives of the potential

$$G_x = -\frac{\partial \phi}{\partial x}, \quad G_y = -\frac{\partial \phi}{\partial y}, \quad G_z = -\frac{\partial \phi}{\partial z}. \tag{4.32}$$

Obviously, similar relations exist between gravitational potential and gravitational force of a mass m. It is just a matter of multiplying by m,

$$\mathbf{F} = -\mathbf{grad}\, U_p \tag{4.33}$$

and

$$F_x = -\frac{\partial U_p}{\partial x}, \quad F_y = -\frac{\partial U_p}{\partial y}, \quad F_z = -\frac{\partial U_p}{\partial z}. \tag{4.34}$$

4.9 Galilei and the Jovian System

G. Galilei (1564–1642) was the first human to explore scientifically the sky using the *telescope*, which he had developed. As a matter of fact, combinations of two lenses put one after the other at a certain distance had existed for at least 30 years. The first written mention is in 1589, by Giovanni Battista Della Porta (1535–1615). At the beginning of the XVII century telescopes were built in the Netherlands by eyeglasses manufacturers. They were toys sold in exhibitions at low prices. Galilei new of the Dutch telescope in 1609. He quickly envisioned a way to transform the device into a scientific instrument and immediately started his experimental work, without a solid theoretical basis. What is known today as geometrical optics was developed only in 1611 by Johannes Kepler (motivated by the desire to explain how the telescope works). Lenses had already been produced since the XIII century, but their quality was not adequate for a scientific instrument.

An important property of the telescope is angular magnification, which is the ratio between the angle under which an object is seen through the telescope and the angle under which it is seen with the naked eye. The second property is the resolving power, namely the ability of the telescope to resolve, to see separated, two point images very close one to the other. To increase both properties the diameter of the objective lens (the one farther from the eye) must be increased. However, the larger the lens, the more difficult is its production without any defect. With a series of improvements, and the help of the Venetian glass makers, Galilei developed the technique to the point that he could build a telescope with magnification 10 and, some time later, one with magnification 30, with lenses of perfect optical quality. With this magnification the light reaching the eye in $30^2 = 900$ times as with the naked eye.

Galilei published his first observation in the booklet "Sidereus nuncius" (astronomical notice) in 1610. In addition, the logbooks of his observations have come down to us. One of his great discoveries was that around Jupiter's four satellites orbit, making a small replica of the solar system. A view of the system with a modern telescope is shown in Fig. 4.18. Let us see how he describes his discovery in the Sidereus Nuncius.

On the night of the 7th of January 1610, looking to Jupiter, Galilei observed three small "starlets". They attracted his attention because they were perfectly aligned between them and with Jupiter and on the ecliptic. He did not correlate the

Fig. 4.18 Jupiter and his satellites. Image © NASA

Fig. 4.19 Sketches of the Galilei observations in January 1610 in the nights of **a** 7th, **b** 8th, **c** 10th, **d** 13th

starlets with Jupiter, thinking they were fixed stars in the background. He took note of their positions in the logbook, as we try to reproduce in Fig. 4.19a.

The following night he repeated the observations and noticed that the relative positions had changed, as in Fig. 4.19b. He thought the change to be due to the movement of Jupiter relative to the stars, that he believed to be fixed, with some doubts, because the motion did not match the calculations. He anxiously waited, as he writes, the following night, but his hope was frustrated, because all the sky was cloudy. The night of the 10th the stars were only two and had again changed position, but still on a line, as in Fig. 4.19c. The third one, he thought, should be hidden by Jupiter. Galilei had no more doubts. He writes (translated by the author):

> my perplexity changed to astonishment and I became sure that the apparent movement was not of Jupiter, but in the stars I observed; hence I decided to continue my investigation with increased attention and scrupulosity.

The 13th he saw for the first time the fourth satellite, which had entered the field of view of the telescope, as in Fig. 4.19d.

After several more nights of observations he published the discovery, together with other important ones on the moon and the Milky way in the above quoted book in March 1610.

The next task was the measurement of the periods. The measurement was extremely difficult, as much that Kepler had declared it impossible, because the images of the four starlets were indistinguishable. Galilei understood that the precision on his measurements of the angular distances from the center of Jupiter had to be improved. He had measured them "by eye" with a precision of better than one

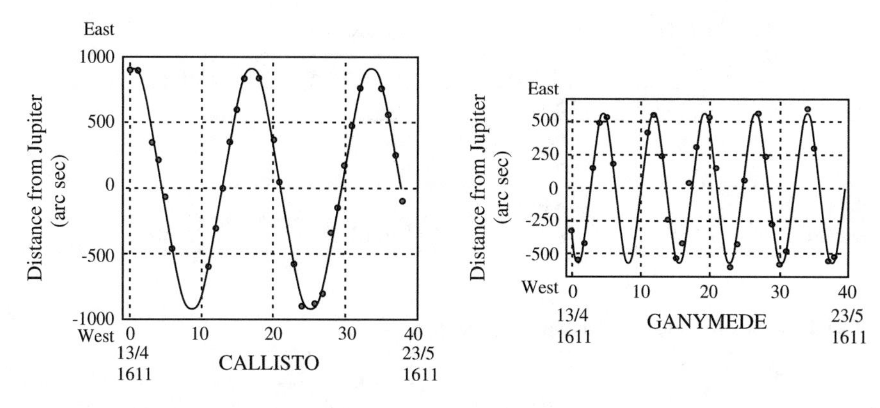

Fig. 4.20 The distances from Jupiter of his two farther satellites as measured by Galilei in spring 1611. The sinusoids are from my calculations

arc minute (1/60°). It was not enough. He developed the micrometer, with which he was to measure the positions with a precision "better than very few arc seconds" (one arc sec = 1/3600°).

Galilei continued his systematic measurements for several years, but already in 1611 he had been able to identify each of the satellites and to calculate their period and the apparent diameters of the orbits. In Fig. 4.20 we report a subset of his measurements made in spring 1611, as taken from his hand notes. For simplicity they are for the two more external ones, Callisto and Ganymede. The planes of the orbits are almost on the line of view from earth. Consequently, if the orbit is an ellipse (or, in particular a circle) the motion appears as sinusoidal functions of time. With a computer it is today easy to find the sinusoid that best interpolates the data, the ones shown in the figure. Clearly, the data are in agreement with the hypothesis. The procedure also gives us a value for the amplitude and the period. Galilei had no computer and made his calculations by hand.

Table 4.3 reports the periods as measured by Galilei and how they are known today. One sees that his measurements were quite good.

Accurate measurements of the apparent amplitudes are more difficult. Notice that these quantities are measured relative to the apparent diameter of Jupiter, namely they are, say, $n = r/r_J$. Table 4.4 reports the values of n as measured by Galilei in subsequent years, showing how the precision is increasing, approaching the presently known values.

The Jovian system is a small solar system. Is the third Kepler law verified? Galilei did not check that, but Newton did. From the data in the two tables, we can do it ourselves obtaining the following table (Table 4.5).

Table 4.3 Periods of the Jupiter satellites (in days)

	Io	Europa	Ganymede	Callisto
Galilei	1.76	3.53	7.16	16.3
modern	1.77	3.55	7.17	16.75

Table 4.4 Angular radii of the orbits relative to the radius of Jupiter

	Io	Europa	Ganymede	Callisto
1610?	3.5	5.7	8.8	15.3
1611	3.8	6.2	8.4	15
1611?	4	7	10	15
1612	5.7	8.6	14	"almost 25"
modern	5.58	8.88	14.16	24.90

Table 4.5 The 3rd Kepler law in the Jovian system

	Galilei			Modern		
	T (d)	$n = r/r_J$	n^3/T^2	T (d)	$n = r/r_J$	n^3/T^2
Io	1.76	5.7	59.8	1.77	5.91	65.8
Europa	3.55	8.6	50.5	3.55	9.40	65.9
Ganymede	7.16	14.0	53.5	7.16	14.97	65.8
Callisto	16.3	24.9	58.1	16.69	26.33	65.5

The 3rd Kepler law is satisfied, better obviously by the modern data, for which the experimental uncertainties are smaller.

We can finally check the universality, namely if the gravitational constant has the same value in the Jovian and in the solar systems. We check if Eq. (4.13), namely, $K_S = T^2/r^3 = 4\pi^2/(G_N M)$, is valid with the same G_N, where now M is the mass of Jupiter, r and T are orbit radium and period of any of the satellites. For that we need absolute values. We now know the distance of Jupiter and then the radii r. The Jupiter mass has been evaluated from his perturbing effects on the other planets. With these values we find that, indeed, the gravitational constant is the same.

4.10 Galaxies, Clusters and Something Else

In this section we shall give two examples of structures of larger scales than the solar systems. The Newton law is valid also at the largest scales. However, we shall also see that the same law gives us evidence that the mass of the Universe is made for its largest fraction of components that are not visible, because they do not emit or absorb light. This is the so-called dark matter, whose nature we do not know.

A first example is shown in Fig. 4.21. It is a *globular cluster*, a system containing millions of stars, which are very old, having an age comparable with the Universe itself. The effect of the gravitational force keeping those stars together is spectacular.

Figure 4.22. shows the image of a spiral galaxy, a system of hundreds and millions of stars kept together by the gravitational attraction. All this enormous

Fig. 4.21 The global cluster NGC 2808. Image © ESA

Fig. 4.22 The galaxy M74 from the Hubble Space Telescope. Image © NASA

system is rotating, as evident by the image. The angular momentum of the huge gas cloud from which the galaxy originated billions of years ago remained constant.

Let us more closely to the rotation. Let us start by considering how the orbital velocity $v(r)$ of a body of mass m orbiting around a central body of mass M (like a planet around the sun) varies with the distance from the center r. Assume for simplicity a circular orbit. We state that the centripetal force must be equal to the gravitational attraction

$$G_N \frac{mM}{r^2} = \frac{mv^2}{r} \tag{4.35}$$

or

$$v = \sqrt{G_N M}/\sqrt{r}. \tag{4.36}$$

The velocity is inversely proportional to the square root of the distance from the center. The validity of the law can be tested on the planets of the solar system.

Five planets are visible with the naked eye and have been known since ancient times. In order of distance from the sun, including earth, they are: Mercury, Venus, Earth, Mars, Jupiter and Saturn. In 1781, William Herschel (1738–1822) discovered a "star", the image of which in the telescope had a non-zero diameter. It was the seventh planet, Uranus. The object had been already observed by Galilei and by more astronomers in the following years. They had not recognized it as a planet, due to the limitations of their telescopes, but had measured its coordinates. On the basis of these measurements, Herschel could reconstruct the parameters of the orbit of Uranus. The motion of Uranus showed some anomalies, when compared to the Newton law predictions. These were interpreted in 1846, independently by Urbain LeVerier (1811–1877) and by Johan Couche Adams (1819–1891), as possibly due to an eighth planet. When his calculations were complete, LeVerrier sent a letter, with the calculated coordinates, to the astronomer Johanne Grottfried Galle (1812–1910) in Berlin, asking him to verify. The following night, Galle found Neptune within 1° of the predicted position. Similarly, in 1930 Pluto was discovered, having its existence predicted from the anomalies of the Neptune motion.

Figure 4.23 shows the orbital velocity of the planets as a function of their distance from the sun. Equation (4.36) is fully satisfied.

Consider now the galaxy, a typical one, shown schematically in Fig. 4.24. The image shows that its luminosity decreases for increasing distance r from the center,

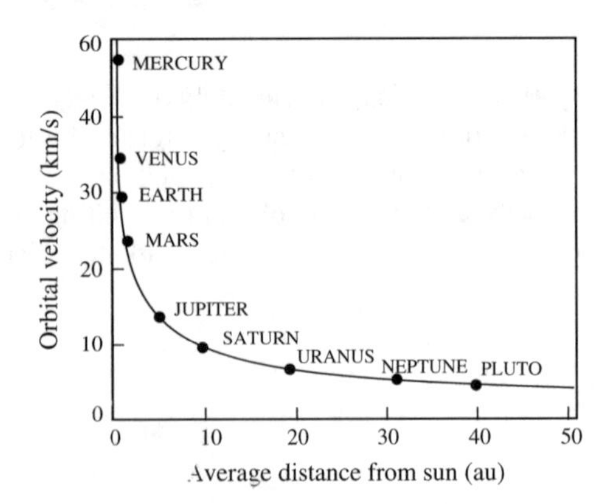

Fig. 4.23 Inverse square root dependence of orbital velocities of the planets

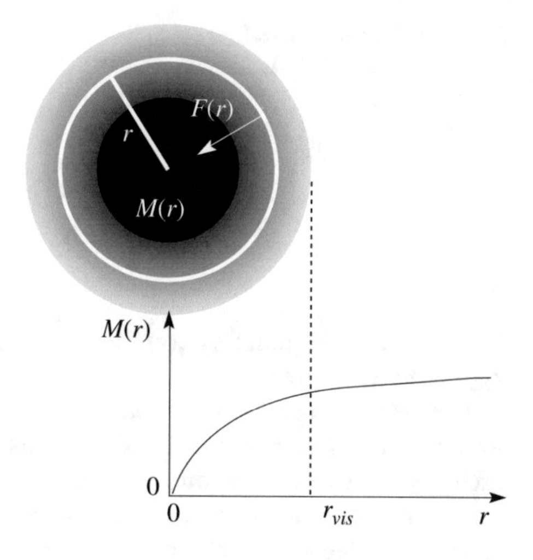

Fig. 4.24 A spherical mass distribution. *M(r)* is the mass in a sphere of radius *r*

till it disappears. This means that the star density decreases departing from the center. We indicate with $M(r)$ the total mass contained in a sphere of radius r. We would guess it having the same behavior as the luminosity. But it is not so. Let $\upsilon(r)$ be the (average) velocity of the points of the galaxy at the distance r from the rotation axis. We can consider with a reasonable approximation the mass distribution as spherically symmetrical. Then, the gravitational force acting on a body, a star or a gas particle, at the distance r is the same as the force of all the mass inside r, concentrated in the center, exactly as for the weight of an apple. Differently from the apple, there is now a lot of mass outside r, but, as we have proven in Sect. 4.6, its gravitational force inside a spherical shell is zero.

The force at the distance r is

$$F(r) = G_N \frac{M(r)m}{r^2} \tag{4.37}$$

and the rotation velocity at the distance r is

$$F(r) = \sqrt{G_N M(r)} \frac{1}{\sqrt{r}}. \tag{4.38}$$

The image of the galaxy shows that the luminosity ends at a certain distance. The visible part of the galaxy has a radius that we call r_{vis}. Typical values vary from 10 kpc to 100 kpc (1 pc, parsec,[1] is 3×10^{16} m = 3.3 light years) from the center. We then expect the function $M(r)$ to increase with r and to become constant at about r_{vis}, because there is no more mass after that, as represented in Fig. 4.24.

[1]A parsec is the distance at which the diameter of the earth orbit is seen under the angle of a second.

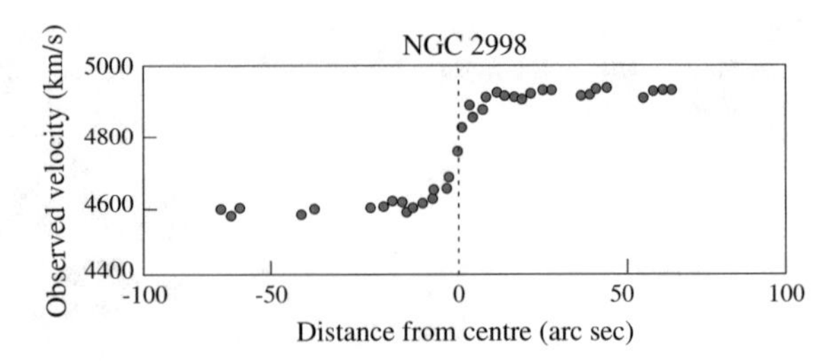

Fig. 4.25 Rotation curve for the galaxy NGC 2998

Consequently, the function $v(r)$ for values of r larger than the radius of the galaxy r_{vis} should decrease as $1/\sqrt{r}$.

How can we measure the rotation velocities of the galaxies at different distances from the axis? The motion of the single stars is not observable from earth. However, each of the elements in nature emits light having a well-defined spectrum, which is characteristic of the element. If the source is moving, the spectrum is shifted in a known way dependent on the relative velocity between source and observer (it is called the Doppler effect).

Consequently, we measure the velocities of the different elements of a galaxy by measuring the spectra of the light they emit. In practice the light emitted by the huge clouds of gases, such as hydrogen and helium that extend farther than the stars from the axis, but do not contribute substantially to the mass.

Figure 4.25 shows the velocities relative to us of the galaxy NGC 2998 as functions of the apparent distance from its center. We can deduce that the galaxy has an average velocity (the velocity of its center) of about 4700 km/s. However, on the left the velocities are systematically smaller, higher on the right. This is because we are observing the rotation of the galactic disk at an angle different from 90°. Consequently the disk is approaching on one side, withdrawing on the other. To have the *rotation curve* of the galaxy, namely the orbital velocities at different distances from its center, we subtract the average velocity. The distance of the galaxy being known, we can convert the apparent distances from axes in absolute distances. We obtain the diagram in Fig. 4.26.

We would expect the orbital velocity to decrease as $1/\sqrt{r}$ at distances larger than the visible radius, which is in this case about 8 kpc. It is not so; the velocity remains practically constant up to the maximum distance explored, much beyond the distance at which no more stars are present.

The behavior of NGC 2998 is not an exception, rather is the norm. The same phenomenon was found in all the spiral galaxies. We need to conclude that either

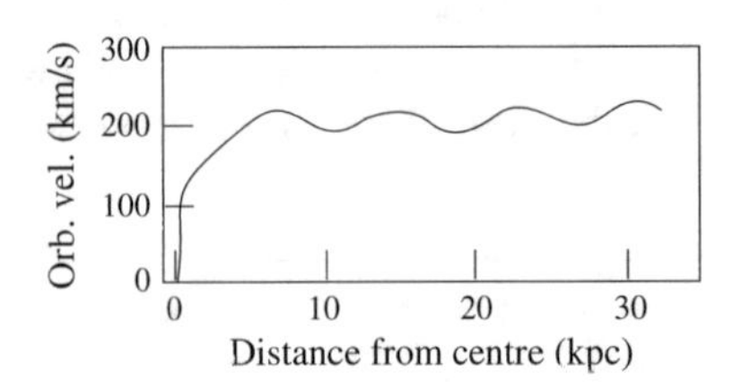

Fig. 4.26 The rotation curve of the galaxy NGC 2998, the orbital velocity versus distance from center

the Newton law is no longer valid in these circumstances, or that there is much more matter in the galaxies than the visible one, which extends much beyond the visible one. It has been called *dark matter* (but invisible matter would be a better name). We now know that the right alternative is the latter. The conclusion comes from a large number of observations, at different length scales, for phenomena ruled by different physics, at different eras of the Universe. All point consistently to the conclusion that dark matter is about five times more abundant than the matter we know. The search for dark matter is one of the frontiers of today's physics.

4.11 Elliptic Orbits

In Sect. 4.4 we have seen the solution of the so-called direct Kepler problem, namely how to find the force from knowledge of the orbit. We have done that however, in the particular case of circular orbits. It is instructive to solve the problem in general, for elliptic orbits. We shall do that in this section using the modern calculus. In the next section we shall show the same, following the Newton demonstration.

We start finding the expressions of velocity and acceleration of a generic material point P, moving an arbitrary plane curve, in polar co-ordinates. We introduce a polar co-ordinate frame with origin O and polar axis x (see Fig. 4.27). We call θ the azimuth of the position vector $\mathbf{r}$, and $\mathbf{u}_\theta$ and $\mathbf{u}_r$ the unitary vectors respectively. The time derivatives of the latter is given by the Poisson formula (1.59)

$$\frac{d\mathbf{u}_r}{dt}=\frac{d\theta}{dt}\mathbf{u}_\theta,\qquad \frac{d\mathbf{u}_\theta}{dt}=-\frac{d\theta}{dt}\mathbf{u}_r. \tag{4.39}$$

We now find the velocity, which is the time derivative of the position vector $\mathbf{r} = r\mathbf{u}_r$

$$\mathbf{v}=\frac{d\mathbf{r}}{dt}=\frac{dr}{dt}\mathbf{u}_r+\frac{d\mathbf{u}_r}{dt}r.$$

which, for the first of Eq. (4.39) is

$$\mathbf{v}=\frac{dr}{dt}\mathbf{u}_r+r\frac{d\theta}{dt}\mathbf{u}_\theta=\upsilon_r\mathbf{u}_r+\upsilon_\theta\mathbf{u}_\theta.$$

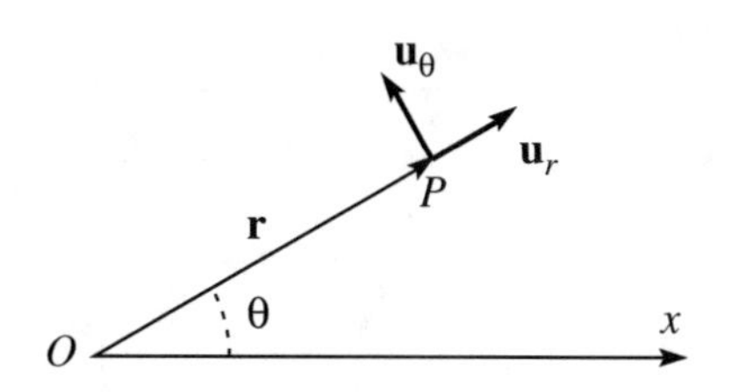

Fig. 4.27 The unit vectors of the polar co-ordinates

We now derive once more to have the acceleration

$$\mathbf{a}=\frac{d\mathbf{v}}{dt}=\frac{d^2r}{dt^2}\mathbf{u}_r+\frac{dr}{dt}\frac{d\theta}{dt}\mathbf{u}_\theta+\frac{dr}{dt}\frac{d\theta}{dt}\mathbf{u}_\theta+r\frac{d^2\theta}{dt^2}\mathbf{u}_\theta-r\frac{d\theta}{dt}\frac{d\theta}{dt}\mathbf{u}_r$$
$$=\left[\frac{d^2r}{dt^2}-r\left(\frac{d\theta}{dt}\right)^2\right]\mathbf{u}_r+\left[2\frac{dr}{dt}\frac{d\theta}{dt}+r\frac{d^2\theta}{dt^2}\right]\mathbf{u}_\theta$$

and finally

$$\mathbf{a}=\left[\frac{d^2r}{dt^2}-r\left(\frac{d\theta}{dt}\right)^2\right]\mathbf{u}_r+\left[\frac{1}{r}\frac{d}{dt}\left(r^2\frac{d\theta}{dt}\right)\right]\mathbf{u}_\theta=a_r\mathbf{u}_r+a_\theta\mathbf{u}_\theta. \tag{4.40}$$

We have now the kinematic expressions we need. Pay attention to the fact that υ_r and a_r are the components of the vectors on the position vector **r** from the focus, not from the center of the ellipse.

We now consider the motion of the planet. The 1st Kepler law states that the orbit is an ellipse with the sun in one of the foci.

We start by recalling the main properties of the ellipse (one of the conic sections, together with the hyperboles and the parabola). We choose the polar co-ordinate frame shown in Fig. 4.28 with the origin in the focus where the sun is and the major axis as polar axis. (Notice that there are also polar co-ordinates with the origin in the center O). The angle θ is called *anomaly* (to be precise, it is sometimes called *true* anomaly, to distinguish it from the case in which the origin is in the center), a and b the semi-major and semi-minor axes.

The equation of the ellipse, in its "canonical" form, is

$$r(1+e\cos\theta)=b^2/a=p, \tag{4.41}$$

where e is the *eccentricity* and p is the *semi-latus rectum* which is the position vector for θ = 90°. The relation between eccentricity and semi-axes is

$$e^2=1-b^2/a^2. \tag{4.42}$$

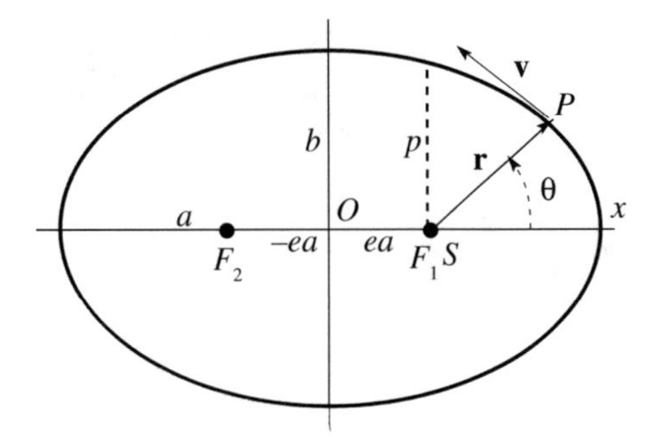

Fig. 4.28 The geometry of the *ellipse* and its main parameters

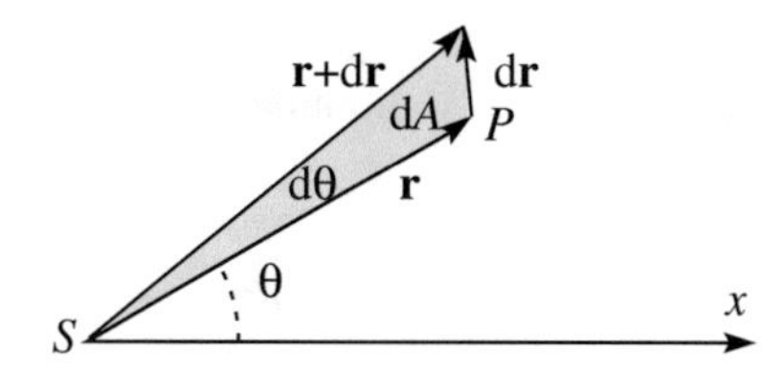

Fig. 4.29 Elementary area swept by the position vector of the planet

The circle can be considered a degenerate ellipse with $e = 0$. The smaller the eccentricity the smaller is the difference of the ellipse from the circle. As a matter of fact, the eccentricities of the planets are in any case quite small, much smaller than in Fig. 4.28.

Consider now the force. First we observe that, being the force directed to the sun, the F_θ component is zero. This statement is equivalent to the 2nd Kepler law and to the conservation of angular momentum. Indeed, from Fig. 4.29 we see that the infinitesimal area swept by the position vector is $dA = (rd\theta)r/2$. Hence the areal velocity is

$$\frac{dA}{dt} = \frac{1}{2} r^2 \frac{d\theta}{dt} \tag{4.43}$$

and

$$F_\theta = ma_\theta = m \frac{d}{dt}\left(r^2 \frac{d\theta}{dt} \right) = 0. \tag{4.44}$$

In addition, calling L the angular momentum and recalling Eq. (4.7) we can write

$$L = 2m \frac{dA}{dt} = mr^2 \frac{d\theta}{dt}. \tag{4.45}$$

This expression will be useful in the following.

We are now ready to go to the acceleration a_r and the force F_r towards the sun. We already found, Eq. (4.40), that

$$a_r = \frac{d^2 r}{dt^2} - r\left(\frac{d\theta}{dt} \right)^2. \tag{4.46}$$

The polar co-ordinates r and θ are not independent, but linked by the ellipse Eq. (4.41). Taking the time derivative of this equation, rearranging the terms and using Eq. (4.45), we have

$$\frac{dr}{dt} = \frac{e}{p} \sin\theta r^2 \frac{d\theta}{dt} = \frac{e}{p} \frac{L}{m} \sin\theta. \tag{4.47}$$

We derive this again, because Eq. (4.46) contains the second derivative, and use again Eq. (4.45), obtaining

$$\frac{d^2r}{dt^2}=\frac{e}{p}\frac{L}{m}\cos\theta\frac{d\theta}{dt}=\frac{e}{p}\left(\frac{L}{m}\right)^2\frac{\cos\theta}{r^2}.$$

We now substitute this in Eq. (4.46), use once more Eq. (4.45) and get

$$a_r=\frac{e}{p}\frac{L^2}{m^2}\frac{\cos\theta}{r^2}-\frac{L^2}{m^2}\frac{1}{r^3}=\frac{L^2}{m^2}\frac{1}{r^2}\left(\frac{e\cos\theta}{p}-\frac{1}{r}\right).$$

Looking back to the equation of the ellipse we recognize that the expression in parenthesis in the last member is just $-1/p$. Finally we have

$$a_r=-\frac{L^2}{m^2}\frac{1}{p}\frac{1}{r^2} \tag{4.48}$$

where the minus sign tells us that the force is opposite to $\mathbf{r}$. We see that the acceleration is inversely proportional to the square of the distance from the sun. The same is true obviously for the force

$$F_r=-\frac{L^2}{m}\frac{1}{p}\frac{1}{r^2}. \tag{4.49}$$

This completes the proof. We have proven that if the orbit is an ellipse with the sun in one of the foci, the force is inversely proportional to the square of the distance. The remaining part of the argument to reach the Newton law is the same we already did for circular orbits, with the conclusion

$$F=\frac{L^2}{m}\frac{1}{p}\frac{1}{r^2}=G_N\frac{mM}{r^2}. \tag{4.50}$$

We did not need the 3rd Kepler law to reach this conclusion, as it had been the case in the particular case of circular orbits. Indeed, in that case Eq. (4.41) reduces to $r = p$ = constant and not all of the arguments of this section any longer hold.

Before concluding we stress once more that there is a unique dependence on r of a central force $F_r(r)$ that produces elliptic orbits with the sun in a focus, $F(r)\propto 1/r^2$. As Newton showed, even the smallest difference in the exponent, $F_r(r)\propto 1/r^{2+\varepsilon}$ would produce an orbit of the type shown in Fig. 4.30, which is, so to say, a slowly rotating ellipse, called a rosette. We shall not reproduce the argument here, but only give a hint. In a motion on an ellipse or on a rosette, both polar co-ordinates, r and θ, vary in time periodically. The period of the latter is in any case the time to increase θ by 2π. The period of r depends on the force. Only if the force is inversely proportional to r^2 is it equal to the period of θ and the

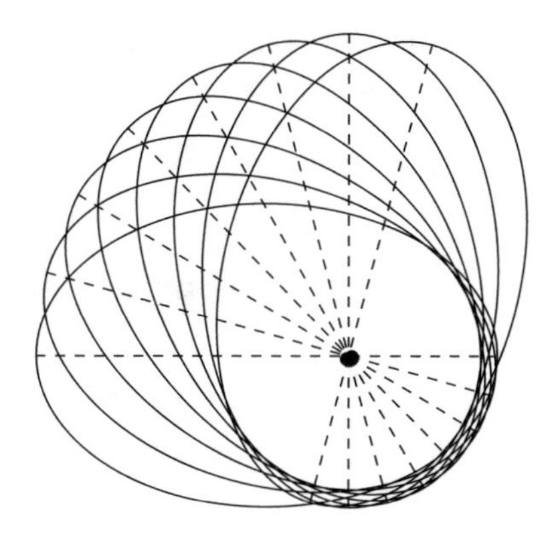

Fig. 4.30 A "rosette" orbit, showing a "snapshot" every 15° of precession

trajectory is closed. If the exponent of $1/r$ is not exactly 2, the two periods are different, the orbit does not close and we have a situation like Fig. 4.30. This effect cannot be seen if the orbit is circular, because a circle rotating on itself is not different from a circle.

Astronomers have observed for centuries the apparent trajectories of the planets in the sky with high accuracy. The absolute trajectories are obtained subtracting the now well-known motion of earth. The aphelia and the perihelia, in particular, can be accurately identified. If the force is proportional to the inverse square distance from the sun, these points should remain fixed. Indeed, this is almost the case, but not quite. Very slow movements, called precessions, of the perihelia are observed. They are the effects of the forces of the other planets, the larger ones in particular, that act on them in addition to the sun. Observations and calculations agree, with an exception, which was found by Le Verrier in 1849. He calculated the precession of the perihelion of Mercury, the nearest to sun, in 10 arc minutes per century. The largest fraction of that is explained by the just mentioned effects of other planets. But not completely; 43 arc sec per century remained unexplained. A number of hypothesis were advanced, but all of them failed. This was the first historical example of the limits of the Newton law. The explanation of the anomalous precession of the Mercury perihelion by Albert Einstein (1879–1955) in 1915 marked the success of general relativity.

The 3rd Kepler law is a consequence of the 1st one. Let us prove that. We start with the consideration that the period T is the area of the orbit A divided by the areal velocity and expressing the latter in terms of the angular momentum L using Eq. (4.45).

$$T = A/\frac{dA}{dt} = \frac{2\pi abm}{L}.$$

We now write the acceleration a_r Eq. (4.48) using this equation and writing the parameter p in terms of the axes, Eq. (4.42)

$$a_r = -\frac{L^2}{m^2}\frac{a}{b^2}\frac{1}{r^2} = -4\pi^2\frac{a^3}{T^2}\frac{1}{r^2}.$$

The force on the planet is the Newton force, and we can write

$$F_r = ma_r = -4\pi^2 m\frac{a^3}{T^2}\frac{1}{r^2} = -G_N\frac{mM}{r^2}$$

and finally

$$\frac{a^3}{T^2} = G_N\frac{M}{4\pi^2}. \tag{4.51}$$

That is the 3rd Kepler law: the squares of the periodic times are proportional to the cubes of the ellipse semi-major axis, for all the bodies orbiting the same central body (of mass *M*).

4.12 The Newton Solution

In this section we shall not introduce any new concept, rather we shall show how Isaac Newton demonstrates some of those we discussed in the previous sections. Reading pages of the giants is, in fact, very instructive, even if, as is the case with Newton, it is not always easy. After having given the necessary preliminary information, we shall read one page of the *Principia*, explaining their meaning line by line. As we shall see, the Newton arguments are mainly geometrical. The novelty, with respect to what was already known to the Greeks, is the final passage to the limit for the length of the considered orbit arc going to zero.

After having stated the laws of motion in Sect. 1 of the *Principia*, Newton dedicates Sect. 2 to "*The determination of the centripetal forces*". Here he considers orbits of various geometrical shapes under the action of a force directed to a immovable center (i.e. *centripetal*). The case we shall take is the ellipse with the center of force in a focus. In the first two "Propositions" of Sect. 2, he shows that, in any case, if the areal velocity is constant the force is directed to the center and vice versa.

Subsequently, in Proposition VI, Newton lays down the basic scheme he shall use to solve the above-mentioned problems. The scheme is shown in Fig. 4.31.

A body moves on the arc *PQ* of its orbit in the short time interval Δt. If there were no gravitational force from the sun, the planet would move of rectilinear uniform motion on the displacement *PR*. On the other hand, if abandoned still in *P* the planet would drop in the time Δt, under the gravitational attraction, by the

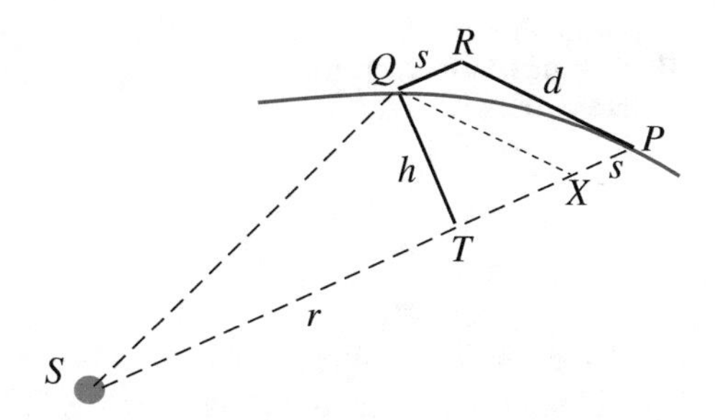

Fig. 4.31 The scheme of Proposition VI

displacement PX. If the force is constant, the motion is uniformly accelerated and PX is proportional to Δt^2. If both conditions are present, the displacement is the diagonal PQ. We now draw the segment QR parallel to PX. QR touches the trajectory in Q.

What we just stated would be true if the force were constant during Δt, which is not true. However, the smaller is Δt the smaller is the variation of the force in that interval. This means going to the limit of $\Delta t \to 0$. The limit geometrically corresponds to approximate the segment of the trajectory with a segment of parabola. The motion is then equal to what was found by Galilei for the projectiles on earth.

On the other hand, QR is also proportional to the acceleration and to the force F we are looking for, namely $QR \propto F \times \Delta t^2$, or $F \propto QR/\Delta t^2$.

For the constancy of the areal velocity, the time interval is proportional to the area swept by the position vector in that interval, which is the area of the triangle SQP. The latter, in turn, is proportional to the product of its base SP and its height QT, and we have

$$F \propto \frac{QR}{QT^2}\frac{1}{SP^2} = \frac{QR}{QT^2}\frac{1}{r^2}. \tag{4.52}$$

This expression is valid for any curve. We shall see how it simplifies in the case of the elliptic orbit, with the center of force in a focus. To do that, we shall need to know some definitions and four properties of the ellipse. We give them here without proof.

A *diameter* is a chord going through the center of the ellipse. Consider the tangent to the ellipse in any given point P on it (see Fig. 4.32). Let be PP' the diameter passing in P and DK the diameter parallel to the tangent in P. The diameters PP' and DK are called *conjugate* diameters.

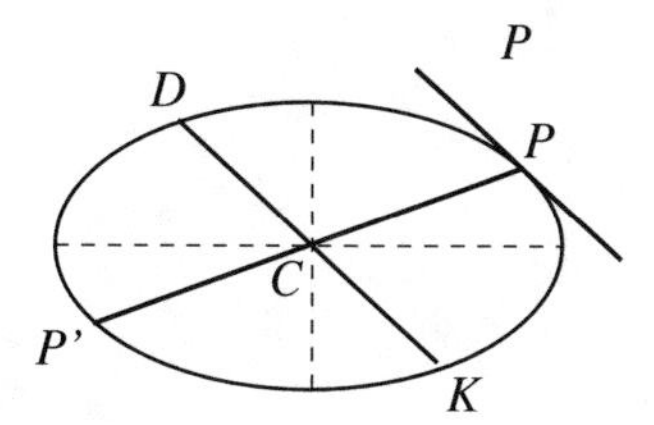

Fig. 4.32 PP' and DK are conjugate diameters

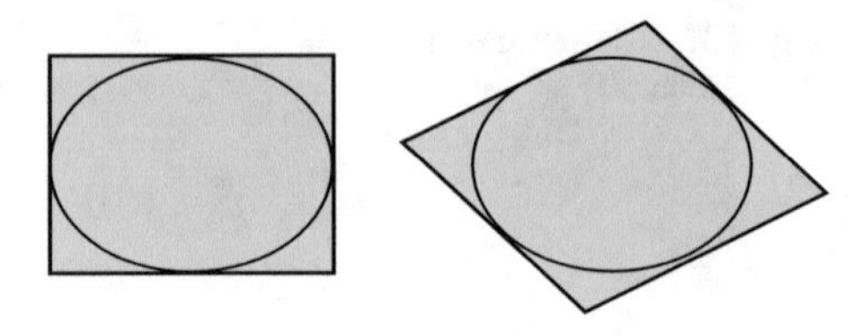

Fig. 4.33 Property 2. Parallelograms with conjugate diameters sides

Notice that the conjugate diameters bisect each other but, in general, do not have equal lengths, neither t0 they cross at right angles.

Property 1. The sums of the distances of any point of the ellipse from the two foci are equal and are equal to the major axis, $2a$.
Property 2. (Fig. 4.33). All the parallelograms having conjugate diameters as sides have the same area. It is equal to the area of the parallelogram having, in particular, the axes as sides, namely $4ab$.
Property 3 (Fig. 4.34). The two focal lines that join any point P of the ellipse form equal angles with the tangent in that point.
Property 4 (Fig. 4.35). Every diameter bisects all the conjugate chords. For any given diameter the ratio between the areas of the rectangles made by the two segments of the diameter and the square of the corresponding semi-chord are equal. Namely

$$\frac{PQ' \times Q'P'}{QQ'^2} = \frac{PR' \times R'P'}{RR'^2} = \frac{PS' \times S'P'}{SS'^2} = \cdots. \tag{4.53}$$

We have now the properties of the ellipse we shall need and we can read Proposition XI.

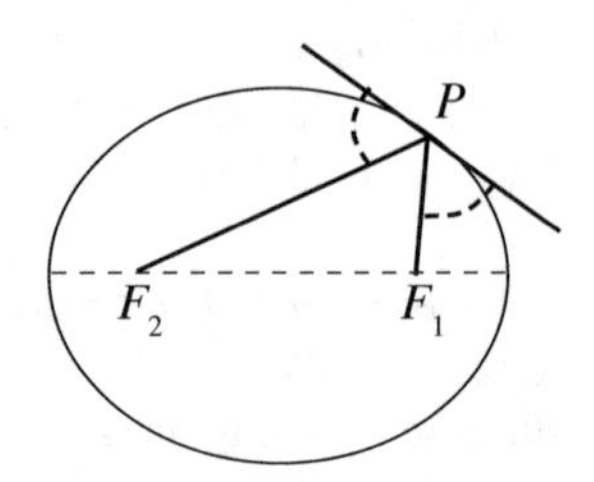

Fig. 4.34 Property 3. Two focal lines and their angles with the tangent

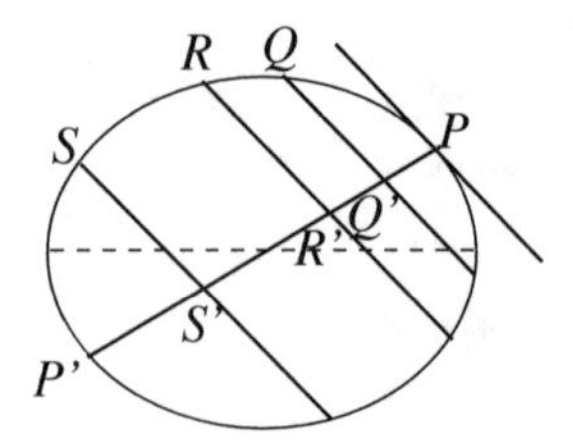

Fig. 4.35 Property 4

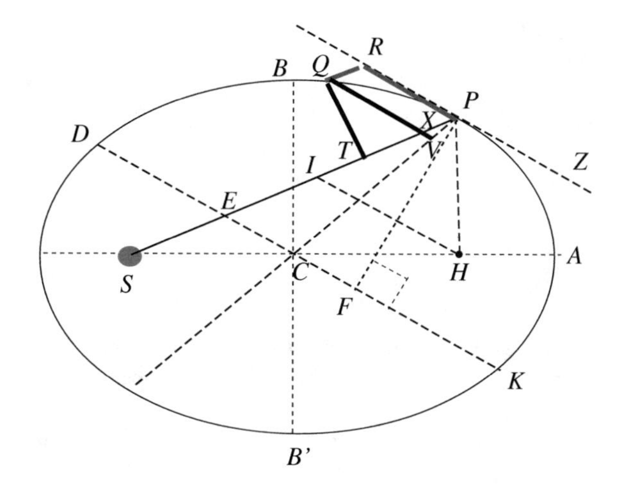

Fig. 4.36 The Newton diagram for Proposition XI

Proposition XI states:

> if a body revolves in an ellipse; it is requested to find the law of the centripetal force directed to the focus of the ellipse.

The proof shows that the ratio QR/QT^2 in Eq. (4.52), in the particular case of the ellipse with the center of force in a focus, is equal to the latus rectum, which we called $2p$ and he calls L. We shall use his symbol in this section (no risk of confusion with the angular momentum).

Figure 4.36 reproduces the diagram on which the theorem is developed. The first lines of the Proposition are:

> Let S be the focus of the ellipse. Draw SP cutting the diameter DK of the ellipse in E, and the ordinate QV in X; and complete the parallelogram QXPR

The sun (the center of force) is in the focus S; H is the other focus, C is the center, $CA = a$ and $CB = b$ are the semi-major and the semi-minor axes respectively. At a certain instant the planet is in P, $SP = r$ is the position vector from the sun. We draw the tangent RPZ to the ellipse in P and the line QV parallel to it. Be X and V the points were it cuts SP and PC respectively. We also draw the lines of $QRPT$ as in Fig. 4.31. To complete the diagram we draw the perpendicular from P to the diameter DK and call F the point in which they meet.

The Newton language is extremely synthetic. What is evident for him is not always evident for us. We shall explain his lines immediately.

> It is evident that EP is equal to the greater semi axis AC: for drawing HI from the other focus H of the ellipse parallel to EC, because CS and CH are equal, ES and EI will be also equal; and hence EP is half the sum of PS and PI, that is (because of the parallels HI and PR, and the equal angles IPR, HPZ) of PS and PH, which taken together are equal to the whole axis 2AC.

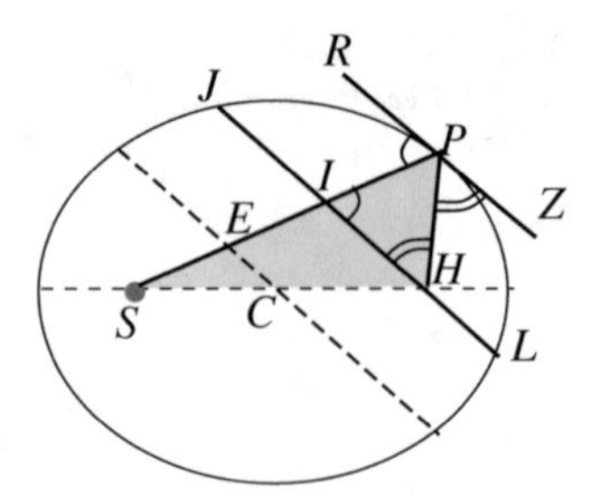

Fig. 4.37 *EP* is equal to the major axis

The geometric elements of Fig. 4.36 that are relevant for this step are redrawn in Fig. 4.37. We start from the equation (Property 1)

$$2a = PH + PS = PH + PI + IE + ES. \tag{4.54}$$

The triangle *IPH* is isosceles with vertex in *P*. This is because:

- the angles *RPI* and *PIH* are equal, as alternate interior angles of the two parallel lines *JL* and *RZ*
- the angles *HPZ* and *IHP* are equal as alternate interior angles of the same lines
- the angles *PIR* and *HPZ* are equal for the Property 3 of the ellipse

Consequently the angles *PIH* and *IHP* are equal, which proves the statement. Hence *PH* = *PI* and we can simplify Eq. (4.54) as

$$2a = 2PI + IE + ES. \tag{4.55}$$

The triangles *ISH* and *ESC* are similar because they have the same angle in the vertex *S* and the sides opposite to it (*EC* and *IH* respectively) are parallel. In addition, *SH* is twice *SC* and consequently *SI* = 2 *ES*, that is also *ES* = *IE*. Substituting in Eq. (4.55) we obtain

$$2a = 2PI + 2IE = 2(PI + IE) = 2PE \tag{4.56}$$

and finally

$$PE = a. \tag{4.57}$$

Now Newton works on *QR*:

> Draw QT perpendicular to SP [*we did that already*], and putting L for the principal latus rectum of the ellipse (or for $2BC^2/AC$ [*see our* Eq. (4.41)]) we shall have $L \cdot QR : L \cdot PV = QR : PV = PE : PC = AC : PC$

Is not so simple to follow the Newton language. He uses proportions, which we shall write as fractions to make them more readable. In addition, when he takes a route he does not tell us the reasons, which are understood only at the end. Let us trust him and follow. He starts from the ratio *QR/PV* with numerator and

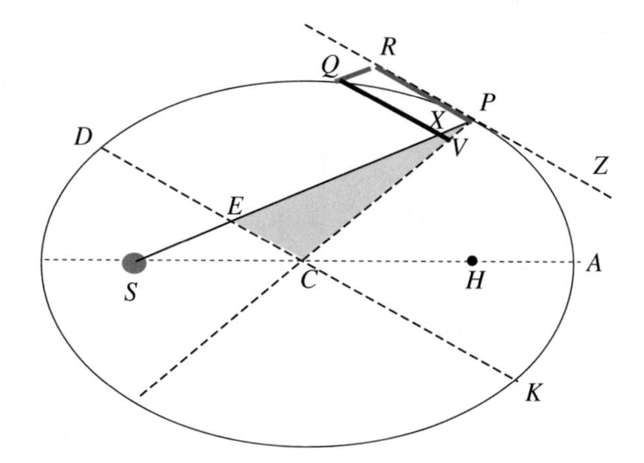

Fig. 4.38 Working on QR

denominator multiplied by L, because at the end this will be useful. The relevant geometrical elements are drawn in Fig. 4.38.

The first step is trivial

$$\frac{L \cdot QR}{L \cdot PV} = \frac{QR}{PV}.$$

But $QR = PX$ by construction. Let us find PX.

The triangles PXV and PEC are similar because they have a common vertex in P and the two opposite sides, XV and EC, are parallel. Consequently $PE/PC = PX/PV$ and also

$$\frac{QR}{PV} = \frac{PE}{PC}.$$

Using Eq. (4.57) namely $PE = AC$ we have

$$\frac{QR}{PV} = \frac{AC}{PC}$$

and in conclusion

$$\frac{L \cdot QR}{L \cdot PV} = \frac{AC}{PC}. \tag{4.58}$$

The next step is working on PV. The single line of Newton is:

also $L \cdot PV : GV \cdot PV = L : GV$ and $GV \cdot PV : QV^2 = PC^2 : CD^2$

Once more Newton works on a ratio, L/GV, and multiplies numerator and denominator by the same quantity, which is PV, the quantity we are now looking for. The relevant geometrical elements are shown in Fig. 4.39.

Fig. 4.39 Work on PV

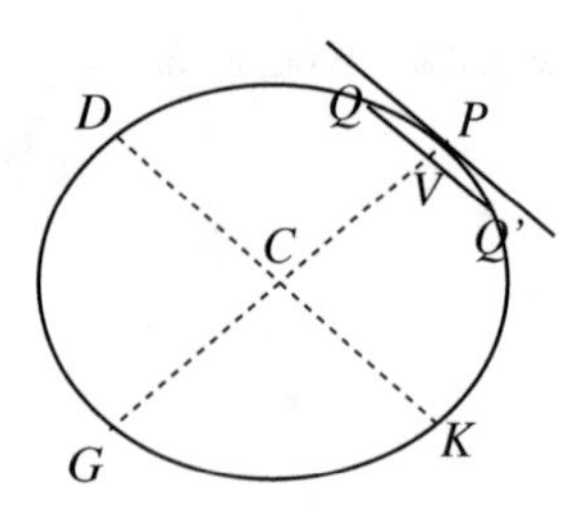

The first step is again trivial,

$$\frac{L \cdot PV}{GV \cdot PV} = \frac{L}{GV}. \tag{4.59}$$

We use the Property 4 of the ellipse applied to the diameter PG and to the semi-chords QV and DC conjugated to it, getting

$$\frac{GV \cdot PV}{QV^2} = \frac{PC \cdot CG}{CD^2} = \frac{PC^2}{CD^2}. \tag{4.60}$$

Newton continues, finding a fourth proportion. Finally he will put the four together. We take a breath, abandon him for a moment and put immediately together the three Eqs. (4.58), (4.59) and (4.60) we found. We multiply them member by member and obtain

$$\frac{L \cdot QR}{L \cdot PV} \frac{L \cdot PV}{GV \cdot PV} \frac{GV \cdot PV}{QV^2} = \frac{AC}{PC} \frac{L}{GV} \frac{PC^2}{CD^2}.$$

Simplifying, but keeping L that will be useful, we have

$$\frac{L \cdot QR}{QV^2} = \frac{L \cdot AC \cdot PC}{GV \cdot CD^2}. \tag{4.61}$$

We need another proportion, the last one.

By Cor. II, Lem. VII [*is the rule for going to the limit*], when points P and Q coincide, $QV^2 = QX^2$ and QX^2 or QV^2: $QT^2 = EP^2$:$PF^2 = CA^2$:PF^2, and (by Lem. XII) = CD^2:CB^2.

We now need to express $1/QT^2$. As usual, Newton works with proportions, and we shall do the same with ratios. This time it is QX^2/QT^2. At the end we shall take the limit for the length of the arc PQ going to zero, namely to have the points P and Q coincident. In this limit, points X and V coincide too and it is then convenient to consider QV^2/QT^2 in place of QX^2/QT^2. The relevant geometrical elements are drawn in Fig. 4.40.

Fig. 4.40 Work on $1/QT^2$

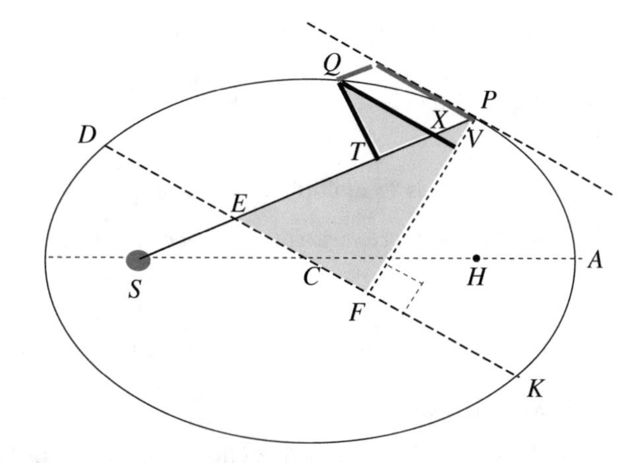

The triangles EPF and XQT are similar, because

- the angles in F and T are equal because are both right,
- the sides of the angles respectively in P and Q (in evidence in the figure) are mutually perpendicular, hence they are equal.

Hence

$$\frac{QV}{QT} = \frac{QX}{QT} = \frac{EF}{PF}$$

and, as $EP = CA$,

$$\frac{QV}{QT} = \frac{CA}{PF}. \tag{4.62}$$

To find PF we use Property 2. Figure 4.41a shows that PF is one half of the height to DK of the drawn parallelogram on conjugate diameters.

Property 2 gives: $PF \cdot CD = BC \cdot CA$, or

$$\frac{CA}{PF} = \frac{CD}{CB},$$

Fig. 4.41 Using Property 2

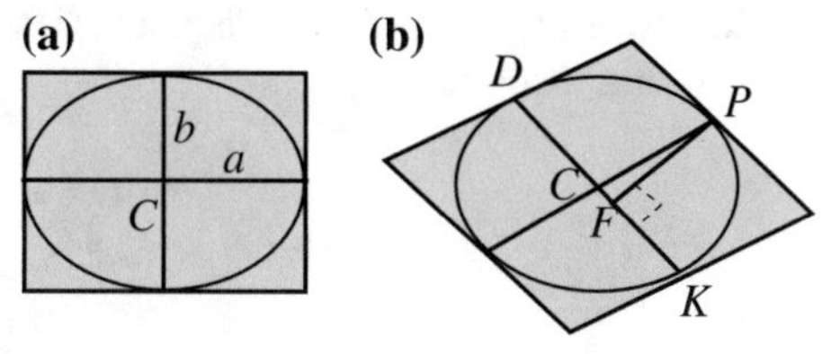

which, substituting in Eq. (4.62) and squaring, gives

$$\frac{QV^2}{QT^2} = \frac{CD^2}{CB^2}. \tag{4.63}$$

The next step is to multiply the four proportions. Newton writes:

Multiplying together corresponding terms of the four proportions, and simplifying, we shall have

$$L \cdot QR : QT^2 = AC \cdot L \cdot PC^2 \cdot CD^2 : PC \cdot GV \cdot CD^2 \cdot CB^2 = 2PC : GV$$

since $AC \cdot L = 2BC^2$''

We have already multiplied the first three ratios obtaining Eq. (4.61). Hence we multiply now its members with those of Eq. (4.63)

$$\frac{L \cdot QR}{QV^2}\frac{QV^2}{QT^2} = \frac{L \cdot AC \cdot PC}{GV \cdot CD^2}\frac{CD^2}{CB^2}$$

and simplify

$$\frac{L \cdot QR}{QT^2} = \frac{L \cdot AC \cdot PC}{GV \cdot CB^2}.$$

Recalling the relation between latus rectum and axes $L \cdot AC = 2CB^2$ and simplifying we finally have

$$\frac{L \cdot QR}{QT^2} = \frac{2PC}{GV}. \tag{4.64}$$

Remember that the factor in Eq. (4.52) we want to express is QR/QT^2. We have it now in Eq. (4.64). The final step is taking the limit for $P \to Q$. Remember that the concept of limit was not known before Newton. He writes:

But the points Q and P coinciding, 2PC and GV are equal. And therefore the quantities $L \cdot QR$ and QT^2, proportional to these, will be also equal. Let those equals be multiplied by $\frac{SP^2}{QR}$, and $L \cdot SP^2$ will become equal to $\frac{SP^2 \cdot QT^2}{QR}$.

In the limit in which the arc PQ becomes infinitely small, point V coincides with P. Consequently, GV becomes equal to $2PC$, and the second member of Eq. (4.64) goes to one, becoming

$$L \cdot QR = QT^2.$$

Now multiply both members by SP^2/QR and get

$$L \cdot SP^2 = \frac{SP^2 \cdot QT^2}{QR}.$$

Finally, Newton concludes:

And therefore (by Cor. I and v, Prop. VI) the centripetal force is inversely as $L \cdot SP^2$, that is, inversely as the square of the distance SP.

Q.E.D.

Namely:

$$F \propto \frac{QR}{QT^2} \frac{1}{SP^2} = \frac{1}{L \cdot SP^2} \tag{4.65}$$

and, given that L, our $2p$, is a constant for a given ellipse,

$$F \propto \frac{1}{SP^2} = \frac{1}{r^2}. \tag{4.66}$$

The force is inversely proportional to the square of the distance from the center. That is what we had to show.

4.13 The Constants of Motion

We now go back to the main stream and consider the potential and the kinetic energy of a body of mass m in the gravitational field of a body of mass M, moving on an ellipse.

We start with its angular momentum, which we call L and go back to our formalism calling p the semi-latus rectum. From Eq. (4.50) we can write

$$\frac{L^2}{m^2} = pG_N M = r^4 \left(\frac{d\theta}{dt} \right)^2, \tag{4.67}$$

which, using Eq. (4.42) is

$$L^2 = \left(1 - e^2\right) a G_N m^2 M. \tag{4.68}$$

In words, the square of the angular momentum is proportional to the major axis. For a given major axis, the angular momentum is the largest for $e = 0$, which is the circle. It decreases for increasing e, i.e. for the ellipse becoming more and more squeezed.

Consider now the potential energy, and make use for r of the ellipse equation

$$U_p = -G_N \frac{mM}{r} = -G_N \frac{mM}{p}(1 + e\cos\theta). \tag{4.69}$$

For the kinetic energy, remember Eq. (4.39)

$$U_k = \frac{m}{2}\left[r^2\left(\frac{d\theta}{dt}\right)^2 + \left(\frac{dr}{dt}\right)^2\right]. \tag{4.70}$$

Using the expression of dr/dt given by Eq. (4.47) and using Eq. (4.52), we have

$$U_k = \frac{G_N Mm}{2} p\left(\frac{1}{r^2} + \frac{e^2}{p^2}\sin^2\theta\right).$$

We use now the Eq. (4.41) of the ellipse to express $1/r^2$ and, taking into account that $\sin^2\theta + \cos^2\theta = 1$, obtain

$$U_k = \frac{G_N Mm}{2p}\left[1 + e^2 + 2e\cos\theta\right]. \tag{4.71}$$

Both potential energy, Eq. (4.69), and kinetic energy Eq. (4.56) depend on the position of the planet and consequently on time. Not so the total energy, which is their sum

$$U_{\text{tot}} = U_p + U_k = G_N \frac{Mm}{2p}\left[e^2 - 1\right], \tag{4.72}$$

which we can also write, in equivalent manner

$$U_{\text{tot}} = -G_N \frac{Mm}{2a}. \tag{4.73}$$

In conclusion, the total energy of the planet depends only on the semi-major axis. Different orbits, such as those in Fig. 4.42, which have the same semi-major

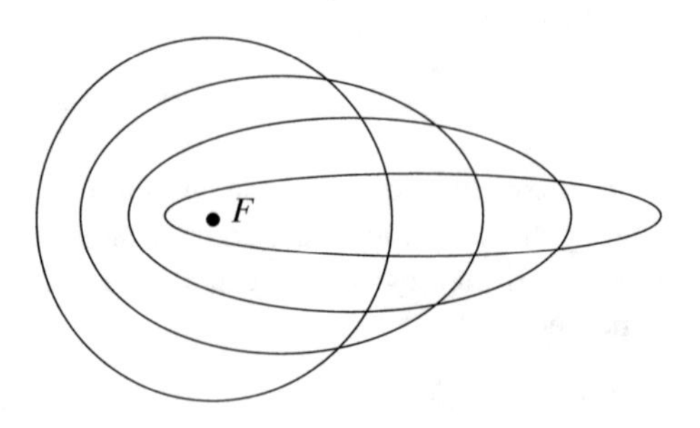

Fig. 4.42 Orbits of the same energy and different angular momenta

but different semi-minor axes have the same total energy. However, as we have seen above, the angular momentum grows for decreasing eccentricity.

Pay also attention to the fact that the total energy is negative. However, this is not the only possibility for a body moving about the sun, or any other source of gravitational force. As a matter of fact, in our demonstration we have used only Eq. (4.41). This is not only the equation of the ellipse, but, more generally of all the conics, ellipse if $e < 1$, parabola if $e = 1$, hyperboles if $e > 1$. The three cases correspond, from the physical point of view, to total energy (4.73) negative, null or positive respectively. The potential energy is always negative, tending to zero at infinite distance from the center. The kinetic energy can be positive or zero. Consequently, at infinite distance the total energy is positive or, as a minimum, zero. If the total energy of a body is negative, it must remain at finite distances. The orbit is said to bound. The ellipse (including the circle as a particular case) is the only conic that does not reach infinity. If on the contrary, the total energy of a body is positive, it will be able to go farther and farther; at infinite distances, or more realistically at distances large enough to have negligible potential energy, all its energy is kinetic, positive in fact. The intermediate case is when the body reaches infinity with zero kinetic (and total) energy. The trajectory is a parabola.

Problems

4.1. A pendulum having 1 s period on the surface of earth is brought on the surface of a planet having the same radius of earth and mass four times larger. What is the period of the pendulum?

4.2. The gravitational potential difference between two points on the earth surface (at the same latitude) is 1000 $m^2 s^{-2}$. What is the difference between the heights were they are located?

4.3. We abandon a body at the distance from earth of the moon orbit with no velocity. Will it fall with constant velocity? With constant acceleration?

4.4. We move a body from the sea level to the top of a mountain 5000 m high (same latitude). How does its mass vary? How does its weight vary?

4.5. Does the velocity at which a satellite moves in a circular orbit around the earth depend on the mass of the earth? on the mass of the satellite? on the radius of the orbit?

4.6. The apparent diameter of the sun as seen from earth is approximately $\alpha = 0.55°$. What would be the period of a hypothetical planet orbiting just out of the sun?

4.7. We want to put an artificial satellite in orbit around the earth having a period of 2 h. Knowing the gravity acceleration g on the surface of earth an its radius R_E, find the height of the requested orbit above the surface.

4.8. Consider a spring (of a ballpoint pen) with rest length 3 cm and elastic constant k = 50 N/m. We fix to its two extremes two equal Pb spheres (density ρ = 11 × 10^3 kg/m^3), of mass m = 10^4 kg each. Assume, unrealistically, that all frictions can be neglected. How much will the spring shrink under the action of the gravitational attraction of the two spheres?

4.9. Knowing the values of g, of G_N and of the radius of earth ($R_E = 6.4 \times 10^6$ m), make an estimate of the mass and of the mean density of earth.

4.10. Knowing the values of G_N and of the radius of earth orbit ($r_E = 1.5 \times 10^{11}$ m) and of its period, make an estimate of the mass of the sun. Knowing that its apparent diameter from earth is 0.55°, estimate its mean density.

4.11. The sun moves on an orbit that we can consider circular about the center of the Galaxy. The radius of the sun orbit is $R_S = 25\,000$ l year $= 2.5 \times 10^{20}$ m, his velocity is $\upsilon_S = 250$ km/s. Compare these data with those relative to the motion of earth about the sun ($r_E = 1.5 \times 10^{11}$ m, $\upsilon_E = 30$ km/s). Make an estimate of the total mass M_{tot} around which the sun orbits; give it as a multiple of the solar mass M_S.

4.12. Io, one of the Jupiter satellites, has the orbital period $T_I = 1.77$ d and the orbit radius $r_I = 4.22 \times 10^8$ m. Compare these data with those of the motion of the earth about the sun ($r_E = 1.5 \times 10^{11}$ m, $\upsilon_E = 30$ km/s). Determine the mass of Jupiter in solar masses.

4.13. Find a procedure to determine the mass of earth.

4.14. Knowing that the earth moves around the sun with the velocity of $\upsilon_E = 30$ km/s, find the gravitational potential of the sun $\phi_S(E)$ in the points of earth orbit. The gravitational potential in a point of the earth is the sum of the just considered $\phi_S(E)$ due to the sun and of the gravitational fields of the earth itself, say $\phi_E(E)$, and of all the Galaxy, say $\phi_G(E)$. Calculate the values of the latter two relative to $\phi_E(E)$, knowing that the masses in the three cases are approximately $M_S = 2\times10^{30}$ kg, $M_E = 6\times10^{24}$ kg, $M_G = 2\times10^{41}$ kg and taking as distances, from earth to sun $r_{ES} = 1.5 \times 10^{11}$ m, radius of earth $r_E = 6.4 \times 10^6$ m, distance from sun to the center of Galaxy $r_{SG} = 2.5 \times 10^{20}$ m

Chapter 5
Relative Motions

In our study of the kinematics of the material point, we have already seen that the equations of motion depend on the reference frame. The law of motions, and more generally all the laws of Physics, transform, as we say, from one frame to another. This chapter is dedicated to the study of these transformations.

Two reference frames may differ in different ways.

The two frames have no relative motion, their co-ordinate homologous axes are parallel, but have different origins; the frames differ for a rigid translation.

The two frames have no relative motion and coincident origins, but the directions of the axes are different; the frames differ for a rigid rotation.

One frame can translate relative to the other in time with uniform or varying velocity, or it can rotate, again with constant or varying angular velocity, or it can translate and rotate contemporarily.

In Sect. 5.1, we shall consider two stationary frames relative to one another, with a relative translation or rotation. We shall see that the laws of Physics have the same form, namely the same mathematical expressions, in both frames. As we say, the laws are covariant under translations and rotations. The meaning of the term will be explained.

We shall then consider frames in relative motion and learn that, when the relative motion is a translation with constant speed, the laws of mechanics are also covariant. This is the relativity principle, a fundamental principle of physics, established by Galilei. For example, experiments done inside a closed room in a ship cannot establish whether the ship is moving in uniform motion or is standing still. One of the consequences is that once we have found an inertial frame, any other frame moving in a uniform translation motion relative to it is also inertial.

In Sect. 5.3, we shall deal with the relative translatory accelerated motion. As already anticipated, in any reference that accelerates relative to an inertial frame, the Newton laws are not valid. For example, a body at rest can start moving without any force acting on it. The motion can be described introducing fictitious forces, which are known by several equivalent names, apparent forces of the relative motions, pseudo-forces and inertial forces. We feel such "force," for example, when

A. Bettini, *A Course in Classical Physics 1—Mechanics*,
Undergraduate Lecture Notes in Physics, DOI 10.1007/978-3-319-29257-1_5

we brake suddenly in a car. In Sect. 5.4, we shall deal with the general case (translation and rotation) and we shall see the relations between velocities and between accelerations in two frames of any relative motion. In Sect. 5.5, we shall discuss several examples of motion in frames rotating relative to an inertial frame.

Any frame at rest in a laboratory on earth does, in fact, move with earth. In initial, and quite good, approximation, these frames can be considered to be inertial. Not completely, however, because earth rotates on its axis and moves along its orbit around the sun, and even the sun moves along its orbit in the galaxy. In Sect. 5.7, we shall study a few effects of the inertial forces in frames at rest relative to earth: the variation with latitude of the magnitude of the weight, the rotation of the oscillation plane of pendulums, the deviation from the vertical of free fall and the circulation of winds.

The inertial forces acting on a body are proportional to its inertial mass, while the gravitational attraction of earth is proportional to its gravitational mass. This observation allows for the realization of very delicate experiments to check whether the two masses are different or equal. We shall describe such an experiment in Sect. 5.8.

5.1 Covariance of the Physical Laws Under Rotations and Translations

Consider two Cartesian reference frames, which are stationary relative to each other, S (coordinates x, y, z, origin O) and S' (coordinates x', y', z', origin O').

A physical law is a mathematical equation between physical quantities. The relation between the two frames can be a rigid rotation or a rigid translation. Let us start with rotations.

We choose the origins of the two frames as coincident. For simplicity, we consider their z-axes also to be coincident. The frames differ for a rotation, by an angle θ, around this axis. The rotation is in the common plane xy, as shown in Fig. 5.1.

Suppose now that an observer in S makes a very simple experiment. He measures, using a balance, the masses of two objects, finding the values m_1 and m_2. He

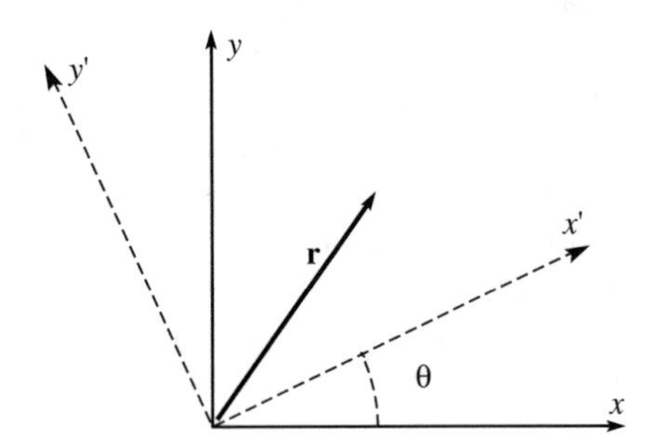

Fig. 5.1 Two reference frames different for a rigid rotation

finds that the second mass is three times the first. He writes the relation (let us call it the “law”)

$$m_2 = 3m_1. \tag{5.1}$$

Another observer in S' performs the same experiment. We indicate with a prime the homologous quantities he finds. In this very simple case, considering that the procedure of measuring the mass with a balance does not depend on the direction of the axes, we can conclude that he will find the same result, namely

$$m'_1 = m_1, \quad m'_2 = m_2. \tag{5.2}$$

The second observer also states that

$$m'_2 = 3m'_1. \tag{5.3}$$

Equation (5.3) has the same *form* as Eq. (5.1). Namely, the two observers describe the same phenomenon with laws of the same form. Indeed, mass is a scalar quantity, which is invariant under rotations of the axes. In general, a relation between scalar quantities, if valid in a frame, is also valid in any frame rotated relative to the first one, because both sides of the equation do not vary going from one frame to the other.

But it is not always so. Suppose that the observer in S measures two components of the velocity of a point, finding the values υ_x and υ_y. He finds, again, the second quantity three times larger than the first and writes the equation

$$\upsilon_y = 3\upsilon_x. \tag{5.4}$$

What would the observer in S' find? We know the answer because we know the relations between velocities in the two frames:

$$\upsilon'_x = \upsilon_x \cos\theta + \upsilon_y \sin\theta; \quad \upsilon'_y = -\upsilon_x \sin\theta + \upsilon_y \cos\theta. \tag{5.5}$$

We calculate the ratio between υ'_x and υ'_y, also employing Eq. (5.5):

$$\frac{\upsilon'_y}{\upsilon'_y} = \frac{-\upsilon_x \sin\theta + \upsilon_y \cos\theta}{\upsilon_x \cos\theta + \upsilon_y \sin\theta} = \frac{-\sin\theta + \frac{\upsilon_y}{\upsilon_x}\cos\theta}{\cos\theta + \frac{\upsilon_y}{\upsilon_x}\sin\theta} = \frac{-\sin\theta + 3\cos\theta}{\cos\theta + 3\sin\theta}.$$

In conclusion, in S', we have

$$\upsilon'_y = \frac{-\sin\theta + 3\cos\theta}{\cos\theta + 3\sin\theta}\upsilon'_x. \tag{5.6}$$

The form of the "law" is different this time in the two frames, being (5.4) in S and (5.6) in S'. This is an obvious consequence of the fact that the components of a vector transform differently one from another.

But wait a moment, a law may be valid in both frames, even if its sides are not invariant, as in the case of the masses; rather, it is sufficient that, if they vary, *in the same way*. Let us see what happens for a law linking vector quantities.

The observer in S', which we assume, for the sake of this example, to be inertial, studies the motion of a material point. He measures the acceleration $\mathbf{a}$ (namely its three components), the force acting on the point $\mathbf{F}$ (again, the three components) and the mass m. He finds the relation

$$\mathbf{F} = m\mathbf{a}. \tag{5.7}$$

More explicitly, this vector relation corresponds to three equations:

$$F_x = ma_x, \quad F_y = ma_y, \quad F_z = ma_z. \tag{5.8}$$

We know how the components of the vectors, such as $\mathbf{F}$ and $\mathbf{a}$ are, transform from one frame to the other, namely

$$\begin{aligned} F'_x &= F_x \cos\theta + F_y \sin\theta; \quad F'_y = -F_x \sin\theta + F_y \cos\theta; \quad F'_z = F_z \\ a'_x &= a_x \cos\theta + a_y \sin\theta; \quad a'_y = -a_x \sin\theta + a_y \cos\theta; \quad a'_z = a_z \end{aligned} \tag{5.9}$$

and we can write

$$\begin{aligned} F'_x &= ma_x \cos\theta + ma_y \sin\theta = m(a_x \cos\theta + a_y \sin\theta) \\ F'_y &= -ma_x \sin\theta + ma_y \cos\theta = m(-a_x \sin\theta + a_y \cos\theta) \\ F'_z &= ma_z \end{aligned}$$

and, for Eq. (5.9),

$$F'_x = ma'_x, \quad F'_y = ma'_y, \quad F'_z = ma'_z, \tag{5.10}$$

which has the same *form* as Eq. (5.8). Both sides of the equations are different, varying from one frame to the other. However, they vary in the same way, because both sides are vectors. Thus, we say that the equation is *covariant*.

In conclusion, the laws of Physics keep the same form under rotations of the axes, or, in other words, are covariant under rotations. And yet, from another perspective, it is impossible experimentally to establish any privileged directions of the reference axes. Space should be considered isotropic, without preferential directions.

The case of the translations is very simple. Scalar quantities obviously have the same values in two frames differing for a translation. This is also valid for vectors, which are simply translated; hence, they are the same vector.

5.2 Uniform Relative Translation. Relativity Principle

Consider now two reference frames, *S* and *S'*, which are in relative motion. We arbitrarily call one of them *S* (origin *O* and coordinates x, y, z) fixed and the other one *S'* (origin *O'* and coordinates x', y', z') mobile. We consider the case of a uniform translation of *S'*. All the points of *S'* move with the same velocity relative to *S*, which is constant in magnitude and direction. The frames, for example, might consist of one fixed on the ground, the other on a carriage moving on straight rails, or a frame fixed at the shore and one on a ship moving straight, in both cases with uniform motion.

The axes of the two frames do not change the relative directions and we can take them as being parallel. Fig. 5.2 shows the two frames at a certain instant. At a later time, the mobile frame will be in a different position, more on the right, but its axes will still be parallel to the axes of *S*. We choose $t = 0$ as the time at which the axes of the two frames overlap.

Figure 5.2 shows the material point *P* and its trajectory. The position vectors **r** and **r'** of *P* in the two frames have the well-known relation

$$\mathbf{r} = \mathbf{r}' + \mathbf{r}_{O'} \tag{5.11}$$

where **r**O' is the position vector of the origin *O'* of the mobile frame *S'* in the fixed frame *O*, namely *OO'*.

A fixed and a mobile observer see the point *P* moving with different velocities, **v** and *v'*. To find their relation, we take the time derivatives of Eq. (5.11), obtaining

$$\mathbf{v} = \mathbf{v}' + \mathbf{v}_{O'} \tag{5.12}$$

where **v***O'* is the velocity of the origin *O'* of the mobile frame, and also of all its points (because the motion is a translation) as seen by *S*. The velocity of an insect flying in the ship in the above example relative to the shore is the vector sum of the velocity of the insect relative to the ship and the velocity of the ship relative to the shore.

A further time derivation gives the relation between accelerations

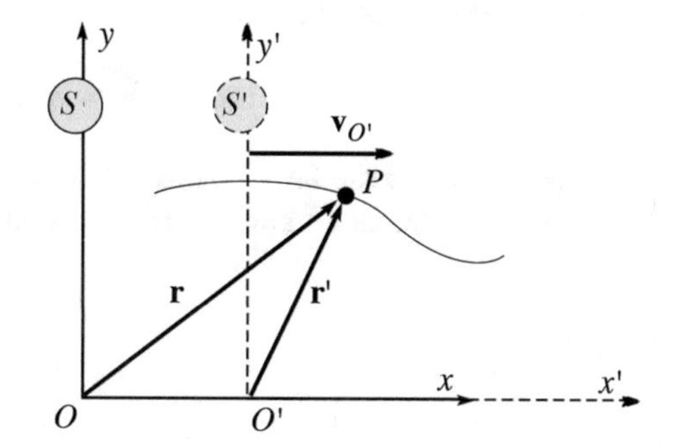

Fig. 5.2 Two reference frames in relative uniform translation motion

$$\mathbf{a} = \mathbf{a}' + \mathbf{a}_{O'} \tag{5.13}$$

where **a**O' is the velocity of the origin *O'* of the mobile frame, and also of all its points.

We now consider the important particular case in which the translation of *S'* relative to *S* is uniform, namely the velocity of its origin, and of all its points, seen by *S* is constant in time

$$\mathbf{v}_{O'} = \text{constant}. \tag{5.14}$$

Then, obviously,

$$\mathbf{a}_{O'} = 0 \tag{5.15}$$

and Eq. (5.13) becomes

$$\mathbf{a} = \mathbf{a}'. \tag{5.16}$$

The accelerations in the two frames are equal. The implications of this simple conclusion are extremely important considering inertial frames.

If *S* is an inertial frame, any material point *P* not subject to forces moves at constant velocity **v** (or remains at rest). In other words, its acceleration is zero, **a** = 0. In the mobile frame, its acceleration **a'**, which is equal to **a**, is also zero. Consequently, *S'* is inertial too.

We conclude that, given an inertial reference frame, any other frame moving relative to it by a uniform translation is also inertial.

What about the second Newton law? It is valid in the frame *S*, which is inertial by assumption. Is it also valid in *S'*? In *S*, we have

$$\mathbf{F} = m\mathbf{a}. \tag{5.17}$$

The observer in *S'* measures the same mass (*m'*=*m*) and the same force (if, e.g., he uses a dynamometer, the spring stretches by the same amount), **F** = **F'**. The acceleration **a'** that he measures is also equal to **a**, but only in the case we are considering of relative translation at constant velocity. Then, in *S'*, the relation between force, mass and acceleration is

$$\mathbf{F}' = m'\mathbf{a}'. \tag{5.18}$$

In other words: the laws of mechanics are covariant under the transformations that link two reference frames in relative uniform translation motion.

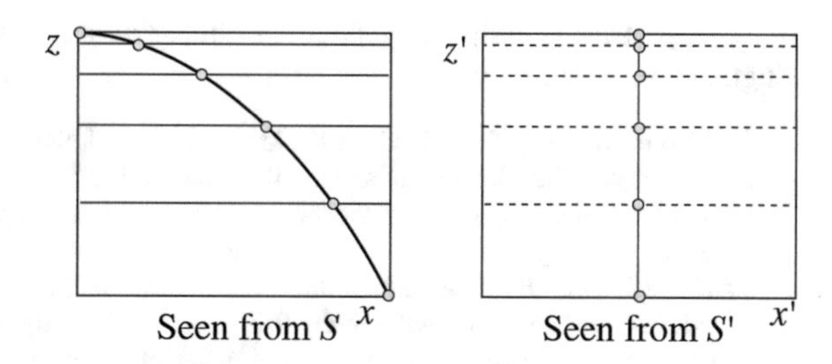

Fig. 5.3 Trajectory of a stone dropped from the top of the mast of a ship, as seen from the ship and the shore

As an example, consider a reference S' fixed on a sailing ship moving on the sea at constant velocity and S a frame fixed to the shore. As above, we choose the axes of the two frames mutually parallel and with coincident origins at $t = 0$. An experimenter climbs on top of the mast and drops a stone. Fig. 5.3 shows the trajectories of the stone as seen by an observer on the shore, a), and on the ship, b).

For the observer in S, the stone falls under the action of its weight, a constant force ($\mathbf{F} = m\mathbf{g}$), directed downwards, opposite to the z-axis (that we have taken to be vertical upwards). The initial velocity of the stone is the velocity of the ship, and we have taken the x-axis in that direction. Hence, the motion of the stone in the z direction is uniformly accelerated, while in the x direction, it is uniform (neglecting the air resistance). The trajectory is a parabola. In the figure, we marked the positions of the stone in time instants separated by the same time interval.

In S', the forces are the same, but the initial conditions are different; the initial velocity of the stone is zero. Hence, it falls vertically along the z'-axis with a uniformly accelerated motion.

Summarizing, in the two frames, the trajectories are different. The reason for the difference is in the different initial conditions of the motion. On the contrary, both observers describe the motion with the same law, $\mathbf{F} = m\mathbf{a}$. The two frames are perfectly equivalent for every dynamic experiment. Each of them can be considered as fixed or movable.

This conclusion is important and is known as the *relativity principle*. The principle does not deal directly with the phenomena but rather with the laws that describe the phenomena. It states that: *the laws of Physics are covariant, namely have the same form, in any reference frame moving of translational uniform relative motion*.

In our discussion, we have seen that the relativity principle is valid for the laws of mechanics, which is the physics chapter we are studying. However, its validity is completely general, including, in particular, all fundamental interactions, gravitational, electromagnetic, nuclear strong and weak interactions. In other words, it is impossible experimentally to establish the relative motion, provided it is as uniform translation. Historically, the principle was established by G. Galilei. He did not use that name, which was given to it by Henri Poincaré (1854–1912) in 1904, but Galilei established it in complete generality, describing, in a beautiful page, a series of experiments, some of which were of an electromagnetic nature, below the deck of a large sailing ship. The page of the *Dialogue* (transalted from Italian into English by the author) is:

> Shut yourself with a few friends in the largest room below decks of some large vessel, and have with you flies, butterflies and similar small flying animals. Let a large bowl of water with several small fish in it be the cabin too. Hang also, at a certain height, a bucket pouring out water drop by drop into another vase with a narrow mouth beneath it. When the ship stands still, carefully observe how those flying small animals fly with equal speed towards all sides of the cabin; you will see the fish swim indifferently in all directions; all the drops will fall into the vessel beneath; and you, when throwing something to a friend, will not need throw it more strongly in one direction than another, when the distances are equal; and jumping up feet together, you will pass equal spaces in all directions.
>
> Once you have observed all these things carefully, though there is no doubt that when the vessel is standing they must happen like that, let the vessel move with speed as high as you like. Then (provided the motion is uniform and not unevenly fluctuating) you will not discover the slightest change in any of the named effects, nor you will be able to understand from any of them whether the ship is moving or standing still. In jumping you will pass on the planking the same spaces as before, nor you will make longer jumps toward the stern than toward the prow, as a consequence of the fast motion of the vessel, despite the fact that during the time you are in the air the planking under you is running in a direction opposite to your jump. In throwing something to your companion, no more force will be needed to reach him whether he is on the side of the prow and you of the stern or your positions are inverted. The drops will fall as before in the lower bowl, without a single one dropping towards the stern, although, while the drop is in the air, the vessel runs many palms. The fish in their water will swim toward the forward part of their vase with no more effort than toward the backward part, and will come with equal ease to food placed anywhere on the rim of the vase. And finally the butterflies and the flies will continue their flights indifferently towards every side, nor will ever happen to find them concentrated close to the wall on the side of the stern, as if tired from keeping up with the course of the ship, from which they, remaining in the air, will have been separated for a long time. And if some smoke will be made burning a bit of incense, it will be seen ascending upward and, similar to a little cloud, remaining still and indifferently moving no more toward one side than the other. The cause of all these correspondences of effects is that the motion of the ship is common to all things contained in it, and to the air also.

We notice here that the development of electromagnetism in the last part of the XIX century led to doubts concerning the general validity of the principle. The process of in depth analysis of the physical laws that followed, leading to the relativity theory, showed that the Galilei relativity principle was, as we have stated, valid in general. However, it was found that the transformations of the co-ordinates, and of the time, between reference frames, valid at a small velocity relative to the speed of light, do not hold at high speeds. We shall discuss that in Chap. 5. Here, we simply anticipate the root of the issue. Consider the transformation equations that link the co-ordinates in S' and in S

$$x' = x - \upsilon_{O'}t; \quad y' = y; \quad z' = z; \quad t' = t, \tag{5.19}$$

where we have included the relation between times t and t' measured by the two observers. Indeed, the measurement of a time interval should be, we think, the same on the shore as on the ship (to continue the example). However, this conclusion, coming from our everyday experience and from experiments at the usual velocities, is wrong at velocities not too small compared to the speed of light. Two observers in two frames moving at those speeds measure different time intervals between the same two events; in other words, t and t' are not equal. A consequence is that the relations between co-ordinates are different from those of Eq. (5.19). The transformation Eq. (5.19), called *Galilei transformations*, fail at high velocities and must be generalized into the *Lorentz transformations*, as we shall see in Chap. 5. But the relativity principle remains completely valid.

5.3 Non-uniform Translation. Pseudo Forces

We now consider the case in which the motion of the reference S' relative to S is still a translation, but with variable velocity. Consider, for example, S' to be fixed on a trolley moving on straight rails with an accelerated (or decelerated) motion relative to S fixed on the ground. We still consider the motion of the point P in Fig. 5.2 as seen by two observers in the two frames. The relation between the accelerations is

$$\mathbf{a} = \mathbf{a}' + \mathbf{a}_{O'}. \tag{5.20}$$

As in the previous section, $\mathbf{a}_O'$ is the acceleration in S of the origin of S' and also of all the points fixed in it (its motion being a translation). Suppose now that S is an inertial frame. If no force acts on P, its acceleration is zero, $\mathbf{a} = 0$. In S', however, $\mathbf{a} = -\mathbf{a}_O' \neq 0$. Namely, in S', a body not subject to forces may accelerate. The law of inertia does not hold. S' is not inertial. Consider the trolley in the above example initially moving at constant speed. If we put a ball on a horizontal plane, it will not move. If the trolley now suddenly slows down, we shall see the ball accelerating forward, without any force acting. This is the interpretation of the observer in S'. The inertial observer in S thinks that there is no force acting on the ball (suppose friction to be negligible), and that it is just continuing its uniform motion (Fig. 5.3).

If a force $\mathbf{F}$ acts on the point P of mass m, the inertial observer in S finds the relation

$$\mathbf{F} = m\mathbf{a}. \tag{5.21}$$

The observer in S' measures the same value of the mass, $m' = m$, the same force, $\mathbf{F}' = \mathbf{F}$, but a different acceleration $\mathbf{a}'$, and finds $\mathbf{F}' = \mathbf{F} = m\mathbf{a} = m'\mathbf{a}' + m'\mathbf{a}_{O'}$, or

$$\mathbf{F}' = m'\mathbf{a}' + m'\mathbf{a}_{O'}. \tag{5.22}$$

We also see that the second Newton law, not only the first one, does not hold for the non-inertial observer S'.

However, the observer in S' can play a trick. Indeed, he is accustomed to thinking that any acceleration will be due to a force and will imagine that a force has suddenly started to act on the ball on the table. Formally, the trick is by Jean Baptiste d'Alembert (1717–1783); we can re-write Eq. (5.22) moving $m'\boldsymbol{a}_{O}'$ to the left-hand side, as

$$\mathbf{F}' - m'\mathbf{a}_{O'} = m'\mathbf{a}' \tag{5.23}$$

and call $-m'\boldsymbol{a}_{O'}$ a force, or, more accurately, a *fictitious force*, or *inertial force*

$$\mathbf{F}_{in} = -m'\mathbf{a}_{O'} \tag{5.24}$$

and Eq. (5.23) becomes

$$\mathbf{F}' + \mathbf{F}_{in} = m'\mathbf{a}'. \tag{5.25}$$

Namely, if we add to the "real" forces the fictitious, or inertial, ones, we re-establish the validity of the 1st and 2nd Newton laws. However, these forces are, as we said, fictitious, not real, because they are not produced by any physical agent. Consequently, there is no corresponding reaction. The 3rd Newton law, the action-reaction law, does not hold for the inertial forces.

Let us go back to the example of a sphere on a table on the trolley. The resultant of the true forces, weight and normal reaction of the plane, is zero. When the velocity of the trolley is constant, the fictitious force $\mathbf{F}_{in}$ is also zero because so is $\mathbf{a}_{O}'$. But when the trolley slows down, the observer in the S' sees the effect of a fictitious force as in Eq. (5.24). It is directed forward, opposite to $\mathbf{a}_{O}'$. He can measure the fictitious force attaching the sphere to a spring and measuring its stretch. In this way, he verifies that Eq. (5.24) is correct.

5.4 Rotation and Translation. Pseudo Forces

Consider now a stationary frame S (origin O and coordinates x, y, z) and a mobile frame S' (origin O' and coordinates x', y', z'), the motion of which is completely general. It may be a translation, with constant or variable velocity, a rotation, again with constant or varying angular velocity, or both of them together. Figure 5.4 represents the two frames at a certain time. At another time, for example, a bit later, both the position of O' and the direction of the axes of S' will be, in general, different.

We begin by finding a formula that will be useful in the following. Consider a vector $\mathbf{A}$, which does not vary with time relative to S'. Examples are the position vector in S' and the velocity of a point moving in rectilinear uniform motion relative to S'. The vector $\mathbf{A}$ is not constant in S. We now find its time derivative. We notice that $\mathbf{A}$ varies relative to S only in direction, not in magnitude. More precisely, $\mathbf{A}$ rotates with the same angular velocity at which, in that instant, the mobile frame S' rotates relative to S. We indicate it with $\boldsymbol{\omega}$. Notice that $\boldsymbol{\omega}$ can vary in time, which

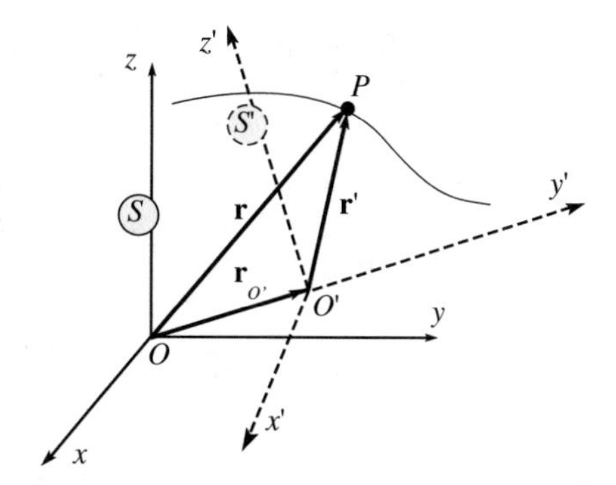

Fig. 5.4 Reference frame S' moves in an arbitrary motion relative to S

is why we specify "in that instant". Under these conditions, the time derivative of **A** is given by the Poisson formula and we have

$$\left(\frac{d\mathbf{A}}{dt}\right)_S = \boldsymbol{\omega} \times \mathbf{A},$$

where the subscript S specifies that it is the rate of change in the reference S.

If the vector **A** also varies in S', we have to sum the rate of change in S', and finally we have

$$\left(\frac{d\mathbf{A}}{dt}\right)_S = \left(\frac{d\mathbf{A}}{dt}\right)_{S'} + \boldsymbol{\omega} \times \mathbf{A}, \tag{5.26}$$

which is the formula we were looking for. Notice that in the preceding sections, we did not take care to specify in which frame we were taking the derivatives. This was allowed because, being the considered transformations translations, the Cartesian components of the vectors were not modified. This can be immediately verified in Eq. (5.26) in which, if $\boldsymbol{\omega} = 0$, the derivatives in the two frames are equal.

We shall now find the relations between the kinematic quantities in S and in S'. We shall call the former *absolute* and the latter *relative*, but we notice that the definition is arbitrary; we could have started calling S' stationary and S mobile.

See that the relation between the position vectors is always

$$\mathbf{r} = \mathbf{r}' + \mathbf{r}_{O'}. \tag{5.27}$$

To obtain the relation between relative (in S') and absolute (in S) velocities, we need the time derivatives. To do that, we need to have on each side of the equation only vectors in one frame. Hence, we re-write Eq. (5.27) as

$$\mathbf{r} - \mathbf{r}_{O'} = \mathbf{r}' \tag{5.28}$$

and derive the vector $\mathbf{r} - \mathbf{r}_O{}'$ using the rule (5.26), obtaining

$$\left(\frac{d(\mathbf{r}-\mathbf{r}_{O'})}{dt}\right)_S = \left(\frac{d(\mathbf{r}-\mathbf{r}_{O'})}{dt}\right)_{S'} + \boldsymbol{\omega}\times(\mathbf{r}-\mathbf{r}_{O'}). \tag{5.29}$$

The meaning of the left-hand side of this equation is clear: it is the difference between the absolute velocities of the point P, say $\mathbf{v}$, and of the point O', say $\mathbf{v}_O{}'$. We substitute Eq. (5.28) on the right-hand side, obtaining

$$\mathbf{v}-\mathbf{v}_{O'} = \left(\frac{d\mathbf{r}'}{dt}\right)_{S'} + \boldsymbol{\omega}\times\mathbf{r}'. \tag{5.30}$$

Now, we see that the first term on the right-hand side is the rate of change in S' of the position vector in S', namely the velocity of P in S', which we call relative and indicate with $\mathbf{v}'$. We then write

$$\mathbf{v} = \mathbf{v}' + \mathbf{v}_{O'} + \boldsymbol{\omega}\times\mathbf{r}' = \mathbf{v}' + \mathbf{v}_t. \tag{5.31}$$

In other words, the velocity $\mathbf{v}$ of the point P in S is the sum of its velocity $\mathbf{v}'$ in S' and of two more terms that we have grouped in $\mathbf{v}_t$. The meaning of the latter is understood considering the case in which the point does not move in S', namely if $\mathbf{v}' = 0$. Then, $\mathbf{v}_t$ is the absolute velocity of the point. We can then state that $\mathbf{v}_t$ is the velocity of the point fixed in the frame S', and call it Q, through which the moving point P passes at the considered time. We can think of $\mathbf{v}_t$ as the velocity of the moving space. It is called the *velocity of transportation*. It contains two terms,

$$\mathbf{v}_t \equiv \mathbf{v}_{O'} + \boldsymbol{\omega}\times\mathbf{r}', \tag{5.32}$$

which we discuss looking at Fig. 5.5. The first one is the velocity of the origin of S' and corresponds to the translational component of its motion relative to S. The

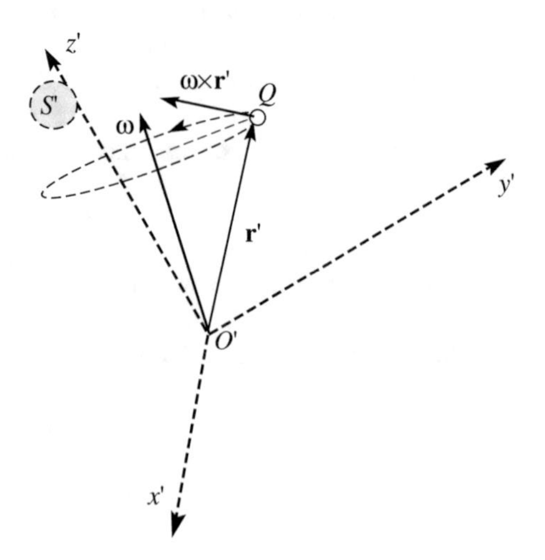

Fig. 5.5 The relative velocity in the rotating frame

second term is due to the rotation of S'. We can think of this as taking place about an instantaneous rotation axis passing through O' with angular velocity, in the considered instant, $\boldsymbol{\omega}$. Indeed, the velocity of the point Q stationary in S' where P is passing is just $\boldsymbol{\omega} \times \mathbf{r}'$.

We shall now find the accelerations, by a further derivative. We shall meet more terms. We start from Eq. (5.31) in the form

$$\mathbf{v} - \mathbf{v}_{O'} = \mathbf{v}' + \boldsymbol{\omega} \times \mathbf{r}' \tag{5.33}$$

and derive the left-hand side using Eq. (5.26), obtaining

$$\left(\frac{d(\mathbf{v} - \mathbf{v}_{O'})}{dt}\right)_S = \left(\frac{d(\mathbf{v} - \mathbf{v}_{O'})}{dt}\right)_{S'} + \boldsymbol{\omega} \times (\mathbf{v} - \mathbf{v}_{O'}). \tag{5.34}$$

Similarly to above, the left-hand side is the difference between the absolute accelerations of P, say $\mathbf{a}$, and O', say $\mathbf{a}_O'$. Still analogously, we use Eq. (5.33) to substitute $\mathbf{v} - \mathbf{v}_O'$ on the right-hand side, obtaining

$$\begin{aligned} \mathbf{a} - \mathbf{a}_{O'} &= \left(\frac{d\mathbf{v}'}{dt}\right)_{S'} + \left(\frac{d(\boldsymbol{\omega} \times \mathbf{r}')}{dt}\right)_{S'} + \boldsymbol{\omega} \times \mathbf{v}' + \boldsymbol{\omega} \times (\boldsymbol{\omega} \times \mathbf{r}') \\ &= \left(\frac{d\mathbf{v}'}{dt}\right)_{S'} + \left(\frac{d\boldsymbol{\omega}}{dt}\right)_{S'} \times \mathbf{r}' + \boldsymbol{\omega} \times \left(\frac{d\mathbf{r}'}{dt}\right)_{S'} + \boldsymbol{\omega} \times \mathbf{v}' + \boldsymbol{\omega} \times (\boldsymbol{\omega} \times \mathbf{r}'). \end{aligned} \tag{5.35}$$

The last term looks a bit complicated, but its terms have well-defined physical meanings. Let us examine them. The first term is the acceleration of P in S, namely the relative acceleration, say $\mathbf{a}'$. In the second term, the angular acceleration of the motion of S' relative to S appears. We shall name it

$$\boldsymbol{\alpha} = \left(\frac{d\boldsymbol{\omega}}{dt}\right)_{S'}. \tag{5.36}$$

The next two terms are equal. We put them together and also group some other terms, writing

$$\mathbf{a} = \mathbf{a}' + [\mathbf{a}_{O'} + \boldsymbol{\alpha} \times \mathbf{r}' + \boldsymbol{\omega} \times (\boldsymbol{\omega} \times \mathbf{r}')] + 2\boldsymbol{\omega} \times \mathbf{v}', \tag{5.37}$$

which expresses the *Coriolis theorem*, after Gustave de Coriolis (1792–1843). We now define

$$\mathbf{a}_t \equiv \mathbf{a}_{O'} + \boldsymbol{\omega} \times (\boldsymbol{\omega} \times \mathbf{r}') + \boldsymbol{\alpha} \times \mathbf{r}', \tag{5.38}$$

which is called the *acceleration of transportation* and

$$\mathbf{a}_{Co} \equiv 2\boldsymbol{\omega} \times \mathbf{v}', \tag{5.39}$$

which is called the *Coriolis*. Finally, we write Eq. (5.37) as

$$\mathbf{a} = \mathbf{a}' + \mathbf{a}_t + \mathbf{a}_{Co}. \tag{5.40}$$

The meaning of the acceleration of transportation $\mathbf{a}_t$ is analogous to that of the velocity of transportation $\mathbf{v}_t$. Indeed, if both velocity and acceleration of P in S' are zero, then its absolute acceleration is $\mathbf{a}_t$, as the other two terms on the right-hand side of Eq. (5.40) are then zero. The term $\mathbf{a}_t$ is the absolute acceleration of the point stationary in S' through which the point P (call it Q again) is passing at the considered instant. It is the sum of three terms. The first is the acceleration relative to S of the origin of the mobile frame S'. The second term is the absolute acceleration of Q due to the rotation of S' relative to S. The situation is shown in Fig. 5.6. Indeed, the velocity of Q (of position vector $\mathbf{r}'$) due to the rotation is $\boldsymbol{\omega} \times \mathbf{r}'$. In turn, this velocity varies in time, and its rate of change is, by the same formula $\boldsymbol{\omega} \times (\boldsymbol{\omega} \times \mathbf{r}')$. This is simply the centripetal acceleration of the point Q. Indeed, as we understand looking at Fig. 5.6, we have

$$|\boldsymbol{\omega} \times \mathbf{r}'| = \omega r' \sin\theta = \omega d,$$

where d is the curvature radius (the radius of the osculating circle) of the curve Q is describing. And further

$$|\boldsymbol{\omega} \times (\boldsymbol{\omega} \times \mathbf{r}')| = \omega^2 d,$$

which is the centripetal acceleration of Q.

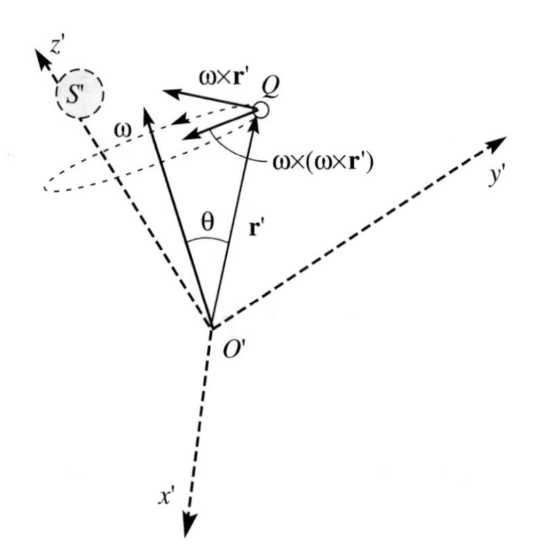

Fig. 5.6 Geometry in a rotating frame

Consider now the third term in Eq. (5.36). If the angular velocity $\boldsymbol{\omega}$ is constant, the absolute velocity of Q varies only in direction, and this term is zero. If $\boldsymbol{\omega}$ is not constant, the magnitude of the absolute velocity of Q also varies. This acceleration is given by the third term, $\boldsymbol{\alpha} \times \mathbf{r}'$.

As for the Coriolis acceleration $\mathbf{a}_{Co}$, we see in Eq. (5.39) that it is zero in three cases: when, in the considered instant, the point P does not move in S' ($\mathbf{v}' = 0$), when the mobile frame does not rotate ($\boldsymbol{\omega} = 0$) and when the velocity of the point P is parallel to the angular velocity. The Coriolis acceleration does not depend on the position of P but does depend on its relative velocity and becomes larger for larger relative velocities. It is always directed perpendicularly to the motion and consequently is a cause of change in its direction, rather than of its magnitude. We shall see examples in the next section.

We shall now assume that S is an inertial frame. As we have seen in the previous section, if S' accelerates relative to S, it is consequently not inertial. In other words, the Newton laws in S do not hold. Let us look at the details.

In the inertial frame S, the law of motion of the mass m under the action of the force $\mathbf{F}$ is $\mathbf{F} = m\mathbf{a}$. This can be written, using Eq. (5.40), as

$$\mathbf{F} = m\mathbf{a}' + m\mathbf{a}_t + m\mathbf{a}_{Co}.$$

The observer in S' measures the acceleration $\mathbf{a}'$ and wants to have that on the right-hand side. We move the other terms to the left-hand side, obtaining

$$\mathbf{F} - m\mathbf{a}_t - m\mathbf{a}_{Co} = m\mathbf{a}'. \tag{5.41}$$

We get the Newton law back formally by defining two fictitious forces

$$\mathbf{F}_t \equiv -m\mathbf{a}_t \tag{5.42}$$

and

$$\mathbf{F}_{Co} = -m\mathbf{a}_{Co}, \tag{5.43}$$

which is called the *Coriolis force*, and we subsequently get

$$\mathbf{F} + \mathbf{F}_t + \mathbf{F}_{Co} = m\mathbf{a}'. \tag{5.44}$$

We can then state that, in a frame mobile with an arbitrary motion relative to an inertial frame, the product of the mass times the acceleration is equal to the resultant of both true and fictitious forces. However, as already stated, the fictitious forces are not real and are not due to any physical agent. Consequently, the action-reaction law is not satisfied.

5.5 Motion in a Rotating Frame

Consider now the simple case in which the reference frame S' rotates relative to the inertial frame S with angular velocity $\boldsymbol{\omega}$ constant in magnitude and direction. For example, S' may be fixed on a rotating platform, for example, a merry-go-round, and S stationary on earth. As we shall see in the next section, such a frame is not exactly inertial due to the rotation of earth on its axis and its revolution around the sun, but the effects of the difference are quite small and we shall disregard them here.

We choose the origin of both frames in the center of the platform, their z and z' axes vertical upwards and, consequently, x, y and x', y' in the horizontal plane of the platform, as shown in Fig. 5.7. The axes x and y are stationary relative to the ground, while x' and y' rotate. With our choice of co-ordinates, the position vectors in the two frames coincide, $\mathbf{r} = \mathbf{r}'$.

In the particular case we are considering, the relevant expressions for velocities and accelerations simplify in

$$\mathbf{v} = \mathbf{v}' + \mathbf{v}_t, \tag{5.45}$$

$$\mathbf{v}_t = \boldsymbol{\omega} \times \mathbf{r}' = \boldsymbol{\omega} \times \mathbf{r}, \tag{5.46}$$

$$\mathbf{a} = \mathbf{a}' + \mathbf{a}_t + \mathbf{a}_{Co}, \tag{5.47}$$

$$\mathbf{a}_t = \boldsymbol{\omega} \times (\boldsymbol{\omega} \times \mathbf{r}') = \boldsymbol{\omega} \times (\boldsymbol{\omega} \times \mathbf{r}), \tag{5.48}$$

$$\mathbf{a}_{Co} = 2\boldsymbol{\omega} \times \mathbf{v}'. \tag{5.49}$$

Let us consider the velocities. In general, the point P is not necessarily on the platform. In Fig. 5.8, we have drawn it somewhat higher up. In general, the vectors $\boldsymbol{\omega}$ and $\mathbf{r}$ are not parallel. Recalling that $\mathbf{v}_t$ is the velocity of the point Q stationary in S' in the instant position of P, we see that it is tangent to the circle thorough

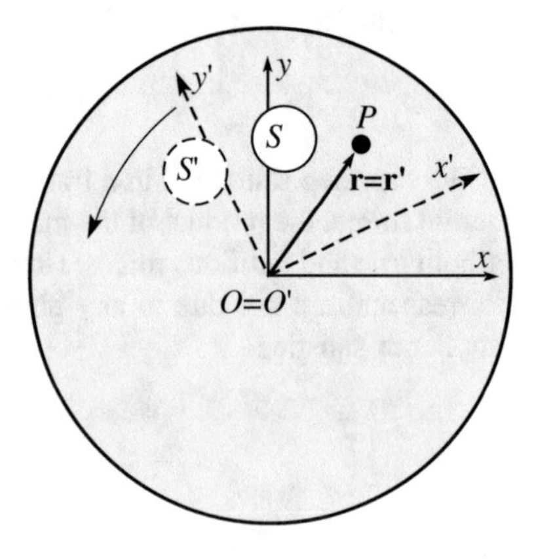

Fig. 5.7 The S reference frame is stationary to the ground, S' rotates with constant angular velocity

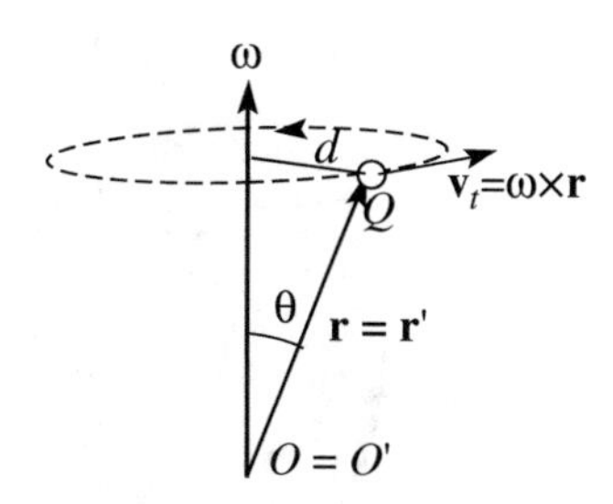

Fig. 5.8 The velocity $\mathbf{v}_t$ of the point Q

P normal to the rotation axis and with its center on the axis. This circle is the trajectory of Q. The magnitude of the velocity of transportation is then

$$\upsilon_t = \omega r \sin\theta = \omega d, \tag{5.50}$$

where d is the radius of the circle, namely the distance from the rotation axis. $\mathbf{v}_t$ is then simply the velocity of Q in its circular motion.

We now consider the accelerations. We immediately see that the $\mathbf{a}_t$ term is simply the centripetal acceleration of the point Q as seen in the inertial frame S.

Let us now consider a point P of mass m to be standing still, relative to S', on the platform at the distance r from the axis. Suppose that the friction is negligible and that P is kept in position by a rubber band attached to a small ring around the axis.

The inertial observer in S sees P moving in uniform circular motion with velocity ωr. He knows that the motion has an acceleration towards the center, the centripetal acceleration, of magnitude $\omega^2 r$ (this is the absolute acceleration in this case). The (centripetal) force causing the acceleration is due to the rubber band. The observer can check that measuring the stretch of the rubber band.

The non-inertial observer in S', on the platform, also sees that the rubber band is stretched, determining that a centripetal force is acting on P. He measures it and finds the same result as the inertial observer. The mobile observer now insists on having the first Newton law be valid and concludes that a second force, equal and opposite to that of the rubber band, must exist. This is the inertial force, due to the acceleration of transportation, $-m\mathbf{a}_t$, the direction of which is opposite to the centripetal force. In this case, the force is *centrifugal*. In this case, and always, the centrifugal forces are not real forces, but pseudo forces of the relative motion. They appear only when we pretend to describe the motion in a non-inertial, rotating frame as if it were inertial. However, the centrifugal force is felt as a real force, such as, for example, in a fast rotating merry-go-round.

We now discuss the Coriolis acceleration (Eq. 5.49) and the effects of the corresponding fictitious Coriolis force

$$\mathbf{F}_{Co} = -m\mathbf{a}_{Co} = -2m\boldsymbol{\omega} \times \mathbf{v}'. \tag{5.51}$$

Consider again the point P lying on the rotating platform. If P does not move relative to the platform, the Coriolis acceleration is null, as in the case just

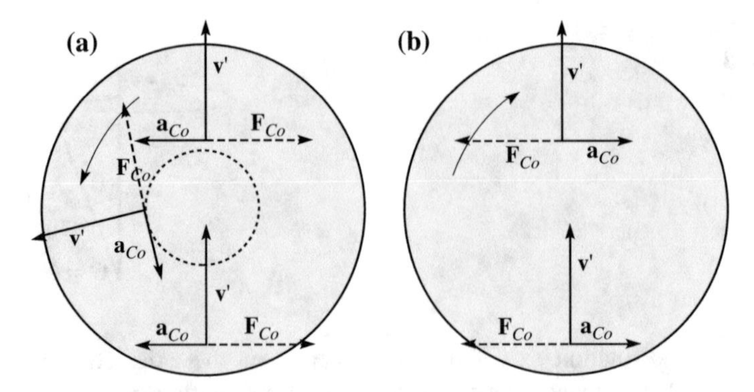

Fig. 5.9 Coriolis acceleration and (pseudo)force on a platform rotating. **a** Counter-clockwise, **b** Clockwise

discussed. Let $\mathbf{v}'$ be this velocity, which we assume, for simplicity, to be parallel to the platform. As we have already noticed, the Coriolis acceleration, and consequently the Coriolis force, does not depend on the position of P on the platform and is in any case perpendicular to the relative velocity. Consider Fig. 5.9. If the angular velocity $\boldsymbol{\omega}$ is directed out of the plane of the figure, as in Fig. 5.9a, we see the platform turning counter-clockwise. In this case, the Coriolis acceleration is directed towards the left of the motion, and the Coriolis force to the right. Suppose you are the point P waking or running on the platform. You will feel a push to the right of your speed, in whatever direction you move. Contrastingly, if $\boldsymbol{\omega}$ is directed inside the drawing, as in Fig. 5.9b, and the rotation is clockwise, the Coriolis force pushes to the left of the speed.

If we were to look at the earth from some distance from its surface on the axis, we would see the northern hemisphere rotating counter-clockwise if we were above the North pole, and the southern one clockwise if we were above the South pole. The Coriolis forces are the dominant causes of the circulation of winds in the atmosphere and cyclonic and anticyclonic phenomena. We shall discuss that in the next section.

Consider now another example, namely a material point P, standing in equilibrium above the platform in a fixed position relative to S, i.e., to the ground. We might think about a fly located just above the platform. The observer in S sees P at rest. Knowing that it is subject to its weight, he understands that another force, equal and opposite to the weight, should exist. The force is exerted by the beating of the fly's wings.

For the observer in S', the description is more complicated. He sees P moving in a circular uniform motion on a circle of radius r with velocity ωr. The motion is accelerated with a centripetal acceleration $\omega^2 r$. He deduces that a force $m\omega^2 r$ should act on the fly. However, he also knows, as the result of experiments he has done in the past, such as the one we just discussed, that a centrifugal force exists on the platform, namely a force of magnitude $m\omega^2 r$ directed outwards. Considering that the point moves on a circle, he concludes that the centripetal force on the fly must be twice as large, namely 2 $m\omega^2 r$. From where is this force is coming? It is the Coriolis force. In this case, $\boldsymbol{\omega}$ and $\mathbf{v}'$ are mutually perpendicular; Eq. (5.49) says that the magnitude of this force is just 2 $m\omega^2 r$ and that its direction is radial, towards the center. Physics is difficult in non-inertial frames, but the factor two is needed!

As a final example, let us go back to the first one, in which the point P is kept still on the platform by a rubber band attached to the axis. The motion seen by S is circular uniform. At a certain instant when we cut the band, S will see P sliding on the platform of a straight uniform motion at the velocity it had at the moment of the cut, directed as the tangent to the circle in that moment. Indeed, there is no net force acting on P.

How does the observer in S' describe the motion? To be concrete, assume the rotation to be counter-clockwise. When the rubber band is cut, the force that is needed in the rotating system to keep the objects standing disappears, and we might expect to see the point P moving outside along the radius of the platform. But this is not what we observe; rather, the point moves outside describing a curve. The reason is the Coriolis force. Before the rubber band was cut, P did not move on the platform, and the Coriolis force was null, but it is not so any longer since P has started moving. The Coriolis force acts, pushing P to the right all along its trajectory. Observing from outside, we can better understand what is going on. When the rubber band is cut, P moves with the same velocity as the point of the platform on which it is seated. While moving outwards, P reaches points of the platform having higher speeds, because they are farther from the axis, and consequently is left behind by them.

5.6 The Inertial Frame

As we have already stated, a reference frame is defined as *inertial*, if in that frame the first Newton law is valid. We have also seen that if a reference frame is inertial, any other one moving in uniform translation motion relative to it is also inertial. Indeed, the *relativity principle* we saw in Sect. 5.2 states that no experiment can distinguish between them. In other words, there is no *absolute* reference frame. Finally, we have seen that the Newton laws are covariant under the Galilei transformations.

However, nature does not necessarily behave according to our definition, and inertial reference frames might just not exist. The answer must come, as always, from the experiment. Basically, we need to check if we can find one reference frame in which material points not subject to forces, or, better yet, subject to forces of null resultant, always move in a rectilinear uniform motion. As a matter of fact, as is often the case in physics, we proceed through successive approximations. We can find reference frames that can be considered inertial, within a certain approximation, namely for experiments of a certain sensitivity or precision. For more precise experiments, we must search for frames that are closer to the inertial one, and we can find them.

Indeed, the largest fraction of the experiments takes place on earth, and is described in a stationary frame relative to the walls of our laboratory. These frames can be considered inertial within a quite good approximation, although not perfect. Indeed, earth rotates on its axis, making a turn (2π angle) in a day (84 600 s). The corresponding angular velocity, directed from the South to the North pole, is $\omega_{rot} = 7.3 \times 10^{-5}\ s^{-1}$. Figure 5.10a, for example, shows a stationary reference frame on earth at a certain latitude λ. In this frame, the transportation and Coriolis acceleration are present.

Let us analyze the first one, to which the centrifugal (pseudo)force corresponds. The magnitude of this force on a point P of mass m is the product of the mass, the square of the angular velocity (equal everywhere on earth) and the radius of the circle on which P moves. The latter is the distance from the axis, $R\cos\lambda$, where R is the earth radius and λ is the latitude of P. Calling a_1 the acceleration, the magnitude of the force is

$$F_1 = ma_1 = m\omega_{rot}^2 R\cos\lambda. \tag{5.52}$$

Let us look at the numbers. Recalling that $R = 6.4 \times 10^6$ m and taking, for example, $\lambda = 45°$, the acceleration, which is also the force per unit mass, is

$$a_1 = 2.4 \times 10^{-2}\ \mathrm{ms}^{-2}, \tag{5.53}$$

which is quite small, less than a per mille of the gravitational acceleration. However, for precise measurements, it can be relevant. The Coriolis force is usually smaller. However, it is important for large-scale phenomena, as we shall see in the next section.

However, a stationary reference frame on earth differs from an inertial frame for a second reason, to even smaller effect. Indeed, earth moves along its orbit, turning around in a year, with an angular velocity of $\omega_{riv} = 2\times10^{-7}\ s^{-1}$ on an orbit of radius $R_{orb} = 1.49 \times 10^{11}$ m (Fig. 5.10b). The centripetal acceleration is

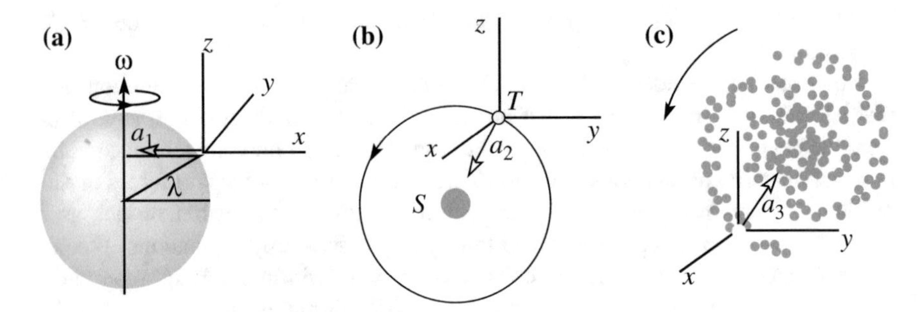

Fig. 5.10 Three reference frames with, acceleration towards **a** the rotation axis of earth $a_1 = 2.4 \times 10^{-2}\ \mathrm{ms}^{-2}$, **b** the sun $a_2 = 5.9 \times 10^{-3}\ \mathrm{ms}^{-2}$, **c** the center of the Galaxy $a_3 = 10^{-10}\ \mathrm{ms}^{-2}$

$$a_2 = \omega_{\text{riv}}^2 R_{\text{orb}} = 5.9 \times 10^{-3}\ \text{ms}^{-2}, \tag{5.54}$$

which is an order of magnitude smaller than a_1. The effects of the corresponding pseudo force are negligible, if not for the most precise measurements. Usually, the Coriolis force is even smaller.

Even these small effects, however, can be eliminated by choosing a reference frame with its origin in the sun and directions of the axes stationary to the fixed stars. This frame is inertial to an extremely good approximation, although not perfect. Indeed, the sun is located at the periphery of our spiral galaxy (10^{11} stars in order of magnitude). The sun turns around the center of the galaxy in an orbit of radius $R_S \approx 2.4 \times 10^{20}$ m over a period of about 150 million years, corresponding to the angular velocity of $\omega_S = 7.9 \times 10^{-16}\ \text{s}^{-1}$. The corresponding centripetal acceleration is

$$a_3 = \omega_S^2 R_S \sim 10^{-10}\ \text{ms}^{-2}. \tag{5.55}$$

This is very small indeed. However, experiments exist that are so sensitive, they are able to detect deviations from the state of inertia even at these extremely small levels. As a matter of fact, our galaxy moves too, in a non-uniform motion. However, when needed, we know how to eliminate the effects.

In conclusion, inertial reference frames exist in nature at every level of approximation we need.

5.7 Earth, as a Non-inertial Frame

As we just saw, the rotation of earth on its axis, with the angular velocity, $\omega_{\text{rot}} = 7.3 \times 10^{-5}\ \text{s}^{-1}$. This implies that in reference frame stationary on earth dynamical effects of the transportation and Coriolis fictitious forces exist. We shall discuss the principal ones in this section.

We take a reference system S with the origin in the center of earth and stationary with it. We shall use the symbols $\mathbf{v}$ and $\mathbf{a}$ for velocities and accelerations in this frame, omitting the prime we used in the previous sections.

The acceleration of transportation is

$$\mathbf{a}_t = \mathbf{a}_O - \omega_{\text{rot}}^2 \mathbf{r}_E,$$

where $\mathbf{a}_O$ is the acceleration of the earth's center. This is the centripetal acceleration of its motion around the sun in a very good approximation, and $\mathbf{r}_E$ is the radius of the orbit, as shown in Fig. 5.11a. The Coriolis acceleration on a point moving with velocity $\mathbf{v}$ in S is

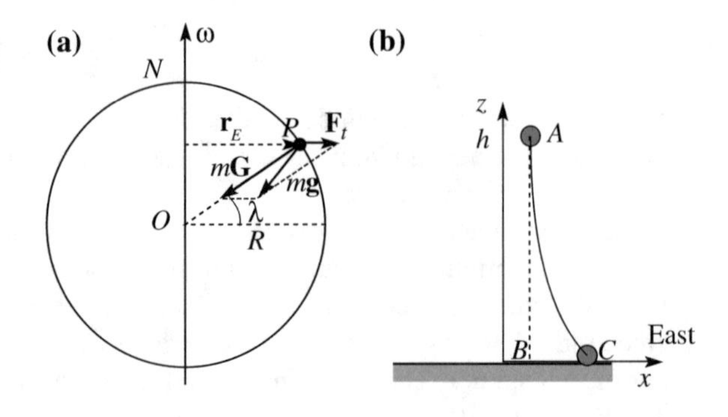

Fig. 5.11 **a** Forces and pseudo forces on matter point P; **b** Displacement to east in the free fall (exaggerated)

$$\mathbf{a}_{Co} = 2\boldsymbol{\omega}_{\text{rot}} \times \mathbf{v}.$$

In S, the equation of motion of a point with mass m subject to the real force $\mathbf{F}_{\text{true}}$ is then

$$m\mathbf{a} = \mathbf{F}_t + \mathbf{F}_{Co} + \mathbf{F}_{\text{true}} = -m\mathbf{a}_O + m\omega_{\text{rot}}^2 \mathbf{r}_E - 2m\boldsymbol{\omega}_{\text{rot}} \times \mathbf{v} + \mathbf{F}_{\text{true}}. \tag{5.56}$$

We can distinguish the following contributions to the true force $\mathbf{F}_{\text{true}}$: the gravitational attraction of earth $\mathbf{F}_E$, the gravitational attraction of all the other heavenly bodies $\mathbf{F}_O$, and of any other force that might be present (air resistance, tension of a wire, etc.), with resultant $\mathbf{F}$. We re-write Eq. (5.56), grouping the terms according to their causes,

$$m\mathbf{a} = \left(\mathbf{F}_E + m\omega_{\text{rot}}^2 \mathbf{r}_E\right) - 2m\boldsymbol{\omega}_{\text{rot}} \times \mathbf{v} + (\mathbf{F}_O - m\mathbf{a}_O) + \mathbf{F}. \tag{5.57}$$

The gravitational force $\mathbf{F}_O$ is due to all the heavenly bodies different from earth, but is largely dominated by the sun. As the diameter of earth is much smaller than the distance from the sun, in a first approximation, we can consider $\mathbf{F}_O$ equal in all the points of the earth. However, the small differences that are present are one of the causes of the tides, as we shall see in Sect. 6.4. The acceleration produced by $\mathbf{F}_O$ on every body is proportional to the mass of the body. Consequently, it is the same on the surface of the earth and in its center. In other words, it is the acceleration $\mathbf{a}_O$, of the earth herself. Hence, $\mathbf{F}_O - m\,\mathbf{a}_O = 0$.

We have reached an important conclusion, which is true as long as $\mathbf{F}_O$ can be considered not to vary on the points of the earth, that *the gravitational forces of the sun, the moon end of the other heavenly bodies do not appear in the equations of motion in reference frames stationary on earth. These forces are exactly balanced by the inertial forces resulting from the acceleration that those agents impart to the earth*.

We can simplify Eq. (5.57) as

$$m\mathbf{a} = \left(\mathbf{F}_E + m\omega_{\text{rot}}^2\mathbf{r}_E\right) - 2m\boldsymbol{\omega}_{\text{rot}} \times \mathbf{v} + \mathbf{F}. \tag{5.58}$$

Now, we are ready to consider several important examples.

The first case is of a body at rest, and **F** is simply its weight. This is the force we measure with a balance and that we have written as

$$\mathbf{F}_w = m\mathbf{g}, \tag{5.59}$$

where **g** is a vector quantity, which is equal for all the bodies in a given position. Up to now, we have talked of it as gravitational acceleration, but we are now ready to see that it is only approximately so. Equation (5.58) indicates that the force pushing a body downwards that does not move ($\mathbf{v} = 0$, $\mathbf{a} = 0$) is $\mathbf{F}_E + m\omega_{\text{rot}}^2\mathbf{r}_E$. We can say that the gravitational force of the earth on the body is

$$\mathbf{F}_G = m\mathbf{G} \tag{5.60}$$

and write

$$\mathbf{F}_w = m\mathbf{g} = m\left(\mathbf{G} + \omega_{\text{rot}}^2\mathbf{r}_E\right), \tag{5.61}$$

where **G** is the gravitational field of earth, and

$$\mathbf{g} = \mathbf{G} + \omega_{\text{rot}}^2\mathbf{r}_E. \tag{5.62}$$

The acceleration is the same for all the bodies in the same location. Equation (5.58) shows that a body dropped in absence of any force other than its weight, from a position of rest, $\mathbf{v} = 0$, moves with an acceleration $\mathbf{a} = \mathbf{g}$. We can say that **g** is the acceleration of the free-fall of any body, provided its velocity is null in the considered instant. If $\mathbf{v} \neq 0$, the Coriolis acceleration is, in general, present too.

In any case, Eq. (5.61) tells us that the weight is the sum of two contributions: the gravitational attraction $m\mathbf{G}$ of the earth, which largely dominates, and the centrifugal force due to the rotation of earth, which is much smaller and varies with the position. We will now discuss the observable consequences of that.

The local value of **g**. Suppose we take a plumb and fix it at a support. In the equilibrium position, its weight $\mathbf{F}_w$, given by Eq. (5.59), and the tension of the wire are equal and opposite. The direction is given by the wire. The distance from the rotation axis of a point P on the surface at the latitude λ is $r_E = R\cos\lambda$, where R is the earth radius (Fig. 5.11a). The weight $\mathbf{F}_w$ can be decomposed in a component, let us call it $F_{w,r}$, directed to the center of earth, and a component, $F_{w,\theta}$, in the direction of the meridian, to the North in the northern hemisphere and to the South in the southern one. The two components are

$$F_{wr} = mG - F_t \cos\lambda = m\left(G - \omega_{\mathrm{rot}}^2 R \cos^2\lambda\right)$$
$$F_{w\theta} = F_t \sin\lambda = m\omega_{\mathrm{rot}}^2 R \sin\lambda \cos\lambda. \tag{5.63}$$

The centrifugal term, the first one, is zero at the poles and maximum at the Equator. The tangential component is zero both at the poles and at the Equator. In these locations, but not elsewhere, the weight is precisely directed to the center of earth. As for the magnitude, the measured values are $g = 9.832\ \mathrm{ms}^{-2}$ at the poles and $g = 9.780\ \mathrm{ms}^{-2}$ at the Equator. If we approximate the shape of the earth surface with a sphere, all its points are at the same distance from the center, and if the mass distribution inside the earth is spherically symmetric, the gravitational term G is equal everywhere. It should be equal to g at the poles, $G = 9.780\ \mathrm{ms}^{-2}$. Let us check by giving an estimate, starting from g at the Equator.

$$G = g + \omega_{\mathrm{rot}}^2 R = 9.780 + \left(7.3 \times 10^{-5}\right)^2 \times 6.378 \times 10^6 = 9.814\ \mathrm{ms}^{-2}.$$

This value is close, but still a bit smaller than what we found from g at the poles. The main reason for that is that earth is not really spherical but somewhat squeezed at the pole, an effect of the centrifugal forces. Consequently, the poles are a bit closer to the center than the Equator.

Notice however, that small differences on the value of g in the different points of the surface are present, due to the local geology.

Absence of weight. If we measure the weight of an object with a balance on the space station, we find it to be zero. Such is also the weight of all the objects in the station, and in every artificial satellite. The arguments we just made are still valid, if we put the station in the place of earth, and consider the earth as an external body, as the sun, the moon and the other planets are. The spaceship is small enough for the gravitational force of those bodies to be considered equal at all the points of the ship. This force is exactly balanced by the inertial force to the acceleration of the spaceship. If its engines are shot, the ship freely falls under the action of gravitation. In this case, the equivalent of the weight on earth, namely the gravitational attraction of the ship on the body inside it, is completely negligible, $\mathbf{F}_w = 0$. The centrifugal term to the weight in the space ship is also negligible because the ship does not rotate appreciably. The weight in the ship is zero.

Eastwards shift in the free-fall. If a material point P of mass m is dropped with null initial velocity at a height h from the ground, it initially falls under the action of the weight, $\mathbf{F}_w$. However, as soon as its velocity, $\mathbf{v}$, is appreciably different from zero, a second inertial force, the Coriolis force, enters into action. It is

$$\mathbf{F}_{Co} = -2m\boldsymbol{\omega}_{\mathrm{rot}} \times \mathbf{v}. \tag{5.64}$$

The velocity $\mathbf{v}$ relative to earth is in the plane containing the earth's axis and point P, namely the plane PON in Fig. 5.11a. Consequently, the Coriolis force is perpendicular to this plane. Considering that the direction of the angular velocity is from South to North, and that $\mathbf{v}$ is downwards, we see that the Coriolis force is toward East in both hemispheres. The situation is shown in Fig. 5.11b, where AB is the direction of the plumb, i.e., the direction of $\mathbf{F}_w$ (no Coriolis force on the plumb that does not move) and C is the point in which the body reaches the ground, falling from the height h. The shift from the vertical BC is very small, and exaggerated in the figure. Let us calculate it.

We take a reference with the *z-axis* vertical, i.e., in the local direction of the plumb, and the x-axis horizontal towards the East. Within a good approximation, we can take the magnitude of the velocity to be $\upsilon = gt$, as in the vertical fall. Its direction is opposite to the z-axis. The equation of the component of the motion on the x-axis is

$$m\frac{d^2x}{dt^2} = 2m\omega_{\text{rot}}gt\cos\lambda.$$

We solve the equation by integrating twice on time and imposing the initial conditions $x(t) = 0$, $(dx/dt(0)) = 0$, obtaining

$$x = \frac{1}{3}g\omega_{\text{rot}}t^3\cos\lambda. \quad (5.65)$$

The time of the fall is, with good approximation, $t = \sqrt{2h/g}$, and we have

$$x = \frac{2\sqrt{2}}{3}g^{-1/2}\omega_{\text{rot}}h^{3/2}\cos\lambda. \quad (5.66)$$

For example, at the latitude of 45° and a fall from $h = 50$ m, the eastward shift is $x \sim 5$ mm, which is quite small, but has been measured, carefully eliminating perturbing effects.

Horizontal wind circulation. As is well known, the earth's atmosphere in a certain instant contains zones of high pressure and zones of low pressure. Naively, one would expect winds to blow from the former to the latter in the direction of the pressure gradient. However, the direction of the winds is substantially perpendicular to that, moving along the isobars, as you can see watching weather forecasts on TV. The effect is due to the Coriolis force.

Figure 5.12 summarizes the situation. H is the pressure maximum, L a pressure minimum, in the Northern hemisphere. Hence, the earth's angular velocity direction is out of the paper and the Coriolis force is directed, perpendicular to the velocity, to the right. Consider, for simplicity, a horizontal wind at constant velocity (in magnitude). Suppose we insulate a small mass of air within an ideal film and follow its motion. Two vertical and two horizontal forces act on our mass. The vertical ones are the weight and the Archimedes force. As the motion is horizontal, they are equal and opposite. The horizontal forces are the pressure (true) force and the Coriolis (pseudo) force. The pressure force acts on the surfaces of our gas mass. The pressure on its left-hand face pushes to the right, while the pressure on the right face pushes to the left. If the pressure were equal on the two sides, the neat force

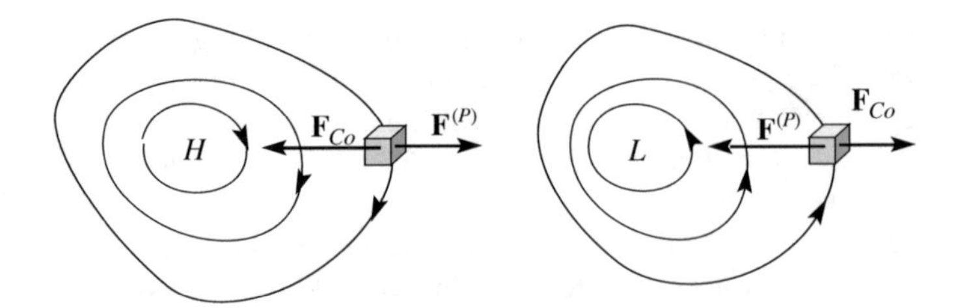

Fig. 5.12 Isobars around pressure maximum (*left*) and minimum (*right-hand*), in the Northern hemisphere and the forces on a mass of air

would be null. However, if there is a pressure maximum on the right of the gas mass we are following, as in Fig. 5.12a, there is a neat pressure force $\mathbf{F}^{(P)}$ pushing to the left. The Coriolis force has an equal and opposite direction. Consequently, the two forces may balance each other, or result in the right value being the centripetal force for the curvature of the wind trajectory. This can happen only if the wind circulates in a counter-clockwise direction around a pressure maximum (anticyclone). Contrastingly, it must circulate clockwise around a minimum (cyclone), as in Fig. 5.12b. The two situations are inverted in the southern hemisphere.

Let us look at the orders of magnitudes. The magnitude of the Coriolis force on an air mass m moving with horizontal speed υ at the latitude λ is

$$F_{Co} = 2m\omega_{\text{rot}}\upsilon \sin \lambda. \tag{5.67}$$

For example, the force on a kilogram of air, which is about 1 m^3, moving at 10 m/s at 45° is about 10^{-3} N. This should be compared to the pressure forces on the same volume. To be of the same order of magnitude, the pressure forces on two opposite sides of our cubic meter volume should be different by 10^{-3} N. This corresponds to a pressure difference of 10^{-3} Pa, being the surface unitary. Hence, the pressure gradient should be of 10^{-3} Pa/m, corresponding, say, to a distance of 100 km between two isobars of 100 Pa difference. This is reasonable (have a look at the weather maps).

The Foucault pendulum. A simple pendulum abandoned in a non-equilibrium position with null velocity oscillates in a vertical plane. However, if we watch carefully for a long enough time, along the order of one hour, we can see that the oscillation plane rotates relative to the laboratory, i.e., relative to a reference fixed on earth. The reason for the rotation is, once more, that the frame is not exactly inertial. As a matter of fact, the oscillation plane is fixed in an inertial frame, relative to which the earth rotates, as in Fig. 5.13.

While the effect has been known since its first observation by Vincenzo Viviani (1622–1703) in 1661, the main experiment and its correct interpretation were done by Léon Foucault (1819–1868) in 1851 in the Pantheon of Paris. His pendulum was 67 m long and had a 28 kg mass.

A similar situation, shown in Fig. 5.14, helps in our understanding. There, we have a pendulum, supported on a turning platform. If we put the pendulum in

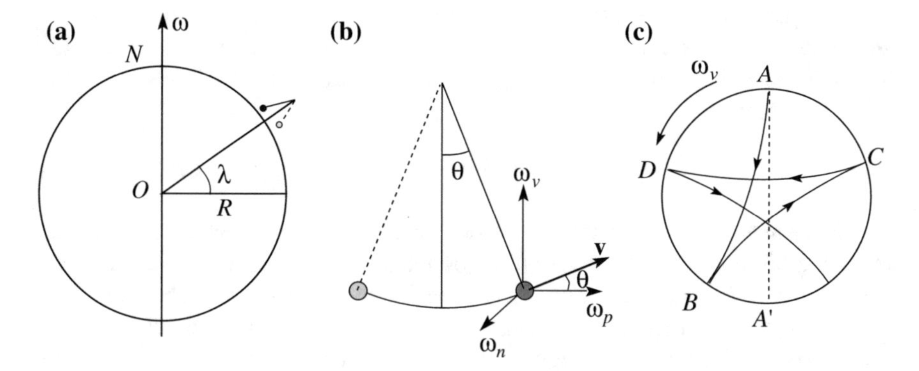

Fig. 5.13 The Foucault pendulum

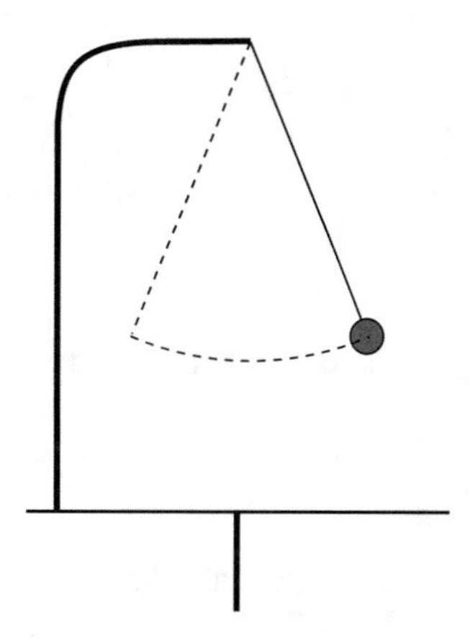

Fig. 5.14 Pendulum on a rotating platform

oscillation and the platform in rotation, we observe the oscillation plane remaining fixed, as expected, and the platform rotating under the pendulum. We can easily imagine what an observer on the platform would see, namely the plane of oscillation rotating in the opposite direction.

In this way, we easily understand what happens on earth if we are on a pole. Here, the angular velocity vector $\boldsymbol{\omega}_{\mathrm{rot}}$ is normal to the "platform", the earth surface, exactly as in that experiment. From the point of view of an inertial observer, the oscillation plane is constant, and he sees the earth turning relative to it. He understands why an observer at the pole sees the oscillation plane rotating and making a complete turn in 24 h. In the reference fixed to earth, the equation of motion is, once more, (5.58), with $\mathbf{F}$ the tension of the wire (neglecting air resistance). The Coriolis force $\mathbf{F}_{Co} = -2m\boldsymbol{\omega}_{\mathrm{rot}} \times \mathbf{v}$ is normal to the oscillation plane, and causes its rotation.

If the experiment is done not at the pole but at a latitude λ, we must pay attention to the vector characteristic of $\boldsymbol{\omega}_{\mathrm{rot}}$. We decompose it in a horizontal component, namely parallel to the ground in our position, $\boldsymbol{\omega}_h$, and a vertical one $\boldsymbol{\omega}_v$: $\boldsymbol{\omega}_{rot} = \boldsymbol{\omega}_h + \boldsymbol{\omega}_v$. We further decompose the horizontal component, which is still a vector, in its components parallel, $\boldsymbol{\omega}_p$, and normal, $\boldsymbol{\omega}_n$, to the oscillation plane and write Eq. (5.56) as

$$m\mathbf{a} = m\mathbf{g} - 2m\boldsymbol{\omega}_v \times \mathbf{v} - 2m\boldsymbol{\omega}_p \times \mathbf{v} - 2m\boldsymbol{\omega}_n \times \mathbf{v} + \mathbf{F}. \tag{5.68}$$

The term $-2m\boldsymbol{\omega}_n \times \mathbf{v}$ has the direction of the wire. Its effect is to change the tension a bit, but it has no effect on the oscillation plane. The term $-2m\boldsymbol{\omega}_v \times \mathbf{v}$ is perpendicular to the oscillation plane and causes its rotation. The third term $-2m\boldsymbol{\omega}_n \times \mathbf{v}$ is also perpendicular to the rotation plane, but is very small. Indeed, as we can see in Fig. 5.13b, it is proportional to $\sin\theta$, where θ is the angle between the wire and the vertical and is small, for small oscillations. We can then simplify Eq. (5.68) by writing

$$m\mathbf{a} = m\mathbf{g} - 2m\boldsymbol{\omega}_v \times \mathbf{v} + \mathbf{F}. \tag{5.69}$$

In concussion, the motion is similar to that at the pole with the sole difference being that in place of the angular velocity $\boldsymbol{\omega}$, we must consider its component along the local vertical, of magnitude

$$\omega_v = \omega_{rot} \sin\lambda. \tag{5.70}$$

The oscillation plane makes a complete turn in the period

$$T_{rot} = \frac{2\pi}{\omega_{rot} \sin\lambda} = \frac{24 \text{ hr}}{\sin\lambda}. \tag{5.71}$$

At 45° latitude, in one hour, the plane rotates by 10.6°.

Figure 5.13c shows the projection on the horizontal plane of the trajectory of the Foucault pendulum. The vector $\boldsymbol{\omega}_v$ is normal to the drawing towards the observer. The Foucault force is always directed normally to the velocity to the right of the direction of motion. The force bends the trajectory, as shown with exaggeration in Fig. 5.13c. Suppose that the pendulum is initially in A and abandoned with null velocity. Initially, when the Coriolis force is very small, the pendulum heads to A'. But as soon as the velocity becomes appreciable, the Coriolis force pushes to the right, bending the trajectory. The pendulum reaches point B, where it stops. When the velocity has again sufficiently increased, but in the opposite direction, the Coriolis force pushes in the opposite direction too, although still to the right of the motion. The pendulum reaches C, etc.

In the Foucault experiment, the length of the pendulum was large, l = 67 m, corresponding to a period T = 16.4 s. With such a long period, the lateral shift can already be observed in a single oscillation. The oscillation amplitude was A = 3 m. At the Paris latitude, $\sin\lambda$ = 0.753 and the rotation period is T_{rot} = 3.8 h = 14 480 s = 31,8 h = 14480 s. In an oscillation period T, the plane rotates at the angle

$2\pi T/T_{\mathrm{rot}}$. Hence, the shift of the oscillation extreme in one period is $s = 2\pi AT/T_{\mathrm{rot}} = 2.7$ mm.

Moreover, the length is important for another reason to which we can only hint. In practice, it happens that the stress forces always present in the wire and in the hook supporting the pendulum result in a spurious rotation of the oscillation plane. The effect is slow, but important for observations of several hours. It can be shown, however, that it is smaller for longer lengths.

5.8 The Eötvös Experiment

In Sect. 2.5, we have seen how Galilei and then Newton experimentally established the identity between inertial and gravitational mass. This is a very fundamental issue, and experiments have been done, and are still being done, to increase the precision within which the equality is verified. We discuss here the beautiful experiments conducted by the Hungarian physicist Loránd Eötvös (1848–1919) in the last years of the XIX century.

In this chapter we gave a number of examples of the effects of the inertial forces, the pseudo forces that appear in non-inertial frames. The inertial forces acting on a mass are proportional to the mass, just like the gravitational force. There is an important difference, however, as inertial forces are proportional to the inertial mass m_i, and the gravitational force is proportional to the gravitational mass m_g. Suppose the ratio between the two types of mass to be different for different substances. We could then hang spheres made of the two substances to two wires and look for any small difference in the directions of the wires. In this section, we shall use different symbols, m_i and m_g, for the two types of mass.

Consider a body hanging from a wire fixed in Ω, as in Fig. 5.15, at a point at the latitude λ. The distance from the axis is $r_E R\cos\lambda$, where R is the earth's radius. The centrifugal force has a direction perpendicular to the axis outwards and magnitude

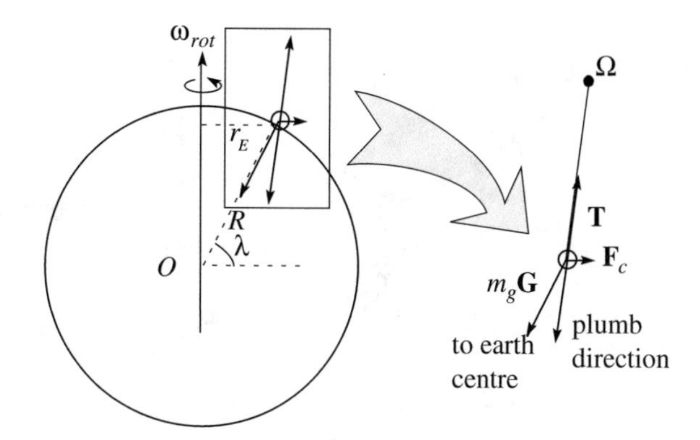

Fig. 5.15 The basis of the Eötvös experiment

$$F_c = m_i \omega_{\text{rot}}^2 R \cos \lambda. \tag{5.72}$$

and the gravitational force

$$F_G = m_g g = G_N \frac{m_G M_E}{R^2}. \tag{5.73}$$

If for two substances, m_i and m_g are different, the angle between the two forces is also different, and so is the direction of the wire. As we saw in Sect. 5.6, the centrifugal acceleration on the earth's surface is of the order of the per mille of the gravity acceleration. Correspondingly, the sought-after effects can be very small.

The Eötvös experiment directly compares the angles of wires to which spheres of different substances are attached. The two wires are attached to the extremes of a rigid bar. The bar is suspended by a metal wire that acts as a torsion balance, as shown in Fig. 5.16, similar to what we described in Sect. 4.7.

Figure 5.16a shows the system in perspective, with Fig. 5.16b looking at it parallel to the bar. If the ratio m_i/m_g is different for the two spheres, the directions α and β of the two tensions are a bit different. This produces a moment on the bar, due to the horizontal components of the two tensions, that rotates it about the wire from which it hangs. Under rotation, the wire develops an elastic moment, which increases with the angle. At the equilibrium angle, the two moments are equal and opposite. Measuring the angle, the torsion balance gives the moment.

The result of the very sensitive Eötvös experiment was null, allowing him to give the upper limit $m_g/m_i - 1 < 5 \times 10^{-9}$, namely that the difference, if any, is less than 5 parts per billion. An experiment of the same type by Robert Henry Dicke (1916–1997) in the 1960s established the even smaller limit of $m_g/m_i - 1 < 3 \times 10^{-11}$.

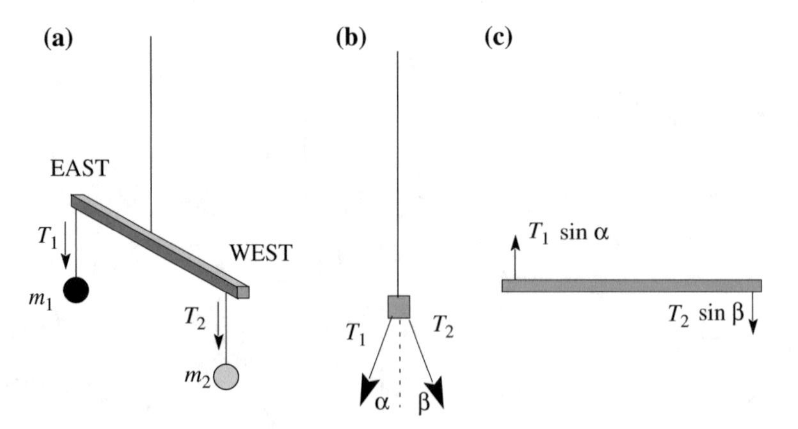

Fig. 5.16 The scheme of the Eötvös experiment

Problems

5.1. A kid sits in a carriage moving on straight rails. (a) If the speed of the carriage is constant, in which direction should he launch a ball to take it back in his hand without moving? In which direction if the carriage accelerates forwards?

5.2. A train travels on straight horizontal rails at the velocity $\upsilon_0 = 30$ m/s. Reaching a station, its stops, with constant acceleration, in $s = 150$ m. A suitcase of mass $m = 10$ kg lies on the floor, with dynamical friction coefficient $\mu_d = 0.20$. During the braking, it slides along the corridor. (a) How much is its acceleration relative to the ground during this time? (b) Which is the velocity of the suitcase when the train stops? (c) After the train stops, the suitcase continues to slide for a while and then itself stops. Which was the total displacement of the suitcase on the floor?

5.3. A man measures his weight in a lift, which is at rest, using a spring and balance, and finds it to be 700 N. With the lift moving, he repeats the measurement and finds it to be 500 N. What can he determine about the lift acceleration? And about its velocity?

5.4. A person sits in a chair standing on the platform of a merry-go-round, which is turning. The person holds a plumb. Draw separate force diagrams for the plumb, the wire, the person, the chair and the platform. Describe each of the forces in words. Identify the action reaction pairs, both for a frame stationary on earth and for one stationary on the platform. In the latter case, specify which of the forces are inertial.

5.5. A kid sits on a merry-go-round that turns at angular velocity $\boldsymbol{\omega}$, while his friend is on the ground. The resultant of the forces on the latter is zero. (a) What is the motion of the second kid seen by the first? (b) What is his acceleration? (c) What are the forces causing it?

5.6. An old vinyl disk rotates at 33 turns per minute. Its radius is $r = 15$ cm. An insect walks from the center towards the border. Will it be able to reach it if the static friction coefficient is $\mu_s = 0.1$?

5.7. A tennis player at 45° latitude is imparting to the ball a speed of 100 km/s, which we assume to be initially horizontal. Willing to hit ground at a distance of 50 m, should he take into account the Coriolis force?

Chapter 6
Relativity

In the preceding chapters we have seen how the Galilei-Newton mechanics is able to describe with simple laws an enormous number of phenomena both at the everyday scale and at cosmic level. Newtonian mechanics is one of the major conceptual constructions of human genius. However, the validity of the theory is limited on two sides, on the side of high velocities and on the side of small dimensions.

Newtonian mechanics is no longer valid for not very small velocities compared to speed of light. The latter is very large, about 3×10^8 m/s. The velocities of all the objects we have to deal with on earth, of planets and of the majority of heavenly bodies are very small in comparison. The velocity of the earth and the planets around the sun, for example, are of the order of one in ten thousand of the speed of light. Consequently, Newtonian mechanics gives an extremely good approximation for these phenomena. In this chapter we shall see how the theory needs to be extended at speeds comparable with the speed of light, in relativistic mechanics.

Classical mechanics, as are called both the Newtonian and relativistic mechanics, cease to apply for objects at molecular or smaller scales, of the order, say, of nanometers. These are one thousand times smaller than microbes. The correct theory, valid at all orders of magnitude is quantum physics. Classical mechanics is the limit of quantum physics for sufficiently large dimensions. In this book we shall only warn the reader of the limits of classical mechanics when needed.

As we have just stated, in this chapter we shall study the fundamental principles of relativistic mechanics and of the high velocity phenomena it describes. In the last part of the XIX century Maxwell formulated a set of equations that completely describe with great accuracy all the electromagnetic phenomena. However, these equations seemed to be in contrast with the relativity principle. If it was so it would have been possible to experimentally find an absolute reference frame. Astronomical observations and accurate experiments, culminated in the experiment by Michelson and Morley in 1887 (described in Sect. 6.2), showed that was impossible. An in depth criticism of the fundamental concepts, in particular on the measurement of

A. Bettini, *A Course in Classical Physics 1—Mechanics*,
Undergraduate Lecture Notes in Physics, DOI 10.1007/978-3-319-29257-1_6

time, followed, leading to development of special relativity mainly by A. Lorentz, H. Poincaré and A. Einstein, which was substantially completed in 1905.

The relativity principle was found to be universally valid, but new transformations of coordinates and time, valid between two inertial frames, the Lorentz transformations were established. The Lorentz covariance, first established for electromagnetism, must be valid for all physics laws. We shall emphasize this point after having followed the historical path.

In the sections from Sects. 6.4 to 6.6 we shall study how the concepts of simultaneity, time interval and distance must be revised. In Sect. 6.7 we shall find the law of addition of velocities, which shows, in particular, that the speed of light is the largest possible one. Space and time become completely correlated concepts and should be considered as a single four-dimensional manifold, space-time, which we shall study in Sect. 6.8.

In Sects. 6.9 and 6.10, we shall discuss relativistic dynamics and see how the fundamental concepts of mass, linear momentum and energy change. In Sect. 6.11 we shall find the relativistic form of the second Newton law.

As already mentioned, all the physics laws should be relativistically covariant. This in fact is the case. We shall give a few hints on that in Sect. 6.12.

Finally, in Sect. 6.13 we shall give a summary of the differences and equalities between Newtonian and relativistic mechanics.

6.1 Does an Absolute Reference Frame Exist?

In Chap. 5 we studied the covariance of physical laws under transformation between two reference frames. We have seen that inertial frames exist, which are by definition the frames in which the inertia law is valid. In such frames also the second and third Newton laws hold. In this chapter we shall consider only inertial frames. Formally, we can state that the physics laws have the same form in two inertial frames in the following cases. The two frames have no relative motion but differ for a translation or a rotation of the axes, or the two frames are in a relative uniform translation motion. In other words we have no means for experimentally observing differences between one location or another (invariance under translations, the space is homogeneous), or between one direction or another (invariance under rotations, the space is isotropic), or, to establish whether a frame is moving in a uniform translation motion or not relative to another inertial system. The latter property is the relativity principle established by Galilei.

We observe now that the set of all translations of the reference frames, all their rotation, and all the transformations between two inertial frames in relative uniform translation motion have the important mathematical characteristics of being a *group*. Let us define what this means.

Consider a set of transformations A, B, C... for which a product operation, $\otimes$, is defined. The set is a group if the following conditions are satisfied:

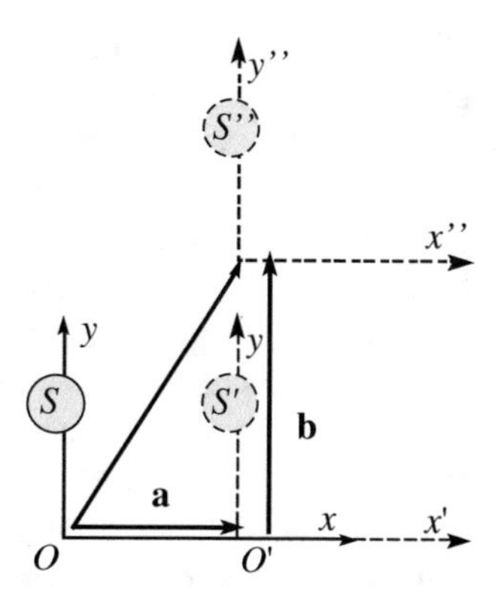

Fig. 6.1 Two translations and their product

1. For any pair of transformations A and B of the set, the product $C = A \otimes B$ is also a transformation of the set. The product is associative, namely $A \otimes (B \otimes C) = (A \otimes B) \otimes C$.
2. The set includes the identity transformation E, such as $A \otimes E = A$.
3. For every transformation A of the set, the inverse transformation, called A^{-1}, exists, such as $A \otimes A^{-1} = E$

With product $B \otimes A$ we mean that we first apply transformation A and then, on the result of that, transformation B.

To better understand that, consider the example of the static translations or displacements, which we take in two dimensions for simplicity. Suppose that the transformation A is the displacement a in the x direction from the coordinates S (x, y) to the coordinates S' (x', y'), as shown in Fig. 6.1.

The transformation A is

$$\begin{aligned} x' &= x + a \\ y' &= y. \end{aligned} \tag{6.1}$$

Let the static transformation B be the displacement b in the y' direction of the result of A, which is $S'(x', y')$, to S'' (x'', y''), namely

$$\begin{aligned} x'' &= x' \\ y'' &= y' + b. \end{aligned} \tag{6.2}$$

The product of the two is the transformation from $S(x, y)$ to S'' (x'', y''). Is it a translation? These relations are

$$\begin{aligned} x'' &= x + a \\ y'' &= y + b, \end{aligned} \tag{6.3}$$

which is the expression of a translation too. It is also easy to see that the associative property holds. Property 2 for being a group is satisfied.

The other two properties are also satisfied. The identity is the translation of null displacement (do nothing). Given a translation by a certain displacement, the static translation of the opposite one is also such a translation. Doing one after the other leads to the identity. In conclusion, static translations form a group.

Particularly important are the rotations. We recall that the covariance of the laws under rotations of the axes correspond to the fact that the quantities appearing in the equations that express the laws (position vector, velocity, acceleration, force, energy, etc.) must have well-defined transformation properties under rotations. They should be scalar, pseudoscalar, vectors or pseudovectors, and both sides of the equation must share the property.

Consider now the time. In Newtonian mechanics time is the same in all reference systems. We need to look at that more carefully. The time interval, as all the physical quantities, must be operationally defined. It is not obvious that the operations to measure the time interval between two events is the same for an observer at rest relative to the events and one moving relative to them. As we shall see, this is not true at high enough velocities, in the domain of relativistic physics.

We state immediately that the covariance properties of physics laws relative to translations and rotations remain equal to those we know, in relativistic physics. The changes are in the covariance properties between two frames in relative uniform translation motion. Let us consider two (inertial) reference frames. The first one has the coordinates x, y, z and time t. We call it S (x, y, z, t). The second frame, $S'(x', y', z', t')$, has axes parallel to the first one. The relative velocity is along the, overlapping, x and x' axes. The constant velocity of S', or of its origin, is $\mathbf{v}_{O'}$, is in the positive direction of x. We choose the origins of the times in both frames in the instant in which O' and O coincide. Figure 6.2 shows the situation.

The covariance of the laws under transformations between two inertial frames is defined once the transformation equations are defined, namely the relations between coordinate and time in S' and in S. The transformation equations we know, including the relation between the times, are

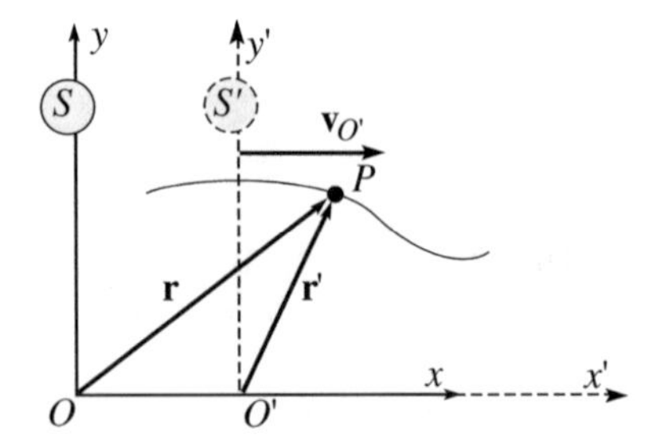

Fig. 6.2 Two reference frames in relative motion

$$\begin{aligned} x' &= x - \upsilon_0 t \\ y' &= y \\ z' &= z \\ t' &= t. \end{aligned} \tag{6.4}$$

More generally, for a generic direction of the velocity $\mathbf{v}_O$, of the points of S' the transformations are

$$\mathbf{r}' = \mathbf{r} - \mathbf{v}_{O'}t, \quad t' = t. \tag{6.5}$$

These are called *Galilei transformations*. An important property of the Galilei transformation is that they form a group.

The time intervals, in particular, are equal in the two frames. In other words, time is absolute, independent of the motion of the observer. This implies that it is possible to synchronize the clocks in S with the ones in S' independently of the relative velocity of the two frames. As is time, simultaneity is absolute. If two events are simultaneous in S they are simultaneous in S' too and in any other (inertial) frame, whatever its velocity.

The Galilei and Newton laws of mechanics that we have studied were established when only one of the fundamental interactions, gravitation, was known. Three more fundamental interactions were discovered in the following centuries. The first one was *electromagnetism*, including electric and magnetic phenomena and will be treated in the third volume of this course. The other two are the *strong interaction* between quarks in the nucleons and the *weak interaction* responsible, in particular, for beta decay. Both of them act at the nuclear and subnuclear scales and are quantum phenomena. Do they obey the invariance principles we have discussed, in particular the relativity principle? Let us see.

The study of electromagnetic phenomena was developed in the second half of the XVII and, mainly, in the XIX Century. In 1820 Hans Christian Ørsted (1777–1851) discovered that electric currents generate magnetic fields, linking for the first time electricity and magnetism. Between 1820 and 1826 André Marie Ampère (1775–1836) completely clarified the relation between electric currents and magnetism with a series of experiments. In 1831 Michael Faraday (1791–1867) discovered the electromagnetic phenomena: magnetic fields varying in time give origin to electric fields. The progress became rapid and in 1865 James Clerk Maxwell (1831–1879) developed the complete theory of electromagnetism. All the electric and magnetic phenomena are described by a set of differential equations, called the *Maxwell equations*. In addition, the theory predicted a new phenomenon. Electric charges in acceleration produce *electromagnetic waves*, which propagate with a well-defined velocity. This velocity can be expressed in terms of two quantities that measure the strength of the force between two electric charges at rest and between two stationary electric currents respectively. Maxwell himself accurately measured these quantities and found the resulting value of the velocity to be, in round figures,

$$c \simeq 3 \times 10^8 \ \mathrm{ms}^{-1}. \tag{6.6}$$

This is just equal to the velocity of light. And Maxwell noticed that

> the only use made of light in the experiment was to see the instruments.

He concluded

> that light is an electromagnetic disturbance propagated through the field according to electromagnetic laws.

The direct experimental confirmation of the existence and of the foreseen properties of the electromagnetic waves was very difficult. Heinrich Rudolf Hertz finally succeeded in that in 1887.

The Maxwell equations led to the unification of electric, magnetic and optical phenomena. However, soon they showed an unexpected behavior. Their form changes between two inertial frames when coordinates and time are transformed according to the Galilei transformations Eq. (6.5). It looked like Maxwell equations did violate relativity principle. If it were so, it should have been possible to design and perform electromagnetic and optic experiments able to establish an absolute reference frame.

Suppose for example we have a light source emitting a light pulse in the positive x direction in Fig. 6.2. Let c be its velocity in S. Notice that light is a wave phenomenon, similar to sound or sea waves. Consequently, its propagation velocity is independent of the velocity of the source relative to the observer. However, differently from the other mentioned cases, light propagates in a vacuum too. Indeed it comes to us from very distant stars. Consequently there is no substance perturbed by the wave and supporting its motion. This fact, which is clear to everybody now, was not so at the end of the XIX Century, when the existence of a substance pervading all space was assumed, the luminifer (light supporting) ether. The ether hypothesis has been a serious problem in the development of electromagnetic physics.

Anyway, as the speed of light is independent of the motion of the source, it should transform as any other velocity, as we have by derivation of Eq. (6.4) in the case of our example

$$c' = c - v_{O'}. \tag{6.7}$$

As any other velocity, the speed of light should be different for the observer in S and in S'. If we then measure the speed of light in S' and find it different from c we would establish that S is the absolute reference frame, namely the only one, amongst all the inertial frames in which the velocity of light has the value of Eq. (6.6).

More specifically, the non-covariance of Maxwell's equations under the Galilei transformations requires us to establish which of the following alternatives is the right one.

(1) The relativity principle is valid for the Newton laws of mechanics but not for the Maxwell laws of electromagnetism. The Galilei transformations are correct. This implies the existence of an absolute reference frame, which should be experimentally found.
(2) The relativity principle is valid for both the Newton laws and electromagnetism. The Galilei transformations are correct, but the Maxwell equations are wrong. In this case we should find modifications to the Maxwell equations that are necessary to have them covariant under Galilei transformations and then experimentally control whether the predictions of these modifications exist or not.
(3) The relativity principle is valid for mechanics and electromagnetism. The Maxwell equations are correct, but the transformation equations between reference frames are not the Galilei transformations. In this case we must find new transformations, different from the Galilei ones and such as to insure the covariance of the Maxwell equations. In addition, the Newton laws would no longer be any more covariant under the new transformations. We should find the modifications needed to guarantee the covariance also of mechanical laws and experimentally verify whether the consequences of the modifications we made are correct. The historical process leading to the clarification of the problem was not straight, but rather along winding paths. After the important contributions of Hendrik Antoon Lorentz (1853–1928), in 1905 two fundamental articles were separately published, the first by Henry Poincaré (1854–1912), the second a few weeks later by Albert Einstein, that laid down the complete theory. It became known as *special relativity*.

The crucial experiment to choose between the above stated alternatives is the measure of the speed of light in inertial frames in relative motion, allowing us to verify whether it is the same or not. The expected effects however, are extremely small and very difficult to detect. The experiment was done by Albert Abraham Michelson (1852–1931) in 1881 and, in a much more sensitive version, together with Edward William Morley (1838–1923) in 1887. We shall describe the 1887 experiment in the next section. We shall see how it showed that the speed of light is the same in all reference frames, so excluding alternatives (1) and (2).

6.2 The Michelson and Morley Experiment

We start with a bit of history. In 1879 Maxwell studied the possibility of establishing the absolute motion of earth relative to the reference in which the speed of light is c, namely the absolute reference frame, on the basis of astronomical data. The absolute reference is the frame in which the ether, which was thought to exist, is still. If this frame exists, it should be at rest relative to the fixed stars, according to astronomical observations. We do not know the velocity of the earth relative to this hypothetical frame, but we know that it should be at least the velocity of the earth in its orbital motion around the sun. This is about $\upsilon_E \approx 30$ km/s in magnitude and varies in direction throughout the year. Let us assume this velocity to be, in order of magnitude, what we have to detect. Its ratio to the speed of light is

$$\beta_E = \frac{\upsilon_E}{c} \simeq 10^{-4}, \tag{6.8}$$

which is a very small value. Maxwell established that only in astronomical phenomena could one expect effects of the first order in β_E. In laboratory experiments, in which the light leaves from a point, moves to a certain distance and comes back to the starting point, or close to it, only effects of the second order were expected, namely of the order of 10^{-8}. This is really a very small number. Maxwell's argument is the following.

Suppose that in our laboratory, namely in a reference in which the earth moves with speed υ_E, we place a bar of length l in the direction of the motion. At one end of the bar we have a source emitting flashes of light and a detector of light nearby. At the other end there is a mirror sending the light pulses back to the detector. The light pulse travels the distance l from the source to the mirror at velocity $c + \upsilon_E$ and when going back from the mirror to the detector at velocity $c - \upsilon_E$. The total time is then

$$t = \frac{l}{c+\upsilon_E} + \frac{l}{c-\upsilon_E} = \frac{2lc}{c^2 - \upsilon_E^2} \approx \frac{2l}{c}\left(1 + \frac{\upsilon_E^2}{c^2}\right) = \frac{2l}{c}\left(1 + \beta_E^2\right). \tag{6.9}$$

Now, 2 l/c would be the round-trip time if the bar were not moving. This is a very short time. But the time to measure is $\beta_E^2 = 10^{-8}$ of it. Maxwell concluded that such an experiment was impossible.

The young, 25 years old, officer of the USA navy Albert Abraham Michelson, who had already performed an accurate measurement of the speed of light, did not accept as obvious the impossibility of a laboratory experiment sensitive to the second order. Rather he worked on the problem and in 2 years found a solution. In 1881, he had already a first result. The sensitivity of this experiment was enough to detect the effect down to one half of the prediction. The result was null. However, the conclusion was so important that a confirmation was needed. Michelson, now with Morley, designed and performed in 1887 a second experiment sensitive to effects 40 times smaller than the predictions. Again the result was null.

The Michelson-Morley experiment is based on the employment of the interferometer shown in Fig. 6.3, which had been developed by Michelson himself (*Michelson interferometer*).

The source L emits a monochromatic line. This means that the wave is a sinusoid. The distance between two consecutive maxima is the wavelength (λ = 0.6 μm). Each point on the wave moves up and down periodically with a

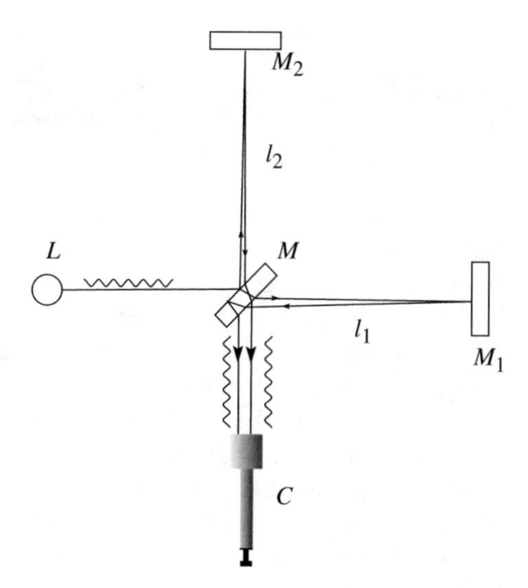

Fig. 6.3 The Michelson interferometer

period T. In an equivalent manner we can say that if we looked at the wave passing on a fixed point, the time interval between the passage of two maxima would be T.

Consequently the ratio between wavelength and period is the speed of the wave. If this is c we have

$$c = \lambda / T. \tag{6.10}$$

In the Michelson interferometer, the light beam is divided in two by a semi-transparent mirror M at 45° with the incident beam direction. One of the two beams after this mirror reaches the totally reflecting mirror M_1, is reflected back, reaches again M, and is reflected towards the telescope C. The other beam on the arm 2 is reflected back by M_2 and, after M, which partially transmits it, rejoins with the first beam. The lengths of the two arms are made as equal as possible. The two light waves are in phase when they leave M for the first time and are also in phase when they recombine, namely in the telescope, provided that the times, call them t_1 and t_2, are identical, or differ exactly by an integer number of periods. This is the situation drafted in Fig. 6.4a. In this situation, the signal they originate when they recombine is a maximum (constructive interference).

(a) In phase waves — Constructive interference

(b) Opposite phase waves — Distructive interference

Fig. 6.4 **a** waves in phase, **b** waves in phase opposition

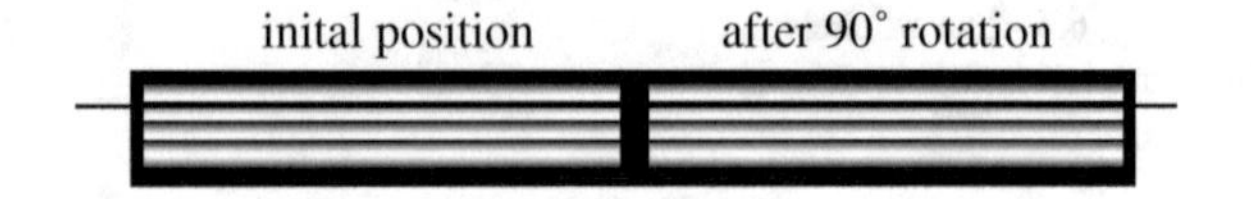

Fig. 6.5 The interference fringes before and after the rotation of 90° of the apparatus

If the travelling times t_1 and t_2 differ by half a period, or an odd number of half periods, as in Fig. 6.4b, the two waves are in phase opposition and cancel each other giving a zero signal (destructive interference). In the intermediate cases, the intensity is intermediate too. If these were the conditions of the field seen by the observer through the telescope, it would be clear in constructive, dark in destructive interference. In practice however, the planes of the two mirrors are never exactly at 90°. Consequently, the conditions of constructive and destructive interference alternate through the width of the beam in the visual field. The observer sees a series of clear and dark bands, called interference fringes. One could change the planes of the mirrors by adjusting screws in order to have the fringes horizontal, as in Fig. 6.5. A reference wire in the eyepiece was used to measure the position of the fringes.

We now evaluate the difference between the times t_1 and t_2. It is due to two causes. The first one is instrumental and due to the fact that the lengths, say l_1 and l_2, of two arms are never exactly equal. Notice that here exactly means to be so within a small fraction of the wavelength, namely a few dozens of nanometers. The other cause is what we want to measure, namely a difference in the light speed, relative to the instrument between the two arms due to the motion of the earth.

Suppose we have aligned the arm 1 parallel to its transportation velocity and evaluate t_1. In the path from M to M_1 the speed of light is $c + \upsilon_E$ and in the path back from M_1 to M is $c - \upsilon_E$. We have already calculated the round-trip time, Eq. (6.9). We can write

$$t_1 = \frac{2l_1 c}{c^2 - \upsilon_E^2} \approx \frac{2l_1}{c}\left(1 + \beta_E^2\right). \tag{6.11}$$

We now calculate the time t_2. If earth moves with velocity υ_E relative to the absolute frame, in the time t_2 is displaced by $\upsilon_E t_2$ as shown in Fig. 6.6. Looking at the figure we write

$$(ct_2/2)^2 = l_2^2 + (\upsilon_E t_2/2)^2$$

and hence

$$t_2 = \frac{2l_2}{c}\frac{1}{\sqrt{1-\beta_E^2}} \approx \frac{2l_2}{c}\left(1 + \frac{\beta_E^2}{2}\right). \tag{6.12}$$

Notice that we have just calculated t_1 in the frame fixed to earth and t_2 in the supposed absolute frame. This was allowed because we have assumed the Galilei

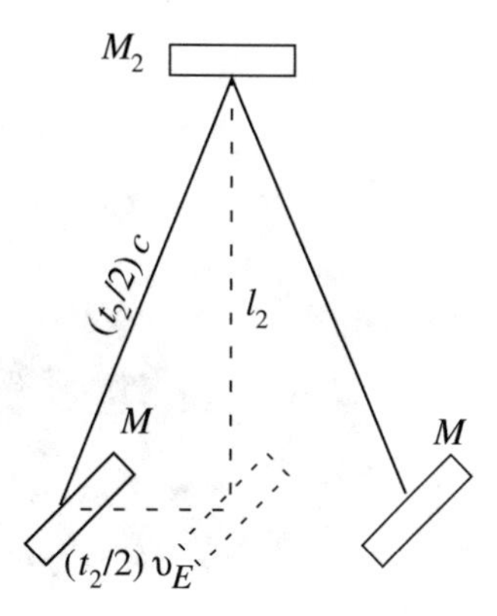

Fig. 6.6 Path of the light in arm 2

transformations to be valid, in particular the time to be absolute. Notice also that, as anticipated, the effect is of the second order, namely as β_E^2.

The difference between the two times is then

$$\Delta t = \frac{2}{c}\left[l_2\left(1+\frac{\beta_E^2}{2}\right) - l_1\left(1+\beta_E^2\right)\right]. \tag{6.13}$$

As we anticipated, the two times differ by the searched for effect, i.e. the term in β_E^2, and for the difference between the arm lengths, $2(l_2 - l_1)/c$. To get rid of the second effect, Michelson employed a measurement method by comparison. The comparison was between a measurement in the just described conditions and one after rotating the whole apparatus by 90°. The time difference, say $\Delta t'$, is Eq. (6.9) with inverted l_1 and l_2, namely

$$\Delta t' = \frac{2}{c}\left[l_2\left(1+\beta_E^2\right) - l_1\left(1+\frac{\beta_E^2}{2}\right)\right]. \tag{6.14}$$

We take the difference between the two differences and obtain

$$\Delta t - \Delta t' = \frac{l_1 + l_2}{c}\beta_E^2. \tag{6.15}$$

If the difference between the differences is zero, the position of the fringes seen by the observer remains fixed relative to the reference wire when we rotate the apparatus. If it is equal to one period the fringe pattern moves by one fringe. In general, the number Δn (not integer in general) of fringes crossing the reference wire during the rotation, is given by

$$\Delta n = \frac{\Delta t' - \Delta t}{T} = \frac{l_1 + l_2}{cT}\beta_E^2 = \frac{2l}{\lambda}\beta_E^2, \tag{6.16}$$

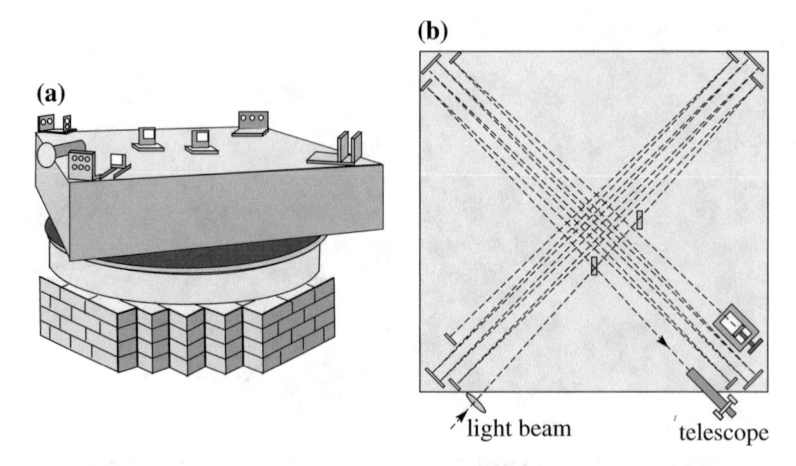

Fig. 6.7 The Michelson Morley experimentMichelson Morley experiment

where, in the last member we have used Eq. (6.15) and introduced the mean value l of the lengths of the two arms.

In the 1881 experiment the length of the arms was $l = 1.2$ m, corresponding to an expected shift of $\Delta n = 0.04$ fringes. Michelson was able to appreciate a shift of 0.02 fringes. He did not observe any and concluded that:

> The consequence of a stationary ether results therefore contradicted by the facts and it must be concluded that the hypothesis of the ether is false.

The second experiment is shown in Fig. 6.7. The optical path, namely the path of the beams, is increased to $l = 11$ m, having the beam going back and forth on its arm eight times with a set of mirrors (Fig. 6.7b). The 90° rotation was an extremely delicate operation. Any vibration even by a small fraction of a wavelength had to be avoided. Everything had to be stable at this level. Michelson and Morley mounted the interferometer on a massive granite bench, which was floating on a mercury bath. The shift expected in the ether hypothesis was now $\Delta n = 0.40$ fringes. Figure 6.4 reproduces the fringes before and after the rotation. No shift can be seen. The sensitivity was one hundredth of a fringe, corresponding to a distance of 6 nm. Figure 6.8 shows the result of the measurements, which are the full lines. The

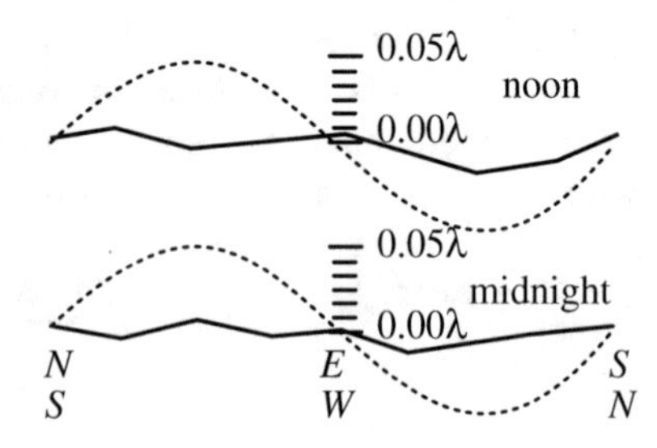

Fig. 6.8 Observed shifts (*continuous lines*) and 1/8 of the expectations (*dotted curves*—to keep them inside the diagram) for Galilei transformations

dotted curves are 1/8th (to enhance the visibility of a possible difference) of the expectations assuming the Galilei transformations. No effect was detected repeating the experiment in day-time and during the night, to check for any effect of the rotation velocity of the earth. The conclusion was definitive: the experiment cannot establish the movement of earth. This is an example of how a null result can give extremely important information.

A first attempt to explain the result was done in 1889 by George FitzGerald (1851–1901) and independently in 1992 by H.A. Lorentz. They advanced the hypothesis that the objects, when in motion, contract, only in the direction of the motion and not in the perpendicular ones. The contraction was able to cancel the effect expected in the ether hypothesis. It was an ad hoc, and wrong, hypothesis but an important step towards relativity theory.

In the following years the Michelson experiment was repeated with increasing precision, always with a null result. Other experiments sensitive to the absolute velocity were done, again with null result. In 1904 H. Poincaré, after a careful analysis of the experimental evidence, drew the conclusion that the relativity principle (so he named it for the first time) holds for all physical laws. His words, similar to those of Galilei three centuries before him, are:

> According the Relativity Principle the laws of the physical phenomena must be the same, whether an observer is fixed, or for an observer moving in an uniform translation motion: so that we have no means, and could not have any, of discovering if are or are not carried along in such a motion.

His second conclusion was that the speed of light is the same in all inertial reference frames, i.e., the speed of light is invariant.

From our side, we concede that only the third alternative of those considered in the previous section can be valid. We must now, first of all, find new transformation laws, in place of the Galilei transformations.

6.3 The Lorentz Transformations

We need now to find new transformation laws of coordinates and of time between two inertial frames in relative uniform translation motion. They must be such, in order to guarantee the relativity principle, that the Maxwell equations are covariant, namely maintain their form, under such transformations. The invariance of the speed of light is an immediate consequence of that. These are the *Lorentz transformations*. After having recalled the important historical elements, we shall give the result without demonstration and shall discuss it. Finally we shall state which are the basic assumptions under which the Galilei and Lorentz transformations are valid.

The Lorentz transformations were, found as the result of a difficult theoretical effort in several subsequent steps of improving precision, by Hendrik Lorentz between 1895 and 1904 and, with a further small correction, in final form, by Henri

Poincaré in 1905, who published the result on the 5th of June 1905. Albert Einstein reached the same result on the 30th June, when his fundamental article was sent for publication.

Consider once more the two inertial reference frames S and S' represented in Fig. 6.2. We have in both frames rulers along the Cartesian axes, to measure the coordinates, and, in every point of the space, we have identical clocks to measure the time. All the clocks in each frame are synchronized with one another. We shall discuss in the next section how this crucial operation can be done. We choose the origins of the times in both frames, $t = 0$ and $t' = 0$, as the time at which the two frames overlap.

We shall call something happening in a definite position an *event*, as defined by the three Cartesian coordinates measured by the rulers in the considered frame, at a definite instant of time, as measured by the clock in that position in the considered frame. There are two relevant parameters, which always are present in relativistic formulas. They are pure numbers and are functions of the velocity $\upsilon_O{}'$ of S' relative to S. The first one is the ratio of this velocity and the light speed

$$\beta_{O'} = \upsilon_{O'}/c, \tag{6.17}$$

and the second is

$$\gamma_{O'} = 1/\sqrt{1-\beta_{O'}^2}. \tag{6.18}$$

The Lorentz transformations are

$$\begin{aligned} x' &= \gamma_{O'}(x - \beta_{O'}ct) \\ y' &= y \\ z' &= z \\ t' &= \gamma_{O'}(t - \beta_{O'}x/c). \end{aligned} \tag{6.19}$$

We immediately see that they are a generalization of the Galilei transformations, tending to them for $\beta_{O'} \to 0$, namely for velocities much smaller than the light speed, $\upsilon_{O'} \ll c$.

The inverse transformations, to go from S' to S, can be found by inverting the system of equations or, in a simpler way, by changing the sign of the velocity. Hence

$$\begin{aligned} x &= \gamma_{O'}(x' + \beta_{O'}ct') \\ y &= y' \\ z &= z' \\ t &= \gamma_{O'}(t' + \beta_{O'}x'/c). \end{aligned} \tag{6.20}$$

The Lorentz transformations show very strange looking aspects. They mix, so to say, space and time. We shall see the consequences in the next sections. Here we shall look at them from a geometrical point of view. Indeed, Eq. (6.20) are similar to the transformations between the coordinates in two frames differing for a rotation of the axes. If the rotation is, for example, around the common z axis, that we can call the height, the transformations are

$$\begin{aligned} x' &= x\cos\theta + y\sin\theta \\ x' &= -x\sin\theta + y\cos\theta. \end{aligned} \tag{6.21}$$

Also in this case, the quantities in the second frame are mixtures, better linear combinations, of the quantities in the first. If we look at an object we refer to one of its dimensions as width, another as thickness. If we now rotate our point of view by an angle around a vertical axis, the new width, namely the angle under which we see the object in the horizontal plane, contains a part of what we called depth before the rotation, and vice versa. It follows that depth and width are not absolute properties, rather they depend on the point of view, namely they are relative to the reference frame. The Lorentz transformations are analogous. They tell us that the length measurements made by a person contain some of the time measured by another person moving relative to the first one. When speeds are high, close to the speed of light, the objects are mixtures of space and time, as usually they are of width and depth. When we turn around an object and we see it from different angles, our brain automatically recalculates depth and width, because it developed under these conditions. If we were living at high speed we might have a brain able to calculate the new mixture of space and time every time we change speed. We do not have this automatic habit and must understand the situation by carefully reasoning.

As we well know, the norm of a vector in our three dimensional space is the sum of the squares of its Cartesian components. In particular the norm of the position vector is

$$r^2 = x^2 + y^2 + z^2. \tag{6.22}$$

If we consider for simplicity a plane, we have $r^2 = x^2 + y^2$, which is the Pythagorean theorem. Notice that the same is not true, for example, on a spherical, rather than plane, surface. The Pythagorean theorem is valid if the two dimensional space is flat. The same is true in three dimensions. A space in which the squares of the distances are given by Eq. (6.22) is said to be an *Euclidean space*.

We also know that a property of the rotation of the axes is to leave the norm of the vectors invariant. We can see the reason for that writing Eq. (6.21) as a product of matrices

$$\begin{pmatrix} x' \\ y' \\ z' \end{pmatrix} = \begin{pmatrix} \cos\theta & \sin\theta & 0 \\ -\sin\theta & \cos\theta & 0 \\ 0 & 0 & 1 \end{pmatrix} \begin{pmatrix} x \\ y \\ z \end{pmatrix}. \tag{6.23}$$

We see that the square matrix in the transformation is orthogonal. Do the Lorentz transformations have the same property?

Consider the following two events in S. The first one is the start of a light pulse from its origin O at the instant $t = 0$, the second is the arrival of the pulse in the point (x, y, z) at time t. We express the fact that the speed of light is c by writing

$$x^2 + y^2 + z^2 - (ct)^2 = 0. \tag{6.24}$$

In S' too the light propagates with the same velocity c and we can write

$$x'^2 + y'^2 + z'^2 - (ct')^2 = 0. \tag{6.25}$$

The quantities in the left-hand side are very similar to the norm of a vector in four dimensions. They are called *intervals*. The difference is the minus sign in front of the temporal term. Technically, the four dimensional space—time is said to be a *pseudo-Euclidean space*. Another way to cope with the issue is to define an imaginary time coordinate, ict. To simplify the expressions we shall use the symbols

$$x_1 = x,\ x_2 = y,\ x_3 = z,\ x_4 = ict. \tag{6.26}$$

An event is a point in space-time. Analogous to the position vector in three dimensions is the four dimensional vector of coordinates given by Eq. (6.26). We shall call these vectors, four-vectors, to distinguish them from the vectors in three dimensions (three-vectors). The Lorentz transformations written as products of matrices are

$$\begin{pmatrix} x_1 \\ x_2 \\ x_3 \\ x_4 \end{pmatrix} = \begin{pmatrix} \gamma & 0 & 0 & i\beta\gamma \\ 0 & 1 & 0 & 0 \\ 0 & 0 & 1 & 0 \\ -i\beta\gamma & 0 & 0 & \gamma \end{pmatrix} \begin{pmatrix} x'_1 \\ x'_2 \\ x'_3 \\ x'_4 \end{pmatrix}. \tag{6.27}$$

As first established by Poincaré in 1905, the Lorentz transformations joined with the space rotations, forming a group which he named the *Lorentz group*. The matrix corresponding to the product of two transformations is the product of their corresponding matrices.

Equations (6.24) and (6.25) tells us two things. Firstly, two events connected by a light signal are separated by a null interval. This does not mean that they coincide but that the norm of the interval between them is zero. This possibility is a consequence of the minus sign in the temporal term. The norm of a four-vector can be positive, zero, or negative in space-time. Secondly, if an interval is null in S it is null in S' too. This is a formal way to state that the speed of light is invariant. As a matter of fact we can state more. The square matrix in Eq. (6.27) is orthogonal. The consequence is that all the intervals, even more the norms of all the four-vectors, are invariant under Lorentz transformations.

In three-dimensional space we dealt with vectors in three dimensions, which we now call three-vectors. As the reader remembers, a three-vector is an ordered triplet of real numbers that transforms under rotation of the axes as the position vector.

In a similar manner in relativistic physics we deal with four-vectors. A four-vector is a quadruple of numbers, real the first three, imaginary the fourth, which transform from an inertial reference to another, in relative uniform translation motion, as the coordinates do. The norms of all the four-vectors are consequently invariant under Lorentz transformations, in other words they are four-scalars. As such, they play very important roles in relativistic physics. We shall see examples later.

In the next sections we shall discuss the deep consequences of the Lorentz transformations on the basic concepts of space and time. Here we notice the following.

Historically, the Lorentz transformations were found, as mentioned, by the three main authors, in temporal order, Lorentz, then Poincaré and then Einstein. Each of them started from somewhat different hypotheses and followed a different logical path. The path we have followed here is to start from the experimental discovery of the invariance of the speed of light. This was indeed a revolutionary discovery. This was also one of the axioms, together with the relativity principle, assumed by Einstein. From the logical point of view, however, this second axiom is not necessary. Indeed, the relativity principle imposes the covariance of the Maxwell equations. Once this is established, with the Lorentz transformations, the invariance of the speed of light is an immediate consequence.

However, the historical approach we have followed, as the vast majority of textbooks do, tends to hide the logical structure of special relativity and to overemphasize the role of electromagnetism in the foundations of the theory. After more than one century from its creation we know that all the fundamental interactions, not only the electromagnetic one, but also the gravitational, the strong and the weak interactions obey the relativity principle. All the laws that govern them are covariant under Lorentz transformations. The fields of the fundamental interactions, which are analogous to the gravitational field we studied in Chap. 4, in quantum mechanics, are mediated by "quanta". The quantum of light is the photon. Its velocity is the velocity of light. As we shall see in Sect. 6.10 this implies that the mass of the photon is zero. However, it would be logically possible that the photon would be massive. In this case, the Lorentz transformations would not change, but the parameter c appearing in the equations would not be the speed of light and the latter would not be invariant. Indeed, this is the case of the weak interaction, the quanta of which, called Z^0 and $W^{\pm}$ bosons, have mass and do not move at the speed of light. If that was the case, the demonstration based on the invariance of the speed of light would not hold. But the final result would still be valid.

From the logical point of view, we must ask ourselves the following questions. Can we establish the relativity theory independently on electromagnetism? What are the assumptions needed for that? The answer is yes; only a few hypotheses on the basic structure of the space-time are needed. These are the following:

1. Space-time is isotropic and homogeneous.
2. A class of inertial reference frames exist, namely frames in which the inertia law holds.
3. The relativity principle is valid, namely there is no privileged reference frame.
4. The transformations form a group.
5. A class of events exists for which the causality principle holds. In this class the sign of the time differences between events, that is the nature of a possible causal relation, is the same in all the inertial frames.

It can be demonstrated that only two transformations exist under these hypotheses, the Galilei and the Lorentz transformations.[1] The quantity c in the latter has the dimension of a velocity and enjoys the properties of being invariant and being the maximum possible velocity. Galilei transformations are the limit of the Lorentz ones for $c \to \infty$. There is no need to rely on electromagnetism. The electromagnetism enters the game only to give to c the physical meaning of speed of light.

6.4 Criticism of Simultaneity

The most important difference between Galilei and Lorentz transformation is on time measurements. In the former the result of the measurement of a time interval is the same in S and in S'. Time is absolute, independent of the reference frame, in the Galilei and Newton physics. On the contrary, the last Eq. (6.19) states, in particular, that the instant at which an event happens in S' depends not only on the time in which it happens in S (as expected) but also on its position in S, as not expected. Hence, two events happening in two different points that appear to be simultaneous to an observer in S do not appear to be so to an observer in S'. The simultaneity of two events is not an absolute concept, but rather it is frame dependent. The in-depth criticism of the simultaneity concept and of the time intervals measurement was made by H. Poincaré in 1898. We shall explain the argument considering the ideal experiment represented in Fig. 6.9.

We suppose to have fixed in the S' frame a rigid bar parallel to the x' axis. In the middle point of the bar we have installed a light source, which emits a light flash at a certain instant. The flash propagates in all directions, in particular towards two

[1]For an elementary proof of this result, see J-M. Lévy-Leblond "One more derivation of the Lorentz transformation" American Journal of Physics **44** (1976) 271 and A. Pelissetto and M. Testa "Getting Lorentz transformations without requiring an invariant speed" American Journal of Physics **83** (2015) 338.

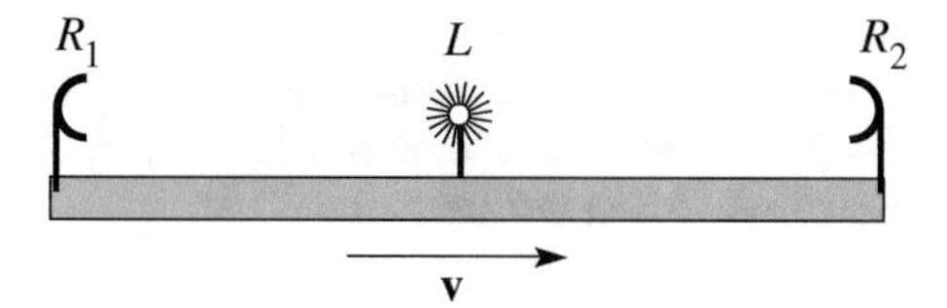

Fig. 6.9 A light flasher and two detectors at equal distances

detectors R_1 and R_2 at the two extremes of the bar. The observer in S' considers the two events of arrival of the flash at the two detectors as simultaneous. Notice that this conclusion can be reached only assuming that light propagates with the same velocity in both directions, namely that space is isotropic. Notice that the assumption is different from the invariance of the speed of light.

For the observer in S the two events are not simultaneous. Suppose that the velocity $\mathbf{v}$ of the bar in S has the direction from R_1 to R_2. One flash travels towards R_1 that is approaching, the other towards R_2 that is receding. The former will then take a shorter time than the latter to reach its detector. The two events are not simultaneous.

The fact that the simultaneity of two events happening in two different points is not absolute is a consequence of the existence of a maximum velocity for the propagation of the signals. This in turn has deep consequences on the measurement of time. We have defined an event as the set of the three spatial coordinates and the temporal one that characterize a phenomenon happening at a certain time in a certain point. To give a physical meaning to this definition, we need to define the sets of operations to be done to measure the space and time coordinates. In particular, to measure the time of the events we need to have identical clocks in all the points of the reference frame. All the clocks must be synchronized. This means that the arms of all the clocks must reach the same position simultaneously. As simultaneity is frame dependent, an observer moving relative to a frame, the clocks of which have been synchronized by the observer at rest in that frame, sees those clocks as not synchronized. The consequence of the frame dependence of simultaneity is the frame dependence of the time measurements. Let us see that in the details.

6.5 Dilation of Time Intervals

Consider two events happening in the same point x_1 of the frame S in two different instants t_1 and t_2. In these conditions we can measure the time with a single clock in x_1. In other words, we have no need to synchronize clocks in different positions. The two events have the space and time coordinates $(x_1, 0, 0, t_1)$ and $(x_1, 0, 0, t_2)$. They are separated by the time interval

$$\Delta t_0 = t_2 - t_1,$$

where the subscript $_0$ is to recall that the time interval is measured in the frame in which the object is at rest. Such intervals are said to be of *proper time*. The observer in S' obviously does not see the two events in the same point of his frame, but, say, in x_1' and x_2'. If he wants to measure the times t_1' and t_2', in which the events happen he needs two clocks, one in x_1' and one in x_2', which must be synchronized. Equation (6.19) tell us that

$$t_1^{'} = \gamma_{O'}\left(t_1 - \beta_{O'}\frac{x_1}{c}\right), \quad t_2^{'} = \gamma_{O'}\left(t_2 - \beta_{O'}\frac{x_1}{c}\right).$$

The time interval in S' is then

$$\Delta t' = t_2^{'} - t_1^{'} = \gamma_{O'}(t_2 - t_1)$$

or

$$\Delta t' = \gamma_{O'}\Delta t_0. \tag{6.28}$$

Consider for example a clock producing periodic ticks. The period, namely the time interval between two consecutive ticks, in the frame in which the clock is at rest, is, say Δt_0. An observer moving with velocity $v_{O'}$ the clock appears emitting ticks with the period

$$\Delta t' = \gamma_{O'}\Delta t_0 = \frac{1}{\sqrt{1 - \frac{v_{O'}^2}{c^2}}}\Delta t_0, \tag{6.29}$$

which is longer than the proper time Δt_0.

It is useful to show this result also with a physics argument. Suppose that the observer in S and in S' have two identical clocks, built as in Fig. 6.10a. The light source L emits a flash at a certain instant, which reaches the mirror R at the distance l and is reflected back to the detector R. When the light pulse reaches R a tick is emitted and the source L emits anther flash, and so on. Let us see now how the two

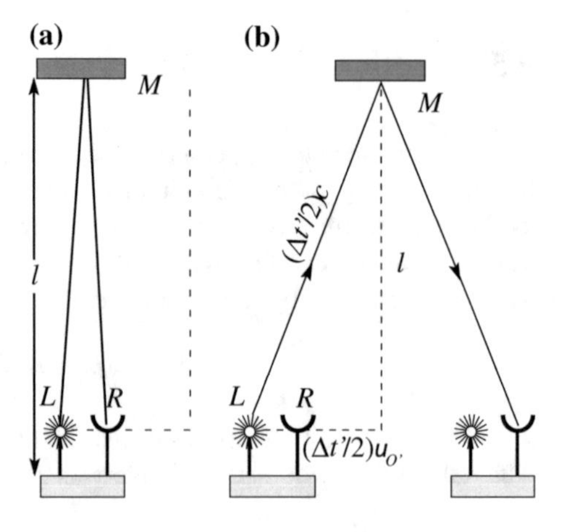

Fig. 6.10 A clock in, **a** seen in its rest frame, **b** seen from a moving observer

observers see the period of the clock. For both observers the period of their clocks is the time to go twice through the distance l, as in Fig. 6.10a, namely the proper time $\Delta t_0 = 2l/c$.

Also, to both observers the clock of the other one appears to move with velocity $\upsilon_{O'} = \upsilon_O$. Suppose that both clocks are oriented perpendicularly to the relative motion. In these conditions, the path of light that the observer S' sees in the clock in S is as represented in Fig. 6.10b, and reciprocally. Light takes half a period $\Delta t'/2$ to go from L to M, and the other half a period to go from M to L. The distance travelled by the flash in half a period is then $(\Delta t'/2)c$. In the same time interval the clock has moved a distance of $(\Delta t'/2)\upsilon_{O'}$. Hence (see figure)

$$(\Delta t'/2)^2 c^2 = (\Delta t'/2)^2 \upsilon_{O'}^2 + l^2 = (\Delta t'/2)^2 \upsilon_{O'}^2 + (\Delta t_0/2)^2 c^2,$$

from which

$$\Delta t' = \frac{1}{\sqrt{1 - \frac{\upsilon_{O'}^2}{c^2}}} \Delta t_0,$$

which is Eq. (6.29).

In the just made argument we implicitly assumed that the length l of the clock is independent of its motion, namely the same for both observers. As we shall prove in the next section, this is true because it is perpendicular to the relative motion. We can reach the same conclusion by observing that it is a consequence of the relativity principle. Indeed, both observers may agree to cut two notches in the positions of the extremes of the moving clock, respectively on the y and y' axis, when this passes by. Such notches must result in the same values of y and y', otherwise the results would be able to distinguish which is moving and which is still. Notice also that the hypothesis of the independence of light speed of the direction is once more necessary.

The phenomenon of time dilation is observed every day in elementary particles laboratories. Protons and electrons are accelerated in accelerators to speed very close to the speed of light. If time dilation were not taken into account, these machines would not work.

Consider as another example a natural phenomenon, the cosmic rays. These are particles, mainly protons and atomic nuclei, accelerated in the galaxy, and above it, to speed close to that of light and constantly entering the earth's atmosphere. In the atmosphere, sooner or later, one of these particles hits a nucleus of the air producing a shower of secondary particles. Some of them are unstable. Among them are the μ leptons, or *muons*, which are very similar to electrons, if not for the mass that is about 200 times larger. Their lifetime is 2.2 μs. In the decay, a muon produces an electron, a neutrino and an antineutrino.

The following experiment was done with didactic aims. Charged particles can be detected using a block of transparent plastic material, doped with substances that emit a flash of light when a charged particle goes through. The small flash of light is converted into an electric current pulse, which is sent to an electronic circuit. When a cosmic ray enters the block a pulse is observed. If it is a muon and if it stops in the block, after a time of the order of the lifetime it dies and the newborn electron gives a second pulse. This signature allows us to discriminate the stopping muons from other events induced by cosmic radiation. The apparatus was first used on Mount Washington (New Hampshire) at 1800 m height. In 1 h 568 stopping muons were counted.

How much time is needed for the muons to travel from 1800 m height to sea level? Obviously that depends on their speed. However, a lower limit is given by assuming they move with the speed of light. This lower limit is 6.3 μs. The experimenters then counted how many muons had lived more than 6.3 μs of the 568 detected on the Mount Washington. They found 27 of them. They then moved their detector to sea level. In absence of time dilation, about 27 events had to be detected. They found 412. This number agrees with Eq. (6.29) if the average muon speed is $\beta = 0.99$.

Now consider an observer sitting on a muon. In this frame the lifetime is not dilated and the muon survives only a few microseconds. How can so many reach sea level? The reason is that, as we shall see in the next section, the distance from the top of Mount Washington to sea level does not appear to the muon to be 1800 m, rather it is contracted by the same factor, the Lorentz γ parameter, as the time dilation. For $\beta = 0.99$, we have $\gamma = 6.1$ and the distance to travel is only 257 m.

6.6 Contraction of Distances

A second consequence of Lorentz transformations is the *contraction of lengths*. We start by observing that the operational definitions of the length of an object at rest and of an object in motion are not the same. The operations to be done in the two cases are indeed completely different. Consequently, there is really no a priori reason for which the two lengths should be equal. It is just every-day experience with objects moving at relatively low velocities that makes us believe in this equality. The lengths are equal, as is easily seen, for the Galilei transformations, not, as we shall now see, for the Lorentz transformations. According to the latter, when a body moves with velocity **v** relative to the observer, its dimension parallel to **v** appears contracted by a factor $1/\gamma$ relative to its value measured at rest. The transverse directions are equal for both observers.

To demonstrate these statements we imagine a rule fixed to the frame S lying on its x-axis. The observer in the moving frame S' determines the length of the ruler by measuring the coordinates of its extremes x_1' and x_2' at the same instant t'. The two corresponding events have the space-time coordinates $(x_1', 0, 0, t')$ and $(x_2', 0, 0, t')$. The length found by the observer is $l' = x_2' - x_1'$. The coordinates in S of the two events are

$$x_1 = \gamma_{O'}\left(x_1' + \upsilon_{O'}t\right), \quad x_2 = \gamma_{O'}\left(x_2' + \upsilon_{O'}t\right)$$

and their distance in S *is*

$$l_0 = x_2 - x_1 = \gamma_{O'}\left(x_2' - x_1'\right).$$

In conclusion, the relation between the length parallel to the relative velocity of an object at rest and moving with velocity $\upsilon_{O'}$ is

$$l' = \frac{l_0}{\gamma_{O'}} = l_0\sqrt{1 - \beta_{O'}^2}, \tag{6.30}$$

where the subscript $_0$ recalls that this is the length at rest. This is called the *proper length*. In any other moving frame the length appears contracted by the factor $1/\gamma$.

As for the dimension of the ruler, or any object, along y and z, perpendicular to the motion, the fact that they do not vary follows immediately from the second and third Eq. (6.19).

In this case too, let us demonstrate the result also with a physics argument. This will show that the contraction of the length is a logical consequence of the time dilation.

We still consider the ruler fixed along the x-axis of S. The observer in S measures the length l, and establishes that the observer in S', which is travelling at speed $\upsilon_{O'}$, crosses the distance l in the time interval $\Delta t = l/\upsilon_{O'}$. This time is not a proper time, because it is between two events happening in different locations, the passage of the mobile observer at one extreme and at the other. As such it is measured with two different clocks. On the other hand, for the observer in S' the two events happen in the same point and he can measure the time interval, $\Delta t'$, with the same clock. $\Delta t'$ is a proper time interval and, for what we saw in the last section, $\Delta t' = \Delta t/\gamma_{O'}$, and, as $\Delta t = l/\upsilon_{O'}$, it is $\Delta t' = \Delta t/(\upsilon_{O'}\gamma_{O'})$. The mobile observer sees the rule moving at the speed $\upsilon_{O'}$ and consequently evaluates its length to be $l' = \upsilon_{O'}\Delta t' = l/\gamma_{O'}$, which is the result that had to be demonstrated.

6.7 Addition of Velocities

In this section we shall find the rule of addition of velocities in relativistic physics. We just recall that for the Galilei transformations, if, for example, a ship moves relative to shore with velocity $\mathbf{u}$ and on the ship a passenger moves with velocity $\mathbf{v}'$, relative to the ship, the velocity of the passenger relative to shore is $\mathbf{v} = \mathbf{u} + \mathbf{v}'$. This is the Galilean composition rules of velocities. We shall now find the corresponding rule for Lorentz transformation, still in the particular case in which the two frames S and S' are those in Fig. 6.11.

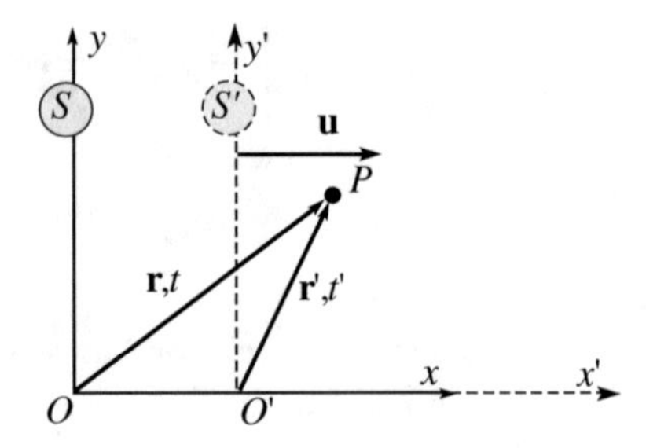

Fig. 6.11 Two frames in relative motion

The velocity $\mathbf{v}'$ of a point in S' is $\mathbf{v}' = \left(\frac{dx'}{dt'}, \frac{dy'}{dt'}, \frac{dz'}{dt'}\right)$ and the corresponding in S $\mathbf{v} = \left(\frac{dx}{dt}, \frac{dx}{dt}, \frac{dx}{dt}\right)$. Notice that each derivative in each frame is with respect to the time in that frame. We shall now use the Lorentz transformations Eq. (6.19) with $\beta = u/c$ and $\gamma = \left(1-\beta^2\right)^{-1/2}$. We have

$$\begin{aligned} dx &= \gamma(dx' + \beta c dt') \\ dy &= dy' \\ dz &= dz' \\ dt &= \gamma\left(dt' + \frac{\beta}{c}dx'\right). \end{aligned}$$

By dividing the first three equations by the fourth we have

$$\begin{aligned} \upsilon_x &= \frac{dx}{dt} = \frac{dx' + \beta c dt'}{dt' + \frac{\beta}{c}dx'} = \frac{\upsilon'_x + u}{1 + \beta\frac{\upsilon'_x}{c}} \\ \upsilon_y &= \frac{dy}{dt} = \frac{dy'}{\gamma\left(dt' + \frac{\beta}{c}dx'\right)} = \frac{\upsilon'_y}{\gamma\left(1 + \beta\frac{\upsilon'_x}{c}\right)} \\ \upsilon_z &= \frac{dz}{dt} = \frac{dz'}{\gamma\left(dt' + \frac{\beta}{c}dx'\right)} = \frac{\upsilon'_z}{\gamma\left(1 + \beta\frac{\upsilon'_x}{c}\right)}. \end{aligned}$$

We then write the conclusion

$$\upsilon_x = \frac{\upsilon'_x + u}{1 + \beta\frac{\upsilon'_x}{c}}, \quad \upsilon_y = \frac{\upsilon'_y}{\gamma\left(1 + \beta\frac{\upsilon'_x}{c}\right)}, \quad \upsilon_z = \frac{\upsilon'_z}{\gamma\left(1 + \beta\frac{\upsilon'_x}{c}\right)}. \tag{6.31}$$

Notice that not only the components parallel to the relative motion, but also the normal ones, are different in the two frames. The complicated behavior of the velocity stems from the fact that its components are not the three components of a four-vector. This is because, while (dx, dy, dz) are such components, dt is not a four-scalar.

It is easy to verify that the Eq. (6.31) tend to the Galilean one for $\beta \to 0$.

Example E 6.1 Consider a particle moving with velocity $\upsilon'_x = c/2$ relative to S', in the positive direction of x'. The reference S' moves relative to S at the speed $u = c/2$ in the same direction. Notice that if the transformation were the Galilean ones the velocity of the particle relative to S would have been equal to c. With the Lorentz transformation we have

$$\upsilon_x = \frac{c/2 + c/2}{1 + (1/2)(1/2)} = \frac{4}{5}c.$$

Example E 6.2 Consider S' to be a (very fast) ship and shooting a ball vertically upwards with velocity υ_z'. Which velocity of the ball is seen from shore? With $\upsilon_x' = \upsilon_y' = 0$ Eq. (6.31) give

$$\upsilon_x = u, \quad \upsilon_y = 0, \quad \upsilon_z = \upsilon_z'/\gamma.$$

Consider now the important case of a light signal propagating along the x' axis of S'. Its velocity relative to S is

$$\upsilon_x = \frac{c + u}{1 + u/c} = c \tag{6.32}$$

Namely, it has the same value in S' and in S, whatever their relative velocity can be. This result was expected considering that the speed of light is invariant under the Lorentz transformations.

A corollary is that combing to velocities smaller than c the resulting velocity is always smaller than c. The speed of light is the maximum possible velocity.

6.8 Space-Time

We have seen in Sect. 6.3 that the Lorentz are, from the geometric point of view, rigid rotations in the space-time, of coordinates (x, y, z, ict).

We cannot represent the four dimensions of the space-time on the two dimensions of a page of a book. However, we can learn a lot considering a particle moving in just one dimension, x. The space-time diagram has then two axes, the space coordinate x and the time, or, better to have the same physical dimensions ct, as shown in Fig. 6.12.

A point on this diagram represents an event happening in the space point x at time t. A particle at rest in the frame is, in space-time, a sequence of events at different times that have all the same coordinates. This is a line parallel to the ct axis, as line 1 in the figure. Such a line, in general, is called the *lifeline* of the particle.

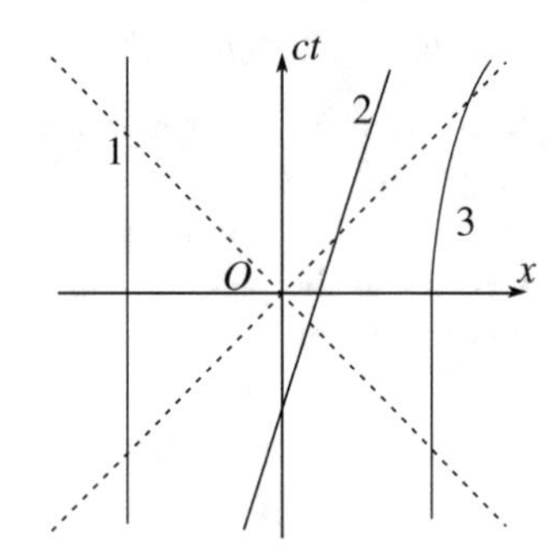

Fig. 6.12 The space-time diagram

If the particle moves with a constant speed υ its lifeline is a straight line, like 2 in the figure, having a slope relative to the *ct* axis equal to υ/c. Notice that the scales of the axes are such that the lifelines of any particle moving at the usual velocities are very near to being vertical ($\upsilon/c << 1$). On the other hand, the lifelines of the light signals are straight lines at +45° or −45° (dotted in Fig. 6.12) depending on the direction of propagation being the same or opposite to the *x*-axis. Line 3 is the life line of a particle that is at rest at a positive value of *x* at time 0, and that later on moves in the positive *x* direction of an accelerated motion, soon reaching speeds close to *c*. Notice that no lifeline can have a slope relative to the *ct* axis larger than one, namely a velocity larger than *c*.

Consider now the event *O* in the origin of the space time reference frame, namely the instant $t = 0$ in the point $x = 0$. Suppose this event being the start of light signals in all directions (the two of the *x* axis in our case). The lifelines of the signals are the bisectors of the axes as shown in Fig. 6.13. In the four dimensional space time, these lines draw a hypercone with vertex in the origin and half vertex angle equal 45°. It is called the *light cone*. The part of the light cone on $ct < 0$ corresponds to a light signal reaching the point $x = 0$ at $t = 0$.

It is not difficult to see that a Lorentz transformation transforms the axes as shown in the figures for x', ct'. The rotation of the axes is different from rotations in space because here the metric is pseudo-Euclidean rather than Euclidean. In the

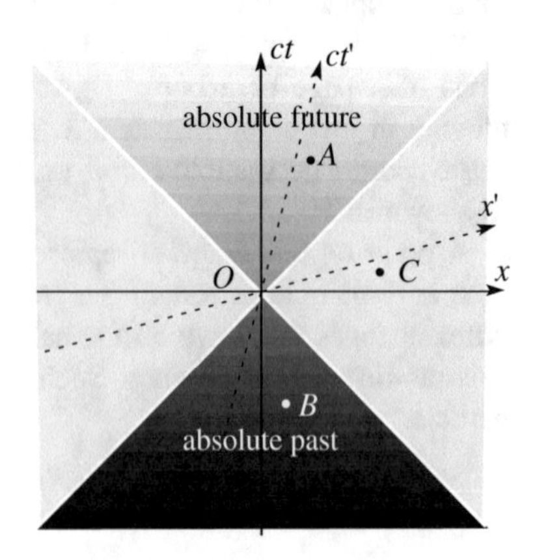

Fig. 6.13 The light cone

space time the x and ct axes rotate in opposite directions, by the same angle, approaching the light cone. The rotation angle is larger for higher relative velocity and tends to 45° when that tends to c. Obviously, the light cones of the two frames coincide, because the light velocity is the same in both.

Consider now events inside the light cone. The intervals between the origin O and each of them, like A and B, are negative. Such intervals are said to be *time-like*, because the purely time intervals are negative. The events outside the light cone are separated from the origin by positive intervals, called *space-like*. The events on the light cone are separated by null intervals and are called *light-like*. The intervals being invariant, these properties are independent of the reference frame.

Two events separated by a time-like interval can be joined by a signal travelling at a speed smaller than light, if the interval is light-like interval, they can be joined by a light signal, but if is space-like, they cannot be joined by any signal. Such a signal should travel faster than light. Consequently, no cause and effect relation can exist between two events at a space-like interval. This conclusion is connected with the fact that the relation past-future is not an absolute one for events outside the light cone, such as the event C in the figure. This event is future relative to O in (it has $t > 0$), while it is past relative to O, the same event, in S' ($t' < 0$) as is clear from the figure.

The events separated from the origin by time-like intervals are, as we have seen, inside the light cone. We can distinguish two parts of the cone. In the upper half cone, with $t > 0$, we have the events future relative to O. In the lower half cone, with $t < 0$, we have the events past relative to O (as B). Consider now, for example, the event A. It is separated from O by a negative interval. As the intervals are invariant there is no reference frame in which A is contemporary to O, because in this case A would be separated from O by a positive, or null interval. We can conclude that there is also no frame in which A is past relative to O, because in this case, for continuity reasons, a frame would exist in which A and O are simultaneous. In conclusion, all the events in the upper half light cone are future to O in any reference frame (*absolute future* of O), those in the lower half cone are past of O in any frame (*absolute past* of O).

6.9 Momentum, Energy and Mass

As we have seen, the Lorentz transformations between inertial frames are such as to guarantee the validity of the relativity principle for the Maxwell equations, the equations that govern electromagnetism. The principle requires however that all the physical laws should be covariant under these transformations. Consequently, we must also find the Lorentz covariant expression that generalizes the second law of Newton. Once we have found the new law, its predictions should be checked against experiments.

We start with the observation that the new law should admit the Newton one as a limit for small velocities. We also notice that, if an equation has to be covariant, all its terms must transform in the same way under Lorentz transformations. All of them must be four-scalar or four-vectors.

We already know a four-vector, the one that identifies the event in space-time, having components (x_1, x_2, x_3, x_4). We have obtained "promoting" the space three-vector $\mathbf{r} = (x_1, x_2, x_3)$ with the addition of the fourth time component. Such a "promotion" is not always possible with every three-vector. As we have already seen, for example, the three components of the velocity three-vector $\mathbf{v} = (dx/dt, dy/dt, dz/dt)$ are not the three space components of a four vector, because such are (dx, dy, dz), but dt is not a four-scalar.

The first step towards using relativistic dynamics is finding the correct expression of linear momentum. As we well know, the linear momentum of a particle of mass m and velocity small relative to c is

$$\mathbf{p} = m\mathbf{v} = m\frac{d\mathbf{r}}{dt}.$$

We can solve the problem of the non-invariance of dt by taking the derivative relative to the proper time t_0, the time in the reference frame moving with the particle, rather than relative to t. Recalling that $dt_0 = dt/\gamma$ we have

$$\mathbf{p} = m\gamma\mathbf{v} = \frac{m}{\sqrt{1-\frac{v^2}{c^2}}}\mathbf{v}. \tag{6.33}$$

We can immediately check that this expression tends to the Newtonian one for small velocities, namely for $\gamma \to 1$. As a matter of fact, γ does not differ much from 1 even at quite large velocities. For example, even at $v = 0.25c$, $\gamma = 1.03$, it has increased by only 3 %. However, when the velocity approaches c, the increase of γ becomes very rapid, for example, for $v = 0.5c$, $\gamma = 1.15$, for $v = 0.75c$, $\gamma = 1.51$, for $v = 0.99c$, $\gamma = 7.09$, to diverge for $v \to c$. If we try to accelerate a particle, when its velocity approaches the speed of light the work necessary to increase the velocity further becomes larger and larger. The work uses a larger and larger fraction of force to increase the γ factor and less and less to increase the velocity. The work to reach c would be infinite.

We have now found the space vector Eq. (6.33) that can be promoted to four-vector, which is called *four-momentum*. What is its fourth component? Taking into account that $dt/dt_0 = \gamma$ it is clearly

$$p_4 = m\frac{dx_4}{dt/\gamma} = ic\gamma m. \tag{6.34}$$

This very important quantity is, as a part of a constant, the energy of a free particle, as will become clear soon after having found the law of motion.

Before doing that we express the norm of the four-momentum $\mathbf{p}^2 - (c\gamma m)^2$. As all the norms of the four-vectors, this is a Lorentz invariant quantity, a four-scalar. Its expression is particularly simple in the rest frame of the particle, in which $\mathbf{p} = 0$, and we have

$$\mathbf{p}^2 - (c\gamma m)^2 = -c^2 m^2. \tag{6.35}$$

The norm of the four-momentum is proportional to the mass squared of the particle.

We now state without demonstration that, once the expression of the momentum is changed according to Eq. (6.33), the expression of the Newton law does not need any further change. However, there are now two time dependent factors in the derivative, the velocity and γ. We have

$$\mathbf{F} = \frac{d\mathbf{p}}{dt} = m\frac{d}{dt}[\gamma(\upsilon)\mathbf{v}] \tag{6.36}$$

Notice that neither the force nor the time derivative of the momentum are the space components of a four-vector. However, such are $\mathbf{F}dt$ and $d\mathbf{p}$, and consequently Eq. (6.36) is Lorentz covariant. Historically, the equation was found for the first time in June 1905 by H. Poincaré, who demonstrated its covariance and, in addition, that it is the unique expression enjoying such a property.

We are now ready to see the physical meaning of the fourth components of the four-momentum and of $\mathbf{F}dt$, namely of $\mathbf{F}\cdot d\mathbf{s}$. We shall proceed in a way quite similar to what we did for the kinetic energy theorem. Let $\mathbf{F}(\mathbf{r})$ be the resultant force acting on the particle at the position vector $\mathbf{r}$. We calculate its work when the particle moves from A to B on a certain trajectory, as shown in Fig. 6.14.

The elementary displacement $d\mathbf{s}$ in the time interval dt is $d\mathbf{s} = \mathbf{v}\, dt$. The work done by $\mathbf{F}$ is $dW = \mathbf{F}\cdot d\mathbf{s} = \mathbf{F}\cdot \mathbf{v}dt = \frac{d\mathbf{p}}{dt}\cdot \mathbf{v}dt = \mathbf{v}\cdot d\mathbf{p}$. To evaluate the last dot product we differentiate Eq. (6.35), obtaining $2\mathbf{p}\cdot d\mathbf{p} - 2m^2c^2\gamma d\gamma = 0$. Substituting $\mathbf{p} = m\,\gamma\,\mathbf{v}$, and simplifying, we have

$$\mathbf{v}\cdot d\mathbf{p} = mc^2 d\gamma \tag{6.37}$$

and the elementary work is then

$$dW = mc^2 d\gamma. \tag{6.38}$$

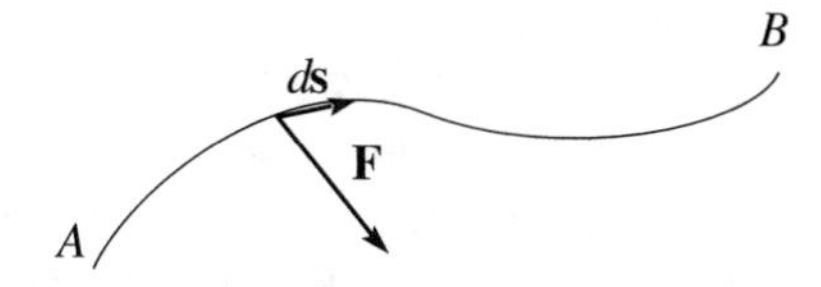

Fig. 6.14 The trajectory of a particle and the force acting on it

The work done by the force when the particle moves from A to B is

$$W_{AB} = mc^2 \int_A^B d\gamma = mc^2\gamma(\upsilon_B) - mc^2\gamma(\upsilon_A) = \frac{mc^2}{\sqrt{1-\frac{\upsilon_B^2}{c^2}}} - \frac{mc^2}{\sqrt{1-\frac{\upsilon_A^2}{c^2}}}. \quad (6.39)$$

Exactly as in Newtonian physics, the work done by the resultant of the forces on the particle is the difference between the values of a function of the velocity only at the end and at the beginning of the considered trajectory. In the following we shall consider only free particles, namely in absence of potential energy. In these conditions, we can say that the energy of the particle is

$$E = m\gamma c^2. \quad (6.40)$$

We see that the fourth component of the four-momentum is just the energy of the particle, divided by c. For this reason, the four-momentum is also called an *energy-momentum vector*. Its components are $(m\gamma\mathbf{v}, im\gamma c)$. Its norm, or better the opposite of its norm is

$$m^2c^2 = (E/c)^2 - p^2. \quad (6.41)$$

The relativistic energy of a free particle, Eq. (6.40), is not only kinetic energy. Indeed, the particle has energy also when it is at rest. It is called *rest energy* and we shall indicate it with

$$E_0 = mc^2. \quad (6.42)$$

We can say that the relativistic kinetic energy of a free particle is its total energy less its rest energy, namely

$$E_K = E - mc^2. \quad (6.43)$$

We see immediately, by developing in series of β^2, that the relativistic kinetic energy tends to the non-relativistic one at low velocities:

$$E_K = E - mc^2 = mc^2\left[\left(1-\beta^2\right)^{-1/2} - 1\right] = mc^2\left[1 + \frac{1}{2}\beta + \ldots - 1\right] \simeq \frac{1}{2}m\upsilon^2.$$

On the other hand, at very high velocities, Eq. (6.40) shows that the energy of the particle grows without limits when its velocity approaches the speed of light. As we have seen for the momentum, this is due to divergence of the γ factor. The particle "accelerators" of the laboratories studying the elementary particles work usually with protons or electrons "accelerated" at a speed very close to c. Accelerators act to increase the energy of the particles, while their velocity may change only by very small amounts. They should be more properly called "energizers". Indeed, particles of non-zero mass can never reach the speed of light. Their energy and momentum would be infinite. We shall come back to massless particles soon.

The fundamental mechanical quantities of a free particle are its mass, its momentum and its energy. These quantities are linked by two fundamental equations, Eq. (6.41) that we shall now write in a bit different form (multiplying by c^2) and a somewhat different expression of Eq. (6.33). They are

$$E^2 = \left(mc^2\right)^2 + (pc)^2, \tag{6.45}$$

$$\mathbf{p} = \frac{E}{c^2}\mathbf{v}. \tag{6.46}$$

We now observe that in nature elementary massless particles exist. Such are the photons, the quanta of light, and also the quanta of the strong interaction binding the quarks in a proton and in a nucleon, which are called *gluons*. When $m = 0$, the expression Eq. (6.33) has no meaning, because it contains the ratio between a null and an infinite quantity. The most general expression of the relativistic momentum is Eq. (6.46) that is valid both for massive and for massless particles.

Let us have a better look at Eq. (6.45) with the help of the "cartoon" of Fig. 6.15. In the general case, Fig. 6.15a, the energy is like the hypotenuse of a right triangle having mc^2 and pc as sides. It is given by the quadratic sum of the two quantities, namely it is the square root of the sum of their squares. One of them, mc^2, is the mass energy, the other one, pc, is the energy of its motion.

If the particle is at rest, its energy is only mass energy, or rest energy

$$E_0 = mc^2 \tag{6.47}$$

Here we must warn the reader that this equation is often written in the press, but also in the scientific literature, as $E = mc^2$, which is not true, because, as we saw in general it is $E = m\gamma c^2$, Eq. (6.40). The confusion is increased by writing $m\gamma$ "relativistic mass" and talking of mass varying with velocity. These are archaic concepts that were introduced when relativity theory was being developed, but should be avoided. Indeed the mass is an invariant quantity and does not vary with velocity. The term $m\gamma$ is apart from a factor c^2 not else than the energy, which is the fourth component of a four-vector.

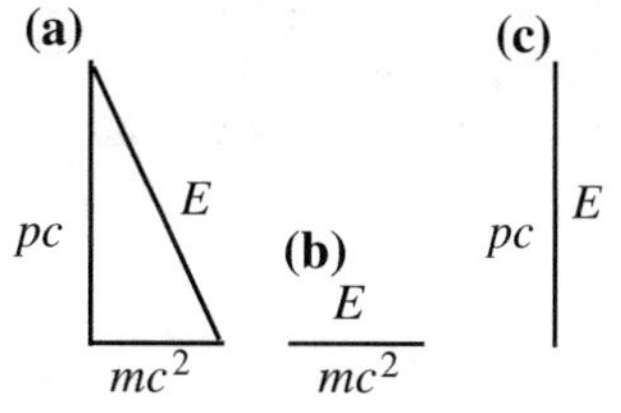

Fig. 6.15 Relation between energy, momentum and mass. **a** Generic, **b** particle at rest, **c** massless particle

Equation (6.46) tells us that the mass energy is enormous, due to the c^2 factor. However matter and energy are not equivalent. Indeed, matter has existed since the origin of the universe and does not convert into energy. The reason is that the matter particles have charges, the electric, the weak and the strong ones. These charges are conserved. We cannot destroy, for example, an electron and get energy from its mass. We can however, annihilate an electron with its antiparticle, the positron that has opposite charge. However, the quantity of antimatter in the universe is very small. We shall come back to the mass and to energy transformations in the next section.

Figure 6.15c shows the case of a massless particle, say a photon. For Eq. (6.45), being massless means that

$$E = pc \tag{6.48}$$

and from Eq. (6.46), for photons

$$\upsilon = c \tag{6.49}$$

a free massless particle can move at only one speed, the speed of light.

6.10 Mass, Momentum and Energy for a System of Particles

We now consider a system of free particles, namely there are no forces, external or internal, acting on them. As in non-relativistic physics, the total momentum and the total energy of the system are the sum of the homologues quantities of the single particles, namely

$$E = \sum_{i=1}^{N} E_i, \quad \mathbf{p} = \sum_{i=1}^{N} \mathbf{p}_i. \tag{6.50}$$

The situation is more complex if the particles interact with internal forces. In particular, Eq. (6.50) are not valid. We do not have the time to discuss the issue here, but only mention that, in addition to the mechanical ones of the particles, there are both energy and momentum distributed in the fields of forces.

Coming back to the system of relativistic non-interacting particles, we shall now look at its total mass. As for the single particle, the total momentum and the total energy of a system are (taking into account the c factors) the four components of a four-vector, of which Mc^2 is the norm.

$$M = \left[\sum_{i=1}^{N} \left(\frac{E_i}{c^2} \right)^2 - \sum_{i=1}^{N} \left(\frac{\mathbf{p}_i}{c} \right)^2 \right]^{\frac{1}{2}}. \tag{6.51}$$

We see here a fundamental difference from the non-relativistic case: the mass of the system is not the sum of the masses of its constituents.

Consider now several examples.

Example E 6.3 Find the expressions for the mass of the system of two photons of the same energy E, if they move in equal or opposite directions.

For the photon that has zero mass, $pc = E$. Consequently the total energy $E_{tot} = 2E$.

If the photons have the same direction, then the total momentum is $p_{tot} = 2E/c$ and therefore the mass is $m = 0$.

If the velocities of the photons are opposite, it is still $E_{tot} = 2E$, but $p_{tot} = 0$, and hence $m = 2E/c^2$.

In general, if θ is the angle between the velocities,

$$p_{tot}^2 = 2p^2 + 2p^2 \cos\theta = 2\left(E/c^2\right)^2(1 + \cos\theta)$$

and hence $m^2 = 2(E/c^2)^2(1 - \cos\theta)/c$.

Example E 6.4 Consider two particles with the same mass m moving with the same initial velocity υ of opposite direction. The two particles collide and stick together. The final kinetic energy is zero. Macroscopically we call the collision completely inelastic. However, the total energy did not vary, because the rest energy has increased by the same amount. In relativistic mechanics the inelastic collisions do not exist. Energy is always conserved

$$2\frac{mc^2}{\sqrt{1 - \upsilon^2/c^2}} = Mc^2.$$

In other words, the mass of the final body is not $M = 2\ m$, but $M = 2m/\sqrt{1 - \upsilon^2/c^2}$, which is larger than 2 m. The mass increase is extremely small at low velocities. As an example, suppose that $\upsilon = 300$ m/s, which is quite large for everyday life, but very small compared to c, being that $\beta = \upsilon/c = 10^{-6}$. Developing the above expression in series we have

$$M = 2m/\sqrt{1 - \beta^2} \simeq 2m\left(1 + \beta^2/2\right),$$

which differs from m by, in order of magnitude, 10^{-12}. This is so small that it cannot be measured. In other words, the rest energy is so large that its increase corresponding to the decrease in kinetic energy is undetectable. The decrease of kinetic energy between initial and final state is on the contrary evident. It looks like energy is not conserved. But, what appears to have been lost is rather hidden in the mass energy.

Example E 6.5 The most massive nuclei, as some of the Uranium isotopes, are often unstable. They can break up in fragments spontaneously, or make them absorb a neutron. Suppose the fragments to be two and m_1 and m_2 their masses, while M is the mass of the mother nucleus. We state that $m_1 + m_2 < M$. Indeed, the energy conservation requires that

$$Mc^2 = m_1c^2 + E_{K1} + m_2c^2 + E_{K2}.$$

The final kinetic energy $E_{K1} + E_{K2}$ is the energy produced for example in a power station. The remaining energy difference $(M - m_1 - m_2)c^2$ may correspond to a small mass difference, but the corresponding energy can be large due to the factor c^2. Let us see a numerical example.

We profit from an example to introduce a measurement unit of mass that is widely used in atomic and subatomic physics. As we have seen, energy can be measured in electronvolt, Eq. (3.78). As the mass is equal to the rest energy divided by c^2, we shall measure it in eV/c^2.

The simplest nucleus, the hydrogen one, is simply a proton, the mass of which is $m_p = 938.27$ MeV/c^2. The mass of the neutron is a bit larger, $m_n = 939.57$ MeV/c^2. The mass of the electron is about 2000 times smaller, $m_e = 511$ keV/c^2. The most massive nuclei have masses of hundreds of GeV/c^2. In a heavy nuclear fission, namely a break up, the released energy is of several MeV. In other words, the mass difference between the initial and the final state is, in relative value, of a few parts in hundred thousandths. These values are small, but can be measured, and the predictions of the theory can be checked.

In the lightest nuclei the opposite process can happen. That process is fusion. For example, two neutrons and two protons can join together to produce a He nucleus. This is because the mass of the latter, $m_{He} = 3.72741$ GeV/c^2, is smaller than the sum of the initial masses. Let us calculate the mass defect, namely

$$m_{He} - 2m_p - 2m_n = 3277.41 - 2 \times 938.27 - 2 \times 939.57 = -28.3 \text{ MeV}/c^2.$$

The mass defect corresponds to the binding energy, namely to separate the four components of a He nucleus we must give it an energy of 28.3 MeV.

Example E 6.6 Consider now the hydrogen atom, which is made of a proton and an electron. Its binding energy, namely the energy to separate the electron from the proton is $\Delta E = 13.6$ eV. The mass difference in relative values is

$$\frac{m_H - m_p - m_e}{m_H} = \frac{\Delta E}{m_H c^2} = \frac{13.6}{9.39 \times 10^8} = 1.4 \times 10^{-8},$$

which is a very small fraction. The atomic energy scale is much smaller than the nuclear one.

Example E 6.7 When energy is measured in eV, the momenta are measured in eV/*c*. Let us see, for example, the value in SI of a 1 meV/*c* momentum. It is

$$p = 1\ \text{MeV}/c = (1\ \text{MeV}/c)\left(\frac{1.6\times10^{-13}\text{J}}{1\ \text{MeV}}\right)\left(\frac{c}{3\times10^{8}\ \text{m/s}}\right)$$
$$= 5.3\times10^{-22}\ \text{kg m s}^{-1}.$$

6.11 Force and acceleration

As we have seen, the relativistic law of motion of a particle of mass m under the action of the force $\mathbf{F}$ states that the force is equal to the rate of change of momentum. This is $\mathbf{p} = m\gamma(\upsilon)\mathbf{v}$. It contains the product of two functions of time. Consequently, the derivative is the sum of two terms

$$\mathbf{F} = \frac{d\mathbf{p}}{dt} = m\gamma\frac{d\mathbf{v}}{dt} + m\frac{d\gamma}{dt}\mathbf{v} = m\gamma\mathbf{a} + m\frac{d\gamma}{dt}\mathbf{v}. \tag{6.52}$$

Taking the derivative of $\gamma(\upsilon)$, we obtain

$$\frac{d\gamma}{dt} = \frac{d}{dt}\left(1-\upsilon^2/c^2\right)^{-1/2} = -\frac{1}{2}\left(1-\upsilon^2/c^2\right)^{-3/2}\left(-2\upsilon/c^2\right)\frac{d\upsilon}{dt} = \gamma^3\frac{\beta}{c}\frac{d\upsilon}{dt}.$$

We substitute this expression in Eq. (6.52) taking into account that $d\upsilon/dt$ is the component of the acceleration in the direction of the velocity, namely that $d\upsilon/dt = \mathbf{a}\cdot\mathbf{u}_\upsilon$, where $\mathbf{u}_\upsilon$ is the unit vector of velocity, obtaining

$$\mathbf{F} = m\gamma\mathbf{a} + m\gamma^3\beta\boldsymbol{\beta}(\mathbf{a}\cdot\mathbf{u}_\upsilon) = m\gamma\mathbf{a} + m\gamma^3\boldsymbol{\beta}(\mathbf{a}\cdot\boldsymbol{\beta}), \tag{6.52}$$

where $\boldsymbol{\beta}$ is the vector $\mathbf{v}/c$.

We see that the force is the sum of two terms, one parallel to the acceleration and one parallel to the velocity. Therefore, we cannot define any 'mass' as the ratio between force and acceleration. At high speeds, the mass is not the inertia to motion.

To solve for the acceleration we take the scalar product of the two sides of Eq. (6.52) with $\boldsymbol{\beta}$. We obtain

$$\mathbf{F}\cdot\boldsymbol{\beta} = m\gamma\mathbf{a}\cdot\boldsymbol{\beta} + m\gamma^3\beta^2\mathbf{a}\cdot\boldsymbol{\beta} = m\gamma\left(1+\gamma^2\beta^2\right)\mathbf{a}\cdot\boldsymbol{\beta} = m\gamma^3\mathbf{a}\cdot\boldsymbol{\beta}.$$

Hence

$$\mathbf{a}\cdot\boldsymbol{\beta} = \frac{\mathbf{F}\cdot\boldsymbol{\beta}}{m\gamma^3} \tag{6.53}$$

and, by substitution into (6.52)

$$\mathbf{F} - (\mathbf{F} \cdot \boldsymbol{\beta})\boldsymbol{\beta} = m\gamma\mathbf{a}. \tag{6.54}$$

The acceleration is the sum of two terms, one parallel to the force, and one parallel to the speed.

Equation (6.52) and its equivalent Eq. (6.54) have been the object of a large number of experimental controls with high energy charged particles like protons, nuclei and electrons under electric and magnetic forces in different configurations. The engineers designing the accelerators at relativistic energies use these formulas in their everyday work.

We notice that force and acceleration have the same direction in two cases only: 1. force and velocity are parallel: $\mathbf{F} = m\gamma^3\mathbf{a}$; 2. force and velocity are perpendicular: $\mathbf{F} = m\gamma\mathbf{a}$. The proportionality constants are different. Consider for example a particle moving with 95 % of light speed, that is $\beta = 0.95$ and $\gamma = 3.2$. If the particle travels on a circle, the centripetal force should be 3.2 times larger than what was foreseen by Newtonian mechanics. However, if it is in a rectilinear accelerated motion the force necessary to give it the same acceleration is $\gamma^3 = 32.8$ times larger than in Newtonian mechanics. We see that, even in these special cases, we cannot consider mass as the inertia to motion.

6.12 Lorentz Covariance of the Physics Laws

We have seen how the relativity principle, originally established by G. Galilei in the XVII century, was found to hold for electromagnetic interactions, provided that the transformations of coordinates and time between two inertial reference frames are Lorentz transformations. This led to special relativity. The theory, however, can work only if all the physics laws turn out to be Lorentz covariant. Indeed, we have already discussed that for the second Newton law.

We have already firmly stated that the Lorentz transformations, while they historically discovered a guarantee for the relativity principle of a specific interaction, can be demonstrated independently of electromagnetism, on the basis of very general assumptions as we saw at the end of Sect. 6.4.

It remains to be seen, however, whether the other forces, or better interactions, of nature satisfy the relativity principle, namely if the equations that rule them behave in a Lorentz covariant form. The answer is yes, but we can give here only a few hints.

The Newton law of the gravitational force,

$$\mathbf{F}(\mathbf{r}) = -G_N \frac{m_1 m_2}{r^2}\mathbf{u}_r \tag{6.55}$$

is clearly not Lorentz invariant. Indeed, this expression implies instantaneous propagation of the effects over any distance. If, for example, our sun would suddenly disappear, the gravitational force on earth would go to zero immediately. But Lorentz invariance requires that all the fundamental interactions propagate with a speed not larger than c, which is the parameter in a Lorentz transformation. Consequently we would be safe still for 8 min, the time taken by the gravitational wave resulting from the explosion to reach us. The relativistic theory of gravity is called *general relativity*, as we have already mentioned. The equations were sent for publication at the end of 1915 independently by David Hilbert (1862–1943) and A. Einstein. We have now an enormous quantity of experimental proofs of its validity. We only mention, as an example, that the data of the global position system, the GPS, which is based on a constellation of artificial satellites, would give wrong information on our position if not elaborated with general relativity.

All the other forces we studied in Chap. 3, the elastic force, the forces of the constraints, the force between molecules, etc. are, at a fundamental level, due to electromagnetic interaction. As such, the laws by which they are governed are Lorentz invariant.

The other two fundamental interactions, the weak interaction and strong interaction, were discovered after the establishment of special relativity and their equations, which are quantum theories, were written in a Lorentz covariant form since the start. Their validity has been proven with a myriad of very high precision experiments on high energy particles both from natural sources, like the radioactive decays and cosmic rays, and, mainly, in the accelerator laboratories.

6.13 What Is Equal and What Is Different

We summarize here the concepts that are different in relativistic mechanics (r.m.) from Newtonian mechanics (n.m) and those that remain unaltered.

1. The relativity principle is valid both in n.m. and in r.m.
2. The coordinate transformations are different, Galilei in n.m., Lorentz in r. m.
3. Time and simultaneity are absolute in n.m., relative in r.m.
4. The law for summing velocity is different.
5. In n.m., velocities can have any value; in r.m. they cannot be larger than c.
6. The expressions of momentum are different.
7. The forces have the same expressions.
8. The force is equal to the time derivative of the momentum in both.
9. The total momentum (and the total angular momentum) of an isolated system are conserved in both cases.
10. The energy has different expressions. The kinetic energy is directly proportional to the square of velocity in n.m., not in r.m. The rest energy does not exist in n.m.

11. The energy of an isolated system is conserved only if all the forces are conservative in n.m., always in r.m.
12. The total momentum of a system of non-interacting particles is the sum of the momenta of the single particles. The same is true for energies. This both in n.m. and in r.m. In r.m. the same is not true for the systems of interacting particles. We can only hint at the reason for that here. It lays in the fact that the field of the interaction force contains both energy and momentum.
13. The mass of a composite body is the sum of the masses of its components in n. m. it is not in r.m.
14. In n.m., force and acceleration are parallel; they are not so, in general, in r.m.
15. In n.m. the proportionality constant between force and acceleration is the mass, which acts as inertia to the motion. In r.m. acceleration is not proportional to the force, there is no "inertial" mass.
16. The mass is invariant both under the Galilei and the Lorentz transformations.

Problems

6.1. Consider two reference frames, S, which we call fixed, and S', which we call mobile as in Fig. 6.2. In the two frames there are clocks as those in Fig. 6.5. Develop the argument analogous to that of Sect. 6.5 if the arms of the clocks are in the direction of the x axis, namely of the relative velocity.

6.2. A muon is produced by cosmic rays in the atmosphere. It travels at $\upsilon = 0.99c$ for 4 km and then decays. (a) How long does it live in our reference? (b) and in its frame? (c) How much is the thickness of the atmosphere it crossed in its reference?

6.3. A particle of mass m moves in a straight motion along the x axis with $x(t) = \sqrt{x_0^2 + c^2t^2}$. Find its limit velocity for $t \to \infty$. Find the expression of the force acting on the point.

6.4. A particle of mass m moving with the speed $\upsilon = (4/5)c$, hits a particle at rest with the same mass. After the collision the two particles form a unique body of mass M. Find M and the velocity of this body.

6.5. The cosmic rays contain protons with 10^{10} GeV energy. Find the time in the reference frame of such a proton to cross the Galaxy.

6.6. Find its momentum (in MeV/c) of an electron of 1 meV kinetic energy.

6.7. Find the momentum, in MeV/c of an electron travelling at $c/2$.

6.8. Find the energy of an electron travelling at 80 % of the speed of light.

6.9. A particle called ρ having mass 770 meV/c^2 decays at rest in two particles called π, which have mass m = 140 meV/c^2. Find their velocity.

6.10. In the LEP accelerator at CERN, electrons were accelerated up to an energy of 50 GeV. Find the relative difference between the velocity of the electrons and light.

6.11. A particle called tau has a lifetime of 0.3 ps. Find the velocity it should have to travel 1 mm in a lifetime.

6.12. A Z° (mass 91.2 GeV/c^2) particle decays at rest in an electron and a positron (they have equal masses). Find the energy and the momentum of the electrons. How much does, in relative terms, the velocity of the electron differ from c?

Chapter 7
Extended Systems

In this chapter we shall discuss the mechanics of extended systems, namely of mechanical systems composed of more than one particle or by bodies of finite extension. As a matter of fact, even in the simplest case of a point-like body under the action of a force, at least another body, giving origin to the force, must exist. Every action is always accompanied by a reaction. In other words, the simplest mechanical system consists of two interacting particles. We have considered, for example, the motion of earth or of a planet around the sun. We had ignored the sun. We could do that without much error because its mass is enormously bigger. These however are particular cases.

In the first three sections we shall study two-body systems. We shall see (Sect. 7.1) how the potential energy, corresponding to the force that one body exerts on the other is, in fact, relative to the pair. In other words, it is an interaction. We shall then introduce in Sect. 7.2 the concepts of center of mass and of reduced mass. In Sects. 7.3 and 7.4 we shall discuss two examples of a two-body system, the double stars and the tides, a phenomenon in another two-body system, earth and moon.

The experimental study of collisions between two bodies had, and still has, an enormous importance in the development of physics. In the sections from Sects. 7.5 to 7.7 we shall see the collision experiments between two pendulums that led Newton to establish the principle of conservation of linear momentum. This is one of the fundamental principles in physics, strictly connected with the action-reaction law.

We shall then move to systems of many particles, introducing the concepts of total linear momentum (or quantity of motion) and total angular momentum of a system. We shall find the fundamental laws giving their rate of change, and study the properties of a privileged point, the center of mass of the system.

In the last two sections we shall come back to the study of collisions between extended bodies.

A. Bettini, *A Course in Classical Physics 1—Mechanics*,
Undergraduate Lecture Notes in Physics, DOI 10.1007/978-3-319-29257-1_7

7.1 Interaction Energy

In our discussions on potential energy in the preceding chapters, we have analyzed the problems as if only one body existed, on which given forces were acting. For example, we said that the potential energy of the weight of a body of mass m at the height h is mgh. This is a perfectly correct statement when the mass of the body under consideration, an apple for example, is much smaller than the body with which it interacts, the earth in the example. In this situation, a reference frame united with the larger object can be considered at rest. As a matter of fact, when the apple falls towards the earth, also the earth falls towards the apple in an accelerated motion. In practice, both earth velocity and acceleration are completely negligible. Rigorously speaking however, we are dealing with a two-body system, the apple and the earth and mgh is the variation of potential energy of the earth-apple *system*, when the distance between the center of the apple and the center of the earth increase by h. In other words, the potential energy is a property of the couple of objects together; it cannot be associated to one or the other individually. Indeed, if the two interacting bodies have comparable masses, both of them accelerate considerably under the action of interaction forces. The kinetic energy of each of them will vary at the variation of the interaction potential energy. Let us now study the issue.

We start with a simple example in Fig. 7.1. It is made of two small spheres of mass m_1 and m_2 joined by a spring. In the upper part of the figure the system is in its configuration at rest. We now move both spheres and call x_1 and x_2 the two displacements, in value and sign, measured each from its equilibrium position, as in the lower part of the figure. Both forces now act, $\mathbf{F}_{21}$ on sphere 1 and $\mathbf{F}_{12}$ on sphere 2. The forces, an action reaction pair, are equal and opposite. These are elastic forces, which are proportional to the stretch that is $\Delta x = x_1 - x_2$ (N.B. x_1 is positive, x_2 is negative).

The elastic potential energy, Eq. (3.2), is

$$U_p = \frac{1}{2}k\Delta x^2 \tag{7.1}$$

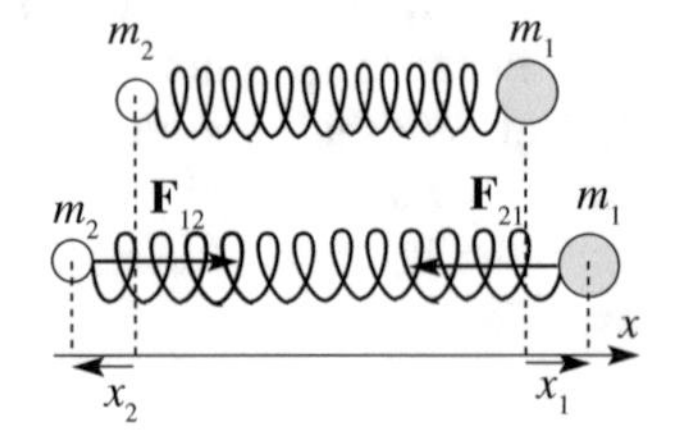

Fig. 7.1 Two masses linked by a spring

where k is the spring constant. Notice that this energy does not belong to one or the other sphere, but to the whole system, in other words is the interaction (through the spring) energy between the spheres.

The potential energy of any system in a given state is always the work that must be done against the forces that the system develops to change its state from the (arbitrarily) defined zero energy state to the given state. In our case the zero energy state is when the spring is not deformed. In the above statement, all the work must go into a change of the potential energy, namely it must be done at constant kinetic energy (zero in particular). Let us check with a direct calculation that our statements are correct.

Suppose we start from the equilibrium position. We first move sphere 1, keeping 2 at rest. Call x the displacement (with sign) of sphere 1 from its equilibrium position. We are moving it from $x = 0$ to $x = x_1$. During the displacement the stretch of the spring is just x. The x component of the force is consequently $F_{21x} = -kx$ and the work to be done is against it,

$$W_1 = -\int_0^{x_1} F_{21x} dx = +k \int_0^{x_1} x dx = \frac{1}{2} k x_1^2.$$

We now move sphere 2, keeping 1 steady. We now call x the displacement of sphere 2 from its equilibrium position. We are moving it from $x = 0$ to $x = x_2$. The stretch of the spring is now $x_1 - x$ and the x component of the force $F_{12x} = k(x_1 - x)$. The work to be done against it is

$$W_2 = -\int_0^{x_1} F_{12x} dx = -k x_1 x_2 + \frac{1}{2} k x_2^2.$$

Finally the total work is

$$W = W_1 + W_2 = \frac{1}{2} k x_1^2 - k x_1 x_2 + \frac{1}{2} k x_2^2,$$

which is clearly Eq. (7.1).

Consider now a second example: the potential energy of the gravitational force. Consider a point-like body of mass m on the surface of the earth (mass M), at the distance R_E from its center. As we know, the potential energy is

$$U_p = -G_N \frac{mM}{R_E}. \tag{7.2}$$

Recalling the arguments of Sect. 2.14 one easily sees that this is the work to be done against the gravitational force to move the mass m, say an apple, at zero kinetic energy, from infinite distance (the state we have defined to have zero potential energy) to the surface of earth. The energy is negative because, from outside of the system, we must work against an attractive force. In other words, the work we are considering is the opposite of the work of the gravitational force. We also see that the energy is not in the apple alone but in the earth and apple system.

As the last example we consider the weight force. The potential energy of a body of mass m at height h over the level we have decided for the potential energy to be zero, say the ground, is

$$U_p = mgh. \tag{7.3}$$

We know that this energy is just Eq. (7.2), apart from an additive constant. Indeed, in the two cases we made a different choice of the zero potential energy state. At first sight the two equations look quite different. However, consider that Eq. (7.3) is an approximate expression, valid for small level differences relative to the earth's radius, $h \ll R_E$. We then start from Eq. (7.2) expanding it in series of h/R_E stopping at the first order. We get

$$U_p(R_E + h) - U_p(R_E) = -G_N \frac{mM}{R_E(1 + h/R_E)} + G_N \frac{mM}{R_E} \simeq G_N \frac{mM}{R_E^2} h.$$

Now consider that $G_N M / R_E^2$ is simply the gravity acceleration on the earth's surface g. Equation (7.3) is valid when taking the potential energy on the earth's surface equal to zero, $U_p(R_E) = 0$. And the last equation becomes Eq. (7.3). In conclusion, the energy of the weight force mgh is not of the body but of the system body and earth.

7.2 Centre of Mass and Reduced Mass

We now come back to the simple mechanical system of two spheres joined by a spring (Fig. 7.2) and consider its motions. We are interested in the motion of one of them, say sphere 1. The sphere is subject to the elastic force $\mathbf{F}_{21}$. In Sect. 3.2 we have already discussed the motion of a material point under the action of an elastic force and found it to be harmonic. In that case, however, the other end of the spring was fixed to a wall and did not move. We can think of the wall as analogous in that

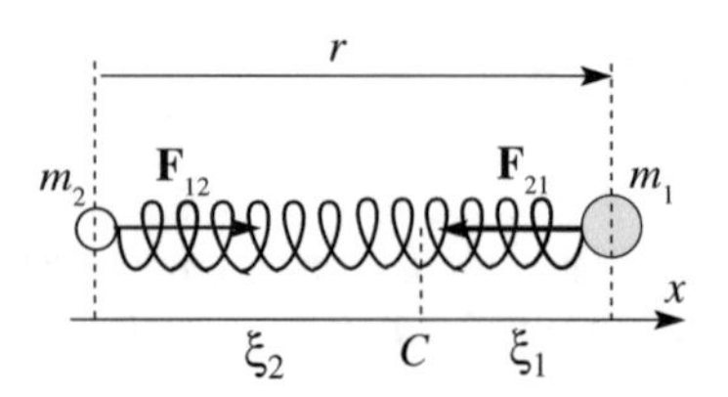

Fig. 7.2 Two spheres connected by a spring

case to sphere 2 in this case. In both cases, the force $\mathbf{F}_{12}$ acts on the second body. But the mass of the wall is so large that its acceleration is completely negligible. In the present case, on the contrary, sphere 2 will accelerate.

The problem we have now is that both points move. As we shall see in this chapter however, for every material system a privileged point, called *center of mass* of the system, exists. It is a geometrical point, not a physical one. In the presence of only internal forces, as in the case under discussion, the acceleration of its center of mass, in an inertial reference frame, is zero. We shall profit from that and describe the motion in a reference frame moving with the center of mass and with its origin in it, called the *center of mass frame*, for a brief *CM* frame. The center of mass of a two point-like bodies system is the point on the segment joining the two points that divide it in parts inversely proportional to the masses at the corresponding extremes.

We shall call C the center of mass, ξ_1 and ξ_2, the distances of the two masses from it and r the coordinate of point 1 measured from point 2. By definition of center of mass

$$\xi_1/\xi_2 = m_2/m_1. \tag{7.4}$$

Considering that $r = \xi_1 + \xi_2$ is the coordinate of point 1, the motion of which we want to study is

$$\xi_1 = \frac{m_2}{m_1 + m_2} r. \tag{7.5}$$

The force $\mathbf{F}_{21}$ acting on point 1 will give it the acceleration a_1 according to the Newton law

$$F_{21} = m_1 a_1 = m_1 \frac{d^2\xi_1}{dt^2} = \frac{m_1 m_2}{m_1 + m_2} \frac{d^2 r}{dt^2}.$$

We have so found, in the last side of this equation, an important quantity, called the *reduced mass* of the system

$$\mu = \frac{m_1 m_2}{m_1 + m_2}. \tag{7.6}$$

We can the write the equation of motion of point 1 as

$$F_{21} = \mu \frac{d^2 r}{dt^2}, \tag{7.7}$$

which is a very simple expression indeed. The equation of motion of point 1 is identical to its equation of motion valid when point 2 is fixed, provided that we are in the CM frame and we substitute for the mass of point 1 the reduced mass of the system.

Let us check if the arguments we made in Sect. 3.2 agree. First, we observe that when m_2 becomes very large compared to m_1, the reduced mass tends to the smaller of the two masses, m_1. To see that, just write Eq. (7.6) as $\mu = m_1/(1 + m_1/m_2)$, from which immediately $\mu \to m_1$ for $m_1/m_2 \to 0$. Clearly, what we said in Sect. 3. 2 is the limit case of what we are discussing here.

We now come back to the problem of the motion of point 1. We call r_0 the length at rest of the spring and s its stretch. Hence $r = r_0 + s$ and $F_{21} = -ks$. But $d^2s/dt^2 = d^2r/dt^2$ and Eq. (7.7) becomes

$$-ks = \mu \frac{d^2 s}{dt^2}, \tag{7.8}$$

which we recognize as the harmonic oscillator equation. We already know its solution

$$s(t) = A\cos(\omega_0 t + \phi) \tag{7.9}$$

where A and ϕ depend on the initial condition and

$$\omega_0 = \sqrt{k/\mu}. \tag{7.10}$$

In the CM frame the motion of point 1 is a harmonic oscillation. The difference with the case when point 2 is at rest is that in place of the mass of the oscillating body we have the reduced mass of the system. Clearly, point 2 moves with a harmonic motion of the same frequency because the reduced mass is the same in both cases.

In Sect. 3.11 we have considered, as an example of mechanical resonance, a diatomic molecule, in particular HCl. The two nuclei are small enough to be considered point-like particles in a very good approximation. Call r_0 their equilibrium distance. When the distance r is different from r_0, the electron cloud that in the molecule surrounds the nuclei exerts a force, which, in a first approximation, is proportional to the displacement $s = r - r_0$. The force is then elastic and the system is quite similar to the one we just discussed. As a matter of fact, the internal motions of molecules are correctly described by quantum mechanics. Our discussion should be considered a first approximation.

The potential energy of the interaction between the two nuclei, which we have already considered in Sect. 3.11, is shown in Fig. 7.3. The dotted parabola around the minimum is an approximation of the potential energy corresponding to the elastic force. In this approximation the potential energy is

$$U_p = \frac{1}{2} k s^2. \tag{7.11}$$

The equation of the parabola is written in Fig. 7.3 in eV units of energy and nanometer units of length. Expressing them in joule and meters respectively we

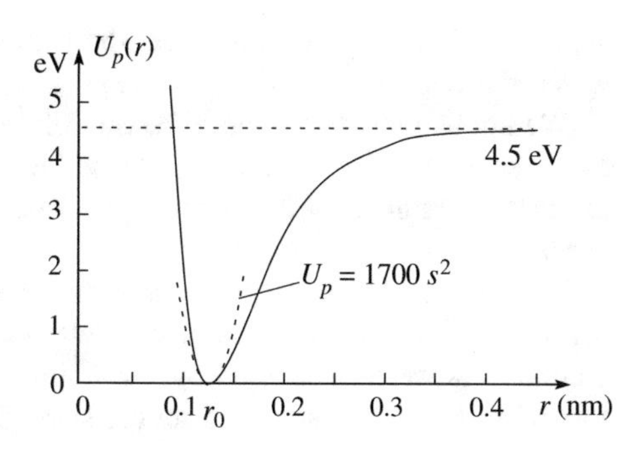

Fig. 7.3 The energy potential of a diatomic molecule

obtain $U_p(r) = 270\ \text{s}^2\ \text{J}$ and the "spring constant" equivalent is $k = 2 \times 270 = 540\ \text{N/m}^2$.

We calculate now the reduced mass. In atomic mass units ($u = 1.66 \times 10^{-27}$ kg) the masses of hydrogen and chlorine are (approximately) equal to 1 u and 35 u. In the same units $\mu = (1 \times 35)/(1 + 35) = 0.97$ u, which is close to the smaller hydrogen mass.

Finally, the proper oscillation frequency is $\nu_0 = \sqrt{k/m}/(2\pi) = 90$ THz, which is the value we used in Sect. 3.11.

As a second example, consider a molecule of carbon oxide (CO). The potential energy is quite similar to HCl, also quantitatively. We then take the same value of the "elastic constant".

As for the reduced mass we must consider that the masses of ^{12}C and ^{16}O are respectively 12 u and 16 u. The reduced mass is then $\mu = (12 \times 16)/(12 + 16) = 6.9$ u $= 1.1 \times 10^{-26}$ kg. Notice that, this time, the two masses are similar and the reduced mass is substantially different from, and smaller than, each of them. The reduced mass of a system of two equal masses is one half of each of them.

Concluding our calculation, we find the oscillation frequency $\nu_0 = 34$ THz, which is not too different, considering our approximations, from the measured value $\nu_0 = 64$ THz

7.3 Double Stars

In this section we shall consider a two-body system moving in two dimensions rather than one dimension as the diatomic molecules. It will also be a much larger astrophysical system. In Chap. 4 we discussed the motion of a planet, of mass m, about the sun, of mass M or of a satellite around its planet, assuming the sun in the first case, the planet in the second, to be at rest. From the discussion of the last sections one clearly understands that the assumption is not rigorously true. Indeed, both bodies move around their center of mass. However, in those cases the mass of the central body is much larger than the one of the orbiting body, and the approximation is quite good. We shall now consider an astronomical system in which the two masses, say m_1 and m_2 are similar.

We know today that a large fraction of the stars are in fact double, or, in several cases, even multiple. To establish these facts, the image of the star system needs to be resolved in those of its components. Telescopes of adequate resolving power are needed.

The first double star system was discovered in 1780 by Sir William Herschel (1738–1822) in the Ursa Major constellation. It is called Xi Ursae Majoris. More double stars were discovered by Sir William and his son John (1792–1871) in the following years. The study of double stars gives a further opportunity to check the Newton theory.

Figure 7.4 shows the apparent positions, namely the angles under which the objects are seen from earth of the Xi, as measured for more than a century. The motion must be studied in the CM frame, as in Fig. 7.5. C is the center of mass of the system, $\mathbf{r}_1$ and $\mathbf{r}_2$ the position vectors of the two stars, which are point-like in a

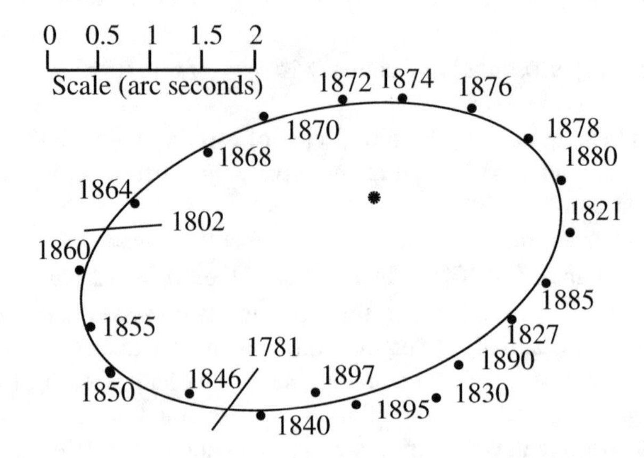

Fig. 7.4 The apparent positions of one star relative to the other (the dot inside the curve) for the Xi Ursae Majoris double star

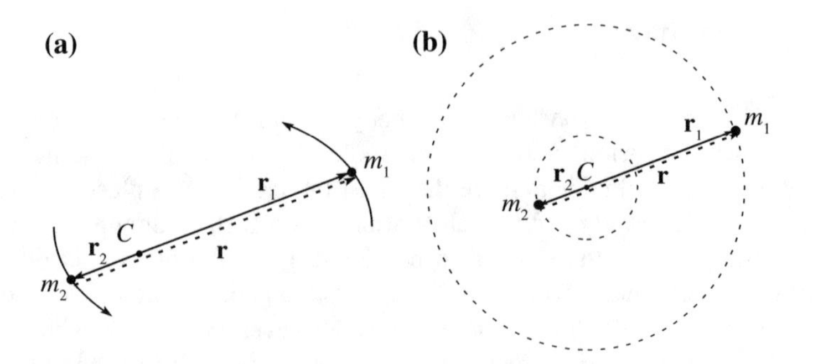

Fig. 7.5 **a** Diagram for the motion of a double star system; **b** case of circular orbits

good approximation, and m_1 and m_2 their masses. Let **r** be the vector from m_2 and m_1.

From the definition of center of mass $m_1 r_1 = m_2 r_2$ and

$$r_1 = \frac{m_2}{m_1 + m_2} r. \tag{7.12}$$

We know that the force, call it $F(r)$, acting on m_1 is the attraction of m_2 and consequently that it is directed as **r**. The acceleration is $a_{r1} = a_r m_2/(m_1 + m_2)$ and the Newton law

$$F(r) = \frac{m_1 m_2}{m_1 + m_2} a_r = \mu a_r. \tag{7.13}$$

In this case too, as in one dimension, we have found that the motion of a body of mass m_1 around another body of mass m_2 when both are moving is the same as when m_2 is at rest if, (a) we substitute for m_1 the reduced mass of the system, (b) we work in the CM taking into account that the center of the forces is the center of mass.

Figure 7.4 shows that the orbit shape is an ellipse. However, one of the stars does not look to be in a focus of the ellipse. This is an optical effect due to the fact that we are not looking normally at the orbit plane, but at a certain angle.

An interesting feature of binary systems is that their period depends only on the sum of the masses and not on their ratio. This is true in general, but, for simplicity, we restrict ourselves to the circular ones, as shown in Fig. 7.5b. The two stars rotate around the center of mass with the common angular velocity ω. The motion of one of them, m_1 for example, is given by the Newton equation $G_N m_1 m_2/r^2 = m_1 \omega^2 r_1$ and hence $\omega^2 = G_N m_2/(r_1 r^2)$ and, for Eq. (7.12)

$$\omega = \frac{2\pi}{T} = \sqrt{G_N \frac{m_1 + m_2}{r^3}}. \tag{7.14}$$

By measuring the period T and the distance r between the stars we can determine the sum of their masses.

7.4 Tides

The level of water contained by the seas and oceans varies during the day. The level grows (*flux*) till it reaches a maximum level (*high tide*) and then decreases (*reflux*) to a minimum (*low tide*) and so on. The phenomenon is periodic with a period (for example between consecutive high tides) of 12 h 25′, which is exactly equal to one half the time taken by the moon to come back to the same position relative to earth, namely its revolution period. Consequently, since ancient times tides were thought to be due to the moon. The explanation of the phenomenon however is not at all simple and had to wait for Newton.

Considering that we observe the phenomenon on earth, we shall describe it in a reference frame fixed on her. The first idea coming to mind is that the moon attracts the parts of the oceans nearest to it more strongly, causing their rise. But it does not work, because after half a period, when the moon is in its farthest position, we observe another rise rather than a lowering. The explanation must be different.

We cannot consider here the earth as point-like. We must take into account that the gravitational field of the moon is different in different points of the earth's surface, that have different distances from the moon. We shall work in a reference frame with the origin in the center of the earth. Notice that it cannot be considered inertial in the present discussion. If the gravitational force was equal in all the points of earth, it would be exactly balanced by the inertial force (centrifugal) due to the accelerated motion of the center of the earth, as we have seen in Sect. 5.7. Actually, the gravitational force is exactly balanced by the centrifugal one only in the earth's center. On the part of the surface nearer to the moon, the moon gravitational force is larger than the centrifugal one. On the opposite part the centrifugal force is larger than the gravitational one.

We underline that the inertia force we are considering is due to the acceleration of the origin of the reference frame (the center of the earth) that is rotating during the day around the center of mass of the earth-moon system. We also observe that we are neglecting the action of the sun on earth, which is much more intense than that of the moon. We can do that, in a first approximation, because what matters here is not the gravitational field itself but its differences in the different points of the earth. As a consequence of the much larger distance (400 times) of the sun than the moon, its field, even if stronger, is much more homogeneous. However, the sun does have an influence. We shall come back to that at the end of the section.

To simplify the problem, we shall consider the earth as a solid sphere with a layer of water of constant depth on the surface. We also assume the moon moving in the plane of the Equator. Figure 7.6 shows a view in this plane. The earth and the moon, a two-body system, rotate about their common center of mass. The

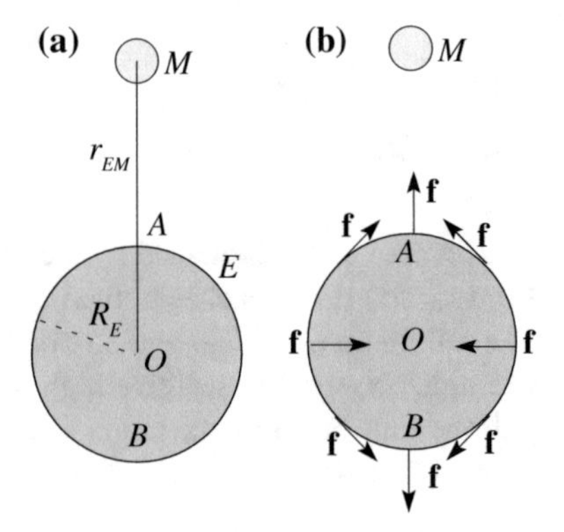

Fig. 7.6 **a** The geometry of the problem, **b** the tide force in different points of the earth surface

accelerations of both are directed towards the center. We can also think that both are continuously falling towards the center of mass.

In the point A, in which the moon is at the zenith, its gravitational attraction is larger than in O, because A is closer to it. As a consequence, the water particles in A fall towards the center of mass, and towards the moon too, with a larger acceleration than the earth's center O. On the contrary, in the point B in which the moon is on nadir, the gravitational attraction of the moon is smaller than in O and the water particles there fall towards the center of mass, and the moon, with an acceleration smaller than O.

We have followed the argument of Newton till here. However, at this point, Newton made a mistake (followed by several authors). The error is to extend what was established for the accelerations of water particles to their displacements. If we could do so, we would say that the water particles in A move towards the moon more than the center O and the sea rises, while those in B move towards the moon less than O. The sea moves away from the moon, and rises here too. The situation is shown in Fig. 7.7a. The ocean presents two bumps, diametrically opposed, on the line joining the moon with the earth's center. The bumps move in phase with the moon. We then expect high tides to take place just when the moon passes at the zenith and at the nadir, the low tides in quadrature, i.e. at a quarter of the period relative to those positions.

The observations, however, do not confirm these predictions. Rather, high tides happen when the moon is about in quadrature, the low tides when she is at the zenith and at the nadir as shown in Fig. 7.7b. The presence of the continents and other effects make the situation more complex. In any case, however, the delay between the passage of the moon on the zenith and nadir and the high tide is always of several hours. The disagreement is a consequence of the above-mentioned mistake. The displacement of a water particle at a certain time is not parallel to the acceleration in that instant.

The correct treatment of the tides can be divided in two parts. In the first part we calculate the *tide-generating force* as a function of a point on the earth's surface. The second part is the calculation of the forced oscillation of the oceans under the action of that force. This calculation is complicated by the presence of continents. We shall show the basic points of the argument.

Fig. 7.7 Schematic view of the earth and of the tides. **a** Phase as foreseen by Newton. **b** Phase as actually observed (approximately)

Let us start with the force. Both the gravitational and the inertia force acting on a water particle are proportional to its mass. We can then consider the force per unit mass. The weight per unit mass, **g**, has no influence on the tides, because it, even if different from point to point, is constant in time in each point. If the moon did not exist, the surfaces of the sea would be in any point perpendicular to **g**. The tide-generating force per unit mass, that we shall call **f**, is, as we already said, the resultant of the gravitational attraction of the moon and of the centrifugal force due to the rotation of the earth's center around the center of mass of the moon-earth system.

We shall not perform the calculation of **f** (which is not difficult). We show the result in Fig. 7.6b. We shall however, evaluate the order of magnitude of f, calculating it in A, where it is particularly easy. In the center O the gravitational attraction of the moon and the centrifugal force are equal. In A the centrifugal force is the same and is larger than in O. In this point they have equal and opposite directions. The magnitude of the sum of the gravitational attraction and the inertia force in A is consequently the difference between the gravitational attraction in A and the gravitational attraction in O (because the latter is equal to the inertia force both in O and in A). In conclusion, with r_{EM} the earth moon distance, R_E the earth radius and M_M the moon mass, we have

$$f = G_N \frac{M_M}{(r_{EM} - R_E)^2} - G_N \frac{M_M}{r_{EM}^2} = G_N \frac{M_M}{r_{EM}^2} \left[\frac{1}{(1 - R_E/r_{EM})^2} - 1 \right].$$

Considering that the radius of the earth is much smaller than the earth-moon distance, $R_E/r_{EM} \sim 1/60$, we can expand this expression in series of this quantity and stop at the first term. We have

$$f = G_N \frac{M_M}{r_{EM}^2} [1 + 2(R_E/r_{EM}) - 1] = 2G_N \frac{M_M R_E}{r_{EM}^3}. \tag{7.15}$$

This is the tide-generating force per unit mass in the point A, which has the dimensions of an acceleration. Let us compare it with the weight per unit mass, $g = G_N M_E / R_E^2$, where M_E is the earth mass. We have, in the right-hand side, with $R_E/r_{EM} = 1/60$ and $M_M/M_E = 1/81$,

$$\frac{f}{g} = 2 \frac{M_M}{M_E} \frac{R_E^3}{r_{EM}^3} = 1.1 \times 10^{-7}. \tag{7.16}$$

First we observe that the tide-generating force is inversely proportional to the *cube* of the earth-moon distance. In fact it depends on the *differences* between the gravitational force in different points, namely the derivative of the gravitational force. The latter varies inversely as the square, its derivative as the cube.

We observe that the tide-generating force is very small, but still enough to be a cause of such important phenomena. As a matter of fact, the height of the tide is of the order of a few to several meters, corresponding to a fraction of 10^{-7} of the earth diameter.

Calculations show that the magnitude of the tide-generating force is the same everywhere, hence is equal to what we calculated. Its direction, as shown in Fig. 7.6b varies as a function of the point.

To be precise, we notice that the moon's orbit is elliptic. Its distance from earth varies between 57 and 63.7 earth radii. Consequently f/g varies from 1.33×10^{-7} to 0.96×10^{-7}.

We now pass to the second part of the theory. Let us look at the situation in a point of the earth's surface. As we have said, the magnitude of the tide generating force is constant in time, but its direction varies. Its variation is a rotation at constant angular velocity. In other words, the components of the force, say the horizontal and vertical ones, vary periodically in time. When the former is a maximum the latter is null and vice versa. The ocean, which we still imagine to cover the entire surface, is subject to a periodic force, varying in time as a circular function. Even if the system is much more complex than a pendulum, it behaves as a *forced oscillator*.

Consider for example a drop of water in the air of a spaceship. Its natural shape is spherical. If we deform it a bit and then we let it go, it will tend to go back to its natural shape. But it cannot do that directly. Rather, like a pendulum, it will oscillate between different shapes and alternate between oblate and prolate. The oscillations have a proper period, which depends on the physical characteristics of the drop, and, if dissipative forces are present, are damped. The same would happen if, in absence of the moon, we would deform the surface of the ocean around the earth and abandon it. The system would oscillate at its proper oscillation frequency or, in other words, with the period, call it T_0, of the free oscillations of the system. Calculating T_0 is extremely difficult due to the complicated shape of the continents and of the sea bottom. Calculations on simplified models lead however, to values of T_0 = 20–30 h.

We can imagine the ocean as an oscillator, with *proper* oscillation period T_0. The oscillator is forced by a periodic force of period $T = 12$ h $25'$, which is much smaller than T_0. In other words, it is an oscillator forced at a frequency substantially larger than the resonance frequency. In these conditions, as we know (see Fig. 3.21b), displacement and force are in phase opposition. Consequently, the correct shape is that of Fig. 7.7b, not that of Fig. 7.7a, in substantial agreement with observations.

We now come back to the action of the sun. The reasoning is exactly the same as for the moon, and the result analogous to Eq. (7.16) is reached, obviously with the mass and the distance of the sun in the place of those of the moon. It is so found that the magnitude of the tide-generating force due to the sun is about half than that due to the moon. The two forces must be obviously summed as vectors. The two forces reinforce one another when the sun and the moon are about on the same line (new and full moon). The tides are then particularly ample (a condition called a syzygy), about one and a half larger than the value for the moon only. On the contrary, when the moon is at the first or last quarter, at 90° with the sun, the two forces partially cancel each other and the tides have small amplitude (quadrature tides), about one half as for the moon alone.

In practice, the height of the tides depends on several other factors, like the shape of the shores of the continents and the islands, the shape of the sea bottom, the oceanic currents, the winds, etc. Near the oceanic islands the height of the tides is typically one meter and near the continental shores it is about twice larger. However, in some sites the tides reach three meters and in a few even six meters. Particularly great tides are observed in deep gulfs or fiords facing the open sea. The greatest tides are in the Bay of Fundy, in Nova Scotia, Canada. Their amplitude is 4 m at the bay entrance, to reach 14 m at its end and even more at the syzygy.

7.5 Impulse and Momentum

Consider a material point of mass m in an inertial reference frame. Let $\mathbf{F}$ be the resultant force acting on the point. The second Newton law can be written in the form

$$\mathbf{F}dt = d\mathbf{p} = d(m\mathbf{v}). \tag{7.17}$$

In words, the effect of a force in the time interval dt is a variation of momentum equal to the product of the force and of the time interval. The vector quantity $\mathbf{F}dt$ is called elementary (meaning infinitesimal) *impulse* of the force in dt. The impulse of a force in a non-infinitesimal time interval from t_1 to t_2 is defined to be

$$\mathbf{i}_{12} \equiv \int_{t_1}^{t_2} \mathbf{F}dt. \tag{7.18}$$

Integrating in that time interval Eq. (7.17) we immediately have

$$\mathbf{i}_{12} = \int_{t_1}^{t_2} d\mathbf{p} = \mathbf{p}_2 - \mathbf{p}_1 = \Delta\mathbf{p}. \tag{7.19}$$

This equation expresses the *impulse-momentum theorem*: the momentum change of a material point under the action of the force $\mathbf{F}$ in the time interval from t_1 to t_2 is equal to the corresponding impulse, whatever is the time variation of the force and whatever is the length of the interval.

The impulse-momentum theorem is useful when the force acts for a short time, like in the collisions, strokes, explosions, etc. In these cases the force is initially null, then it quickly grows and as quickly goes back to zero. In these cases we do not usually know the instantaneous values of the force, but only its average value. The average value of a quantity in a given time interval is, by definition, the integral

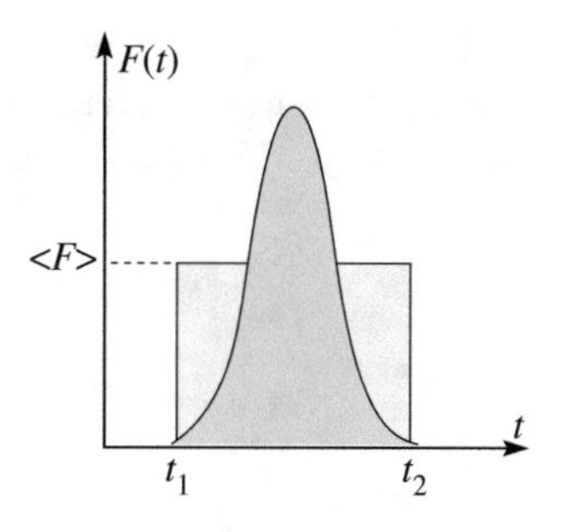

Fig. 7.8 An impulsive force and its average value

of the quantity over that time interval divided by the length of the interval. For the force, as shown in Fig. 7.8,

$$\langle \mathbf{F} \rangle = \frac{1}{t_2 - t_1} \int_{t_1}^{t_2} \mathbf{F} dt.$$

Example E 7.1 The hammer is an instrument used since ancient times to amplify the muscular force. Initially, at time t_1, a hammer of mass m, is at rest. With our arm we apply to it a force of average value $\langle \mathbf{F} \rangle$ till the instant t_2 in which the hammer strikes the head of the nail. In accordance with the impulse-momentum theorem, in this instant the momentum of the hammer is $\mathbf{p}_1 = \langle \mathbf{F} \rangle (t_2 - t_1)$. After that, the hammer slows down and stops (its momentum becomes zero) at time t_3. For the same theorem, the average force on the nail in the interval from t_2 to t_3 is $\langle \mathbf{F}' \rangle = \mathbf{p}_1/(t_3 - t_2)$. In conclusion, $\langle F' \rangle / \langle F \rangle = (t_2 - t_1)/(t_3 - t_2)$. Clearly, $t_3 - t_2$ is much smaller than $t_2 - t_1$ so that we obtain a large amplification of the force, by factors than can well be three orders of magnitude.

7.6 The Action-Reaction Law

Consider again a two-body system, made of two material points, which we call 1 and 2, and later, to follow Newton, A and B, of masses m_1 and m_2. The two points interact, point 1 acting on 2 with the force $\mathbf{F}_{12}$ and point 2 acting on 1 with the force $\mathbf{F}_{21}$. The two forces are an action and reaction pair. The third Newton law states that they are equal and opposite

$$\mathbf{F}_{12} = -\mathbf{F}_{21}. \tag{7.20}$$

We shall assume that no external force exists or, if some do, their resultant is zero. We deal with an *isolated system*.

$\mathbf{F}_{21}$ being the only force acting on point 1, it is equal to the rate of change of its linear momentum, or quantity of motion, $\mathbf{p}_1$, and similarly $\mathbf{F}_{12}$ is equal to the rate of change of $\mathbf{p}_2$. Equation (7.20) immediately gives

$$\frac{d\mathbf{p}_1}{dt} = -\frac{d\mathbf{p}_2}{dt} \tag{7.21}$$

and also

$$\frac{d(\mathbf{p}_1 + \mathbf{p}_2)}{dt} = \frac{d\mathbf{P}}{dt} = 0 \tag{7.22}$$

where we have put $\mathbf{P} = \mathbf{p}_1 + \mathbf{p}_2$. This is *total linear momentum (or total quantity of motion) of the system.* Equation (7.22) implies that

$$\mathbf{P} = \text{const.} \tag{7.23}$$

This equation expresses the *principle of conservation of linear momentum* in the case of a two-particle system. The principle states that total momentum of an isolated system is constant. We shall prove its general validity later in this chapter. In this section we shall use it in an experimental proof of the third law, as Newton himself did. Indeed, we have just seen that, for a two-body system, the principle is a consequence of the action-reaction law. It is also true that, if the total linear momentum of an isolated system is constant, the internal forces must be pairs of equal and opposite ones. Indeed, the most accurate verifications of the action-reaction law are, in fact, verifications of conservation of the total momentum. We observe, however, that in this way we verify the interaction forces to be equal and opposite, not that they have the same application line. We shall come back later on to this point.

Historically, the first experimental checks of the action-reaction law were done by Newton and his contemporaries Christopher Wren (1632–1723), Christiaan Huygens (1629–1695) and John Wallis (1616–1703). Their experiments are very accurate, conceptually simple and elegant. The experiments study the collisions between two spheres of different sizes, measure the momenta before the collision, say $\mathbf{p}_1$ and $\mathbf{p}_2$, and after, say $\mathbf{p}_1'$ and $\mathbf{p}_2'$, as accurately as possible and check if the relation is satisfied or not. The experiments were done by attaching the two spheres to two wires of equal lengths,

$$\mathbf{p}_1 + \mathbf{p}_2 = \mathbf{p}_1' + \mathbf{p}_2' \tag{7.24}$$

thus making two pendulums of the same period. When at rest the two spheres touch each other as in Fig. 7.9a. We move the spheres from equilibrium, each at a certain distance, which we measure. If we let both spheres go at the same instant from rest, they will accelerate, collide with each other in their lowest points, separate and move back together.

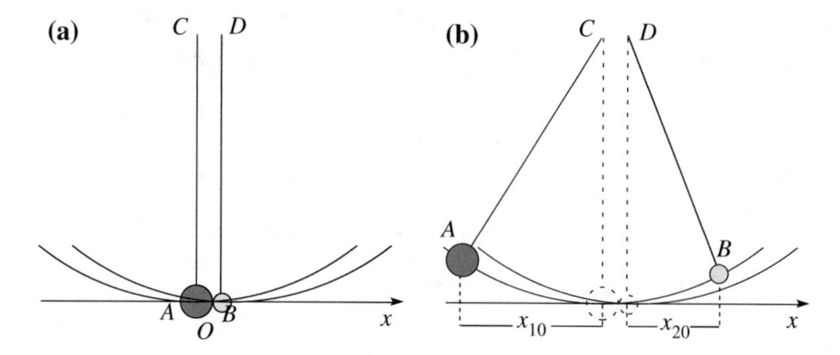

Fig. 7.9 The two-pendulum experiment to verify the momentum conservation. **a** Position at rest, **b** an initial configuration

The experiment profits from two properties of the pendulum. The first property is the isochronism of the (small) oscillations. Having the same lengths, the periods of the two pendulums are equal, independently of the masses of the spheres and of their initial positions (amplitude). Consequently, also the times taken to reach the equilibrium position are equal (a quarter of a period) and they will always collide there, if abandoned at the same time with null velocity. The second property is: the velocity of the pendulum when reaching the equilibrium position starting from a certain distance with null velocity is proportional to that distance. Let us show this property.

Let m be the mass and l the length of the pendulum. Let us remove it from the equilibrium position by x_0 as in Fig. 7.10 and let it go with null velocity. In this position, the pendulum is at a certain height, say h, above the horizontal through the equilibrium position. For small displacement angles we can use for h the approximate expression, Eq. (4.14)

$$h = \frac{x_0^2}{2l}. \tag{7.25}$$

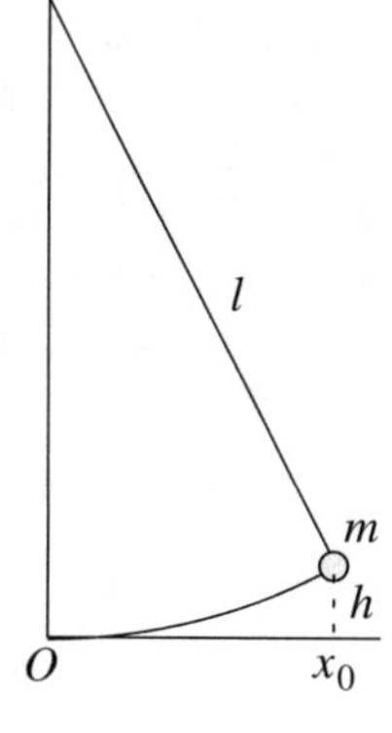

Fig. 7.10 Geometry of the starting configuration of the pendulum

If υ_0 is the velocity of the pendulum in an equilibrium position, the energy conservation law states that $(1/2)m\upsilon_0^2 = mgh$. Hence, using Eq. (7.25),

$$\upsilon_0 = \sqrt{2gh} = x_0\sqrt{\frac{g}{l}}$$

and also if $T = 2\pi\sqrt{l/g}$ is the period of the pendulum

$$\upsilon_0 = (2\pi/T)x_0. \tag{7.26}$$

We conclude that the velocity υ_0 at a collision will be known if we measure the period once and the initial position x_0 for each and every experiment.

We are now ready to read how Newton describes his experiments in the *Principia*. He does that just after having stated the third law to prove experimentally its validity. Newton built two pendulums each 10 ft (about 3.25 m) long, attaching two spheres A and B of the materials to test, and fixing the two wires in C and D as in Fig. 7.9a. We call m_1 and m_2 the masses of A and B respectively and x_1 and x_2 their displacement, measured for each pendulum from its equilibrium position (the position of its center to be precise). We remove both spheres to x_{10} and x_{20} respectively and accurately measure these distances. Notice that x_{10} and x_{20} can be on opposite sides, both on one side or both on the other of O. If we let them go at the very same instant with null velocities, they will collide in O with velocities

$$\upsilon_1 = (2\pi/T)x_{10}; \quad \upsilon_2 = (2\pi/T)x_{20}. \tag{7.27}$$

Let $\upsilon_1{}'$ and $\upsilon_2{}'$ be the velocities immediately after the collision. We can determine them by measuring the maximum distances, $x_{10}{}'$ and $x_{20}{}'$ reached (contemporarily) in their swing back. Indeed, we have

$$\upsilon_1' = (2\pi\ /T)x_{10}'; \quad \upsilon_2' = (2\pi/T)x_{20}'. \tag{7.28}$$

The two particles interact only during the instant of the collision. The external forces acting on them, the weight and the tension of the wire, have zero resultant. However, the system is not exactly isolated because air resistance exists and is an external force. This is small, but it must be taken into account in precision measurements. Newton did that as follows. He started operating with one pendulum only. He removed it from equilibrium at each of the distances that he was going to use in the following experiments. He let it go with zero velocity and observed the position reached after one period, which did not coincide exactly with the original one. He measured the miss. A quarter of that is what is lost in a quarter of a period due to the air resistance.

He made a number of experiments with spheres of different substances. For each of them, he tried different pairs of starting positions x_{10} and $x_{20,}$ measured those reached after the collision x_{10}' and x_{20}' and applied the just described correction. Each of them corresponds to a value of the quantity of motion; he calls that simply "motion", before the collision. The linear momentum conservation law (which is equivalent to the third law) that we need to verify is

$$m_1 \upsilon_1 + m_2 \upsilon_2 = m_1 \upsilon_1' + m_2 \upsilon_2'. \tag{7.29}$$

He writes (in parenthesis some explanations):

Thus trying the thing with pendulums of ten feet (*3.25 m*) in unequal as well as equal bodies, and making the bodies to concur after a descent through large spaces, as of 8, 12, or 16 feet (*2.6, 3.9, 5.2 m*), I found always, without an error of 3 inches (*8 cm*), that when the bodies concurred together directly (*in a straight line*), equal changes towards the contrary parts were produced in their (*quantities of*) motions, and, of consequence, that the action and reaction were always equal.

He continues giving numerical examples of his results. The initial and final momenta are given in "parts of motion", namely in an arbitrary unit. The unit is clearly irrelevant. For clarity, we shall write the values of the two sides of Eq. (7.29) for each quoted result at the beginning of each experiment. For each experiment, he mentions also the changes of the momentum of each body.

In the first experiment *B* is initially at rest (9 + 0 = 2 + 7).

if the body A impinged upon the body B at rest with 9 parts of motion, and losing 7, proceeded after reflection with 2, the body B was carried backwards with those 7 parts.

In the second experiment the initial velocities have opposite directions (12 – 6 = –14 + 8).

If the bodies concurred with contrary motions, A with twelve parts of motion, and B with six, then if A receded (*in its motion after the collision*) with 14, B receded with 8; namely, with a deduction of 14 parts of motion on each side. For from the motion of A subtracting twelve parts, nothing will remain; but subtracting 2 parts more, a motion will be generated of 2 parts towards the contrary way; and so, from the motion of the body B of 6 parts, subtracting 14 parts, a motion is generated of 8 parts towards the contrary way.

In the third experiment the two initial displacements are in the same direction (14 + 5 = 5 + 14).

But if the bodies were made both to move towards the same way, A, the swifter, with 14 parts of motion, B, the slower, with 5, and after reflection A went on with 5, B likewise went on with 14 parts; 9 parts being transferred from A to B. And so in other cases.

Newton then discusses the causes of the errors in the measurements of the distances and, as we have read above, evaluates them less than 3 in., 8 cm. The distances being several meters; this is about 2–3 % error. The relative error on the momenta was similar (masses and periods being known with a much better accuracy).

> It was not easy to let go the two pendulums so exactly together that the bodies should impinge one upon the other in the lowermost place AB; nor to mark the places s, and k, to which the bodies ascended after congress. Nay, and some errors, too, might have happened from the unequal density of the parts of the pendulous bodies themselves, and from the irregularity of the texture proceeding from other causes.

He, and we with him, then observe that the total momentum is conserved both for elastic and non-elastic collisions. A collision is called elastic if energy is conserved. This is an idealization; in practice perfectly elastic collisions do not exist. However, the collision between two steel spheres is close to being so, between two wax ones is not. In an elastic collision the two forces $\mathbf{F}_{12}$ and $\mathbf{F}_{21}$ are conservative. Elastic collisions conserve mechanical energy, inelastic ones do not, but in both cases the total momentum is conserved. Let us go back to Newton.

> But to prevent an objection that may perhaps be alleged against the rule (*the action and reaction law*), for the proof of which this experiment was made, as if this rule did suppose that the bodies were either absolutely hard, or at least perfectly elastic (whereas no such bodies are to be found in Nature), I must add that the experiments we have been describing, by no means depending upon that quality of hardness, do succeed as well in soft as in hard bodies.

Obviously, the relative velocity of the bodies after a collision is smaller for the inelastic than for elastic collisions with the same initial conditions. It may even be null; the two bodies remain attached. The total momentum however is always equal to the initial one.

> This I tried in balls of wool, made up tightly, and strongly compressed.

He compared the results obtained with balls of steel, glass and cork. The Newton conclusion is that

> And thus the third Law, so far as it regards percussions and reflections, is proved by a theory exactly agreeing with experience

In collision experiments the interaction forces act for a very short time, during which they are very intense. We talk of *impulsive forces*. The just described experiments establish that the total momentum is conserved in an isolated system in which the internal forces are impulsive. And if the forces are not impulsive? To answer this question Newton did the following experiment. He fixed a magnet on a piece of wood and a piece of iron on another one. He leaned both of them on the surface of the water in a container, carefully controlling them to be perfectly at rest. He let the two bodies go. The two bodies moved one towards the other, under the attraction of the magnet, attached themselves to each other and remained still. The important observation is that the final body, iron plus magnet, does not move on water, even if there is no impediment to do so. The total final momentum is zero, as the initial one was. In this experiment too the system is isolated. Indeed, the external forces, weight and Archimedes force equilibrate each other.

The conservation of linear momentum in an isolated system is a fundamental law of universal validity.

7.7 Action, Reaction and Linear Momentum Conservation

The conclusions from experiments we have described and many other ones can be summarized as follows. We build an isolated two-body system. The resultant external force is zero. The internal forces are $\mathbf{F}_{21} = -\mathbf{F}_{12}$. We start from an initial state i and measure the two momenta $\mathbf{p}_{i1}$ and $\mathbf{p}_{i2}$. We let the system spontaneously evolve under the action of the internal forces. When the system has reached the state, which we call final, f, we measure again the momenta, $\mathbf{p}_{f1}$ and $\mathbf{p}_{f2}$. We always find out that

$$\mathbf{p}_{1f} + \mathbf{p}_{2f} = \mathbf{p}_{1i} + \mathbf{p}_{2i}. \quad (7.30)$$

The linear momentum is conserved. We stated that this proves the action and reaction law (equality of the application lines apart). Let us look at that more closely. Equivalently, we can write

$$\mathbf{p}_{1f} - \mathbf{p}_{1i} = -(\mathbf{p}_{2f} - \mathbf{p}_{2i}). \quad (7.31)$$

In words, the changes of the linear momentum of the two bodies are equal and opposite. For the impulse-momentum theorem we have

$$\mathbf{p}_{1f} - \mathbf{p}_{1i} = \int_{t_1}^{t_2} \mathbf{F}_{21}(t)dt; \quad \mathbf{p}_{2f} - \mathbf{p}_{2i} = \int_{t_1}^{t_2} \mathbf{F}_{12}(t)dt.$$

The experimental verification of Eq. (7.31) is then a verification of

$$\int_{t_1}^{t_2} \mathbf{F}_{21}(t)dt = -\int_{t_1}^{t_2} \mathbf{F}_{12}(t)dt. \quad (7.32)$$

In conclusion, these experiments verify that the time of the force body 1 exerts on body 2 is equal and opposite to the time integral of the force body 2 exerts on body 1. In absence of any contrary evidence, we assume the instantaneous values of $\mathbf{F}_{21}$ and $\mathbf{F}_{12}$ to be equal and opposite too.

Figure 7.11 shows the time evolution of internal forces in the example of a hypothetical collision. Rigorously speaking, we know from the experiment only that the two areas are equal and assume that the curves have, in addition, mirror shapes, namely that the forces are equal and opposite in any instant.

This assumption is basically a postulate. Moreover, the postulate, namely the action-reaction law, is not true in every circumstance. There is no problem when the two bodies interact through contact forces, as in a collision. Problems arise when the two bodies are separated by a distance. As we have seen in Chap. 6, no effect can propagate over a distance instantaneously. Consequently, when the propagation

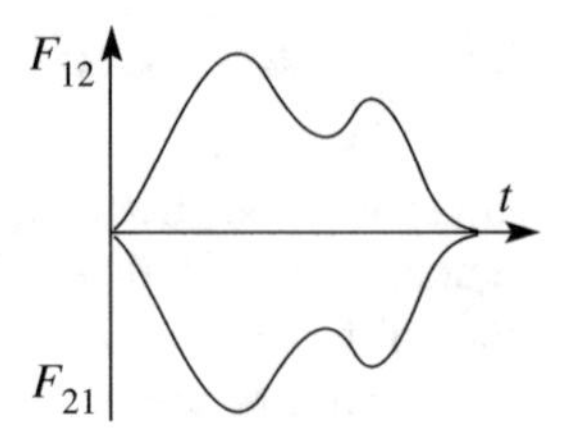

Fig. 7.11 The time evolution of the internal forces during a collision

time is comparable with the time in which the change in motion takes place, the concept of instantaneous equality of action and reaction loses validity.

Figure 7.12 shows a simple mechanical model of a "delayed" interaction. The two bodies are two trolleys moving with negligible friction on straight rails. Trolley 1 carries a gun, trolley 2 a block of material at the height of the gun. At a certain instant the gun shoots a bunch of projectiles. Suppose the projectiles to be invisible, have masses much smaller than the trolleys but very high speeds. Consequently, the bunch carries an appreciable momentum **p**. We observe the system and see, in the instant of the shot, trolley 1 to recoil with a momentum –**p** while trolley 2 remains still. The total momentum of the two trolleys is changed.

If υ is the velocity of the bullets and L the distance between the trolleys, the bullets will reach trolley 2 in a time L/υ and stick to the block. We observe trolley 2 acquiring a momentum **p**. The total momentum of the two blocks is now null, as it was initially. Momentum conservation is restored.

The example looks a bit stupid. The momentum seems not to be conserved during the time the bullets are in flight, just because we did not include their momentum in the total, assuming them to be "invisible". If we include that, as we should, the total momentum is conserved in every instant. However, things are not very different in the cases of actions at a distance as the gravitational and electromagnetic ones. Light, in particular, is an electromagnetic phenomenon. Consider again two trolleys, now very light, again with negligible friction. The first trolley carries a lamp that emits a light flash at a certain instant. Now, light carries momentum, even if in a very small amount. Consequently, the first trolley recoils with an opposite momentum (**p**), while the second is still at rest. Suppose the second trolley carries a black screen, which absorbs the light pulse completely, acquiring the momentum –**p**. The situation is quite similar to the "stupid" mechanical example. However now during the time of flight of the light the total

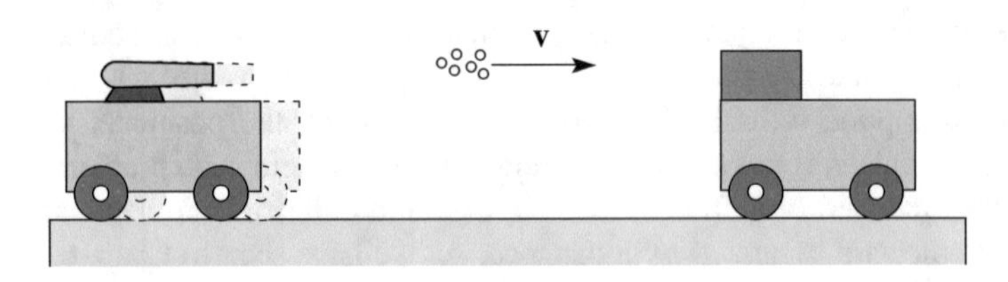

Fig. 7.12 Mechanical model of an action at a distance

mechanical momentum is not conserved. The missing momentum is, during this time, in the electromagnetic field. We shall study this in the 3rd volume of this course. We only notice here that this is basically the reason for which Eq. (6.50) that we found in discussing relativity is not valid for non-interacting particles. In quantum mechanics the analogy is even closer; light is made of "invisible" particles, the photons. A quite similar situation exists for the gravitational interaction. In this case also the gravitational field carries momentum. This is described by general relativity.

7.8 Systems of Particles

We shall now start our study of systems of several, say N, of material points. The relevant physical quantities are shown in Fig. 7.13, in an inertial reference frame. Let $\mathbf{r}_i$ be the position vector of the generic point P_i in a generic instant. We call the set of positions of its constituent material points a *configuration* of the system.

Let m_i be the mass of P_i, $\mathbf{v}_i$ its velocity and $\mathbf{p}_i = m_i\mathbf{v}_I$ its momentum. The forces acting on each point can be usefully divided in internal, due to the other points of the system, and external, due to agents external to the system.

Consider for example the system of Jupiter and its satellites. The forces on one of them, Ganymede for example, are the internal ones due to Jupiter and to the other satellites, Io, Europa, Callisto, and the external ones due to the sun and the other planets. Obviously, being an internal or external force depends on the system under consideration. If the system is the solar system, all the mentioned forces are internal.

We call $\mathbf{F}_i^{(e)}$ the resultant external force and $\mathbf{F}_i^{(i)}$ the resultant internal force acting on P_i. All the forces, both external and internal, according to the Newton law, determine the motion of P_i,

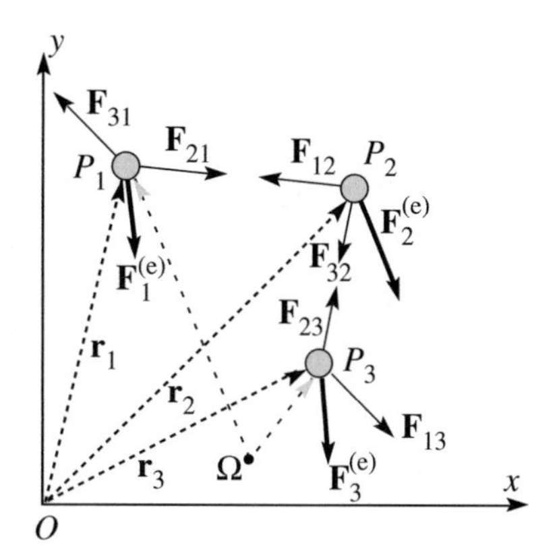

Fig. 7.13 A system of material points

$$\mathbf{F}_i^{(i)} + \mathbf{F}_i^{(e)} = \frac{d\mathbf{p}_i}{dt} = m_i\mathbf{a}_i. \tag{7.33}$$

The motion of a system of N points is described by N independent Eq. (7.33). Their solution is in general quite difficult. Indeed, just think of the fact that the force acting on a certain point at a certain time depends not only on its position, but on those of all the other points too. The problem is so complicated that even in the simplest case $N = 3$ cannot be in general solved analytically. Numerical methods are today available to solve the problem with the help of powerful computers.

We shall not analyze the motions of single points, but rather consider quantities relative to the whole system. We indicate with Ω a geometric point that we choose as the pole of the linear momenta and of the moments of the forces. This point is not necessarily at rest in the reference frame, rather it moves with a velocity that is a function of time, $\mathbf{v}_\Omega$. The angular momentum of point P_i about Ω is

$$\mathbf{l}_i = \mathbf{\Omega P}_i \times \mathbf{p}_i. \tag{7.34}$$

Let $\mathbf{f}_{1,i}$, $\mathbf{f}_{2,i}$,…. be the forces acting on the point P_i and $\mathbf{F}_i = \mathbf{f}_{1,i} + \mathbf{f}_{2,i} + ….$ their resultant. All these forces are applied to the same point and, consequently, their total moment is equal to the moment of their resultant. The external moment acting on P_i is then

$$\boldsymbol{\tau}_i^{(e)} = \mathbf{\Omega P}_i \times \mathbf{F}_i^{(e)}$$

and the internal moment

$$\boldsymbol{\tau}_i^{(i)} = \mathbf{\Omega P}_i \times \mathbf{F}_i^{(i)}.$$

The global quantities of the system that we shall need are the following:

(1) The total linear momentum of the system, which is the vector sum of the linear momenta of the constituent points

$$\mathbf{P} = \sum_{i=1}^{N} \mathbf{p}_i, \tag{7.35}$$

(2) the total angular momentum

$$\mathbf{L}_\Omega = \sum_{i=1}^{N} \mathbf{l}_{\Omega i} = \sum_{i=1}^{N} \mathbf{\Omega P}_i \times \mathbf{p}_i, \tag{7.36}$$

(3) the total kinetic energy

$$U_k = \sum_{i=1}^{N} U_{ki} = \sum_{i=1}^{N} \frac{1}{2} m_i v_i^2, \tag{7.37}$$

(4) the resultant force

$$\mathbf{F} = \sum_{i=1}^{N} \left(\mathbf{F}_i^{(i)} + \mathbf{F}_i^{(e)} \right) = \sum_{i=1}^{N} \mathbf{F}_i^{(i)} + \sum_{i=1}^{N} \mathbf{F}_i^{(e)} = \mathbf{F}^{(i)} + \mathbf{F}^{(e)} \tag{7.38}$$

where the vectors in the last side are the resultants of internal and external forces acting on the system.
We now make a very important observation that will greatly simplify several problems. The internal forces come in pairs; the force exerted on point P_i by another point P_j is equal and opposite to the force that P_j exerts on P_i and their sum is null. Consequently the resultant internal force is zero, $\mathbf{F}^{(i)} = 0$, and Eq. (7.38) becomes

$$\mathbf{F} = \sum_{i=1}^{N} \mathbf{F}_i^{(i)} + \sum_{i=1}^{N} \mathbf{F}_i^{(e)} = \mathbf{F}^{(e)}. \tag{7.39}$$

(5) The total moment about the pole Ω is

$$\mathbf{M}_\Omega = \sum_{i=1}^{N} \left(\tau_{\Omega i}^{(i)} + \tau_{\Omega i}^{(e)} \right) = \sum_{i=1}^{N} \tau_{\Omega i}^{(i)} + \sum_{i=1}^{N} \tau_{\Omega i}^{(e)} = \mathbf{M}_\Omega^{(i)} + \mathbf{M}_\Omega^{(e)} \tag{7.40}$$

where the vectors in the last side are the total moment of the internal and of the external forces respectively.

Notice that we can calculate the total moment of the forces acting on a single point P_i or calculate first the moments of the different forces and then sum them, or sum the forces and then calculate the moment of the resultant. On the contrary, to calculate the total moment acting on the system we must *first* calculate the moments of the forces on the single points and *then* sum those moments. Indeed, in this case the forces are applied in different points.

A second important observation is the following. The internal forces come in pairs that, for the action-reaction law, not only are couples, but also zero arm couples. Consequently, the moment of each couple is null, whatever is the pole. The total internal moment is zero, $\mathbf{M}_\Omega^{(i)} = 0$ and we can write

$$\mathbf{M}_\Omega = \sum_{i=1}^{N} \tau_{\Omega i}^{(i)} + \sum_{i=1}^{N} \tau_{\Omega i}^{(e)} = \mathbf{M}_\Omega^{(e)}. \tag{7.41}$$

7.9 The Center of Mass

We continue with the system of N material points. Figure 7.14 represents the situation.

We define as the *center of mass* of the system the geometric point (it is not a material point) defined by the position vector

$$\mathbf{r}_C = \frac{\sum_{i=1}^{N} m_i \mathbf{r}_i}{\sum_{i=1}^{N} m_i} = \frac{\sum_{i=1}^{N} m_i \mathbf{r}_i}{M} \tag{7.42}$$

where M is the total mass of the system. The coordinates of the center of mass are, clearly

$$x_C = \frac{\sum_{i=1}^{N} m_i x_i}{M}, \; y_C = \frac{\sum_{i=1}^{N} m_i y_i}{M}, \; z_C = \frac{\sum_{i=1}^{N} m_i z_i}{M}. \tag{7.43}$$

It can be shown, but we shall not do so, that the position of the center of mass is independent of the choice of the reference frame. However, obviously, its coordinates depend on that. We already met the center of mass in the particular case of a two-point system. In this case the center of mass is the point of the segment joining the two points at distances from them inversely proportional to the masses. It can be shown that the two definitions agree in this particular case.

We now consider the motion of points of the system. We call $\mathbf{v}_i$ the velocity of P_i (which is a function of time). By deriving Eq. (7.42) we find that the velocity of the center of mass is

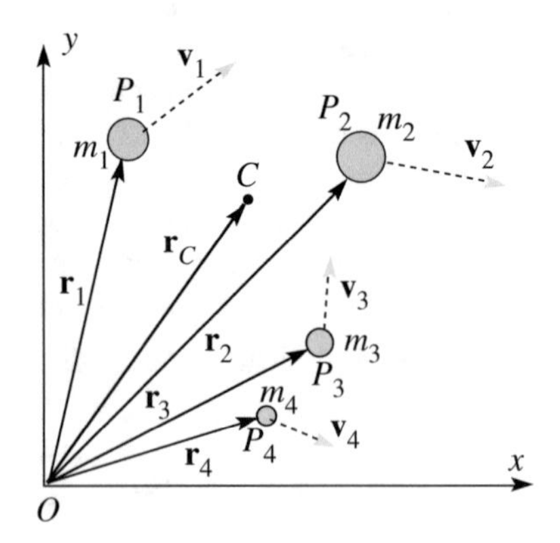

Fig. 7.14 A material system and its center of mass

$$\mathbf{v}_C = \sum_{i=1}^{N} m_i \mathbf{v}_i / M. \tag{7.44}$$

We observe that the sum in the right-hand side of this equation is just the sum of the linear momenta of the points, namely is the total momentum of the system

$$\mathbf{P} = \sum_{i=1}^{N} m_i \mathbf{v}_i = \sum_{i=1}^{N} \mathbf{p}_i. \tag{7.45}$$

We can write Eq. (7.44) as

$$\mathbf{P} = M\mathbf{v}_C, \tag{7.46}$$

which is a very important equation. It states that *the total momentum of the system is equal to the momentum of the center of mass, if considered as a material point in which all the mass of the system is concentrated.*

Consider now how the total momentum varies in time. We work in an inertial reference frame. Taking the derivative of Eq. (7.45) we have

$$\frac{d\mathbf{P}}{dt} = \sum_{i=1}^{N} m_i \mathbf{a}_i, \tag{7.47}$$

but, as we are in an inertial frame, $m_i\mathbf{a}_i$ is equal to the resultant force, both external and internal, acting on P_i.

$$\mathbf{F}_i^{(i)} + \mathbf{F}_i^{(e)} = m_i \mathbf{a}_i. \tag{7.48}$$

Substituting this in Eq. (7.47) we have

$$\frac{d\mathbf{P}}{dt} = \sum_{i=1}^{N} \left(\mathbf{F}_i^{(i)} + \mathbf{F}_i^{(e)} \right) = \mathbf{F}^{(e)} + \mathbf{F}^{(i)},$$

but, as we know, the resultant internal force is zero, a fact that enormously simplifies the equation. It becomes

$$\mathbf{F}^{(e)} = \frac{d\mathbf{P}}{dt}. \tag{7.49}$$

This fundamental equation states that *the rate of change of the total momentum of a mechanical system is equal to the resultant external force acting on the system.* The fact that the internal forces do not contribute to the variation of the total momentum simplifies many problems.

We now go back to Eq. (7.46) and immediately see that

$$\mathbf{F}^{(e)} = M\mathbf{a}_C, \tag{7.50}$$

which is called the *theorem of the center of mass motion*: *the center of mass moves as a material point in which all the mass of the system is concentrated and acted upon by the resultant external force*. Notice that while the motion of the center of mass is determined by the external forces only, the motion of each point of the system depends on both external and internal forces.

As an example, suppose we take in our hand the handle of a hammer, and we launch it in the air. The motion of the hammer will be a complicated combination of rotations and displacements. The motion of its center of mass, on the contrary, will be simply a parabola, with the hammer rotating about it (neglecting air resistance). For that the body does not need to be rigid. If we launch a chain in the air, its center of mass will describe a parabola too. In a similar way, consider the bullet shot by a cannon. It describes a parabola. If at a certain moment the bullet explodes, its pieces will describe complicated trajectories, but their center of mass will continue on the same parabola, as long as the first piece hits the ground. When this happens a new external force, due to the action of ground, starts acting on the system.

The center of mass, as we have seen, is not a material point but behaves as such.

7.10 Linear Momentum Conservation

The *law (or principle) of conservation of linear momentum* states that: *if, in an inertial frame, resultant external force on a system is zero, the total linear momentum is constant in time.* The property is immediately obtained from Eq. (7.50)

$$\mathbf{P} = \text{constant}, \quad \text{if} \quad \mathbf{F}^{(e)} = 0. \tag{7.51}$$

We can also say that, under the same hypotheses

$$\mathbf{a}_C = 0,\ \mathbf{v}_C = \text{constant}, \quad \text{if} \quad \mathbf{F}^{(e)} = 0. \tag{7.52}$$

If the resultant external force is zero in an inertial frame the center of mass remains still if initially still or continues on its rectilinear uniform motion.

In Sects. 7.5 and 7.6, we have already used the center of mass properties and the linear momentum conservation principle in the particular case of two-body systems and discussed the relations with the action-reaction law.

7.11 Continuous Systems

The mechanical systems we have considered so far are discrete, namely composed of a number of point-like particles. We shall now consider continuous mechanical systems. Such are the solid bodies when their physical dimensions cannot be neglected. Figure 7.15 represents a continuous body of mass M and volume V.

We can divide the body into small volumes dV, which we take as cubes with sides parallel to the coordinate axes. Let $\mathbf{r}$ be the position vector of the generic dV and Δm its mass. We define the density $\rho(\mathbf{r})$ of the body in the position $\mathbf{r}$ to be the ratio between the mass and the volume of the element in the limit in which the volume becomes very small, namely

$$\rho(\mathbf{r}) = \lim_{\Delta V \to 0} \frac{\Delta m}{\Delta V} = \frac{dm}{dV}. \tag{7.53}$$

The density can vary from point to point. Think for example of the atmospheric density that decreases with altitude. A body is said to be *homogeneous* if its density does not vary from point to point.

Here we need to specify that the limit $\Delta V \to 0$ should be understood as a physical rather than mathematical limit. Indeed, when seen at a molecular scale, matter is not continuous, but made of small particles, the molecules, separated one from another. Consequently, the limit for volumes going mathematically to zero is not defined. However, the granularity of matter is so small compared to the macroscopic sizes and we can safely state that the limit is taken for volumes very small compared to macroscopic dimensions but still large enough to contain a great number of molecules. Indeed, we can say, for volumes physically tending to zero.

The definition of *center of mass* for a continuous system is completely analogous to that we gave in Sect. 7.9 for a discrete system. We divide the system in N small

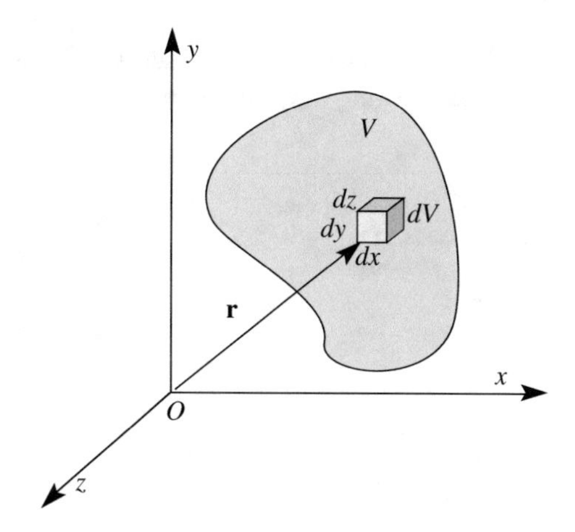

Fig. 7.15 A continuous body and an infinitesimal volume element

volumes ΔV_i, then use Eq. (7.42) to define the center of mass and take the limit for the small volumes tending physically to zero. We obtain

$$\mathbf{r}_C = \frac{1}{M} \lim_{\Delta m \to 0} \sum_{i=1}^{N} \Delta m_i \mathbf{r}_i = \frac{1}{M} \lim_{\Delta V \to 0} \sum_{i=1}^{N} \Delta V_i \rho(\mathbf{r}_i)_i \mathbf{r}_i = \frac{1}{M} \int_V \mathbf{r} \rho(\mathbf{r}) dV.$$

The position vector of the center of mass is then

$$\mathbf{r}_C = \frac{1}{M} \int_V \mathbf{r} \rho(\mathbf{r}) dV \tag{7.54}$$

or, its coordinates are

$$x_C = \frac{1}{M} \int_V x \rho(\mathbf{r}) dV, \; y_C = \frac{1}{M} \int_V y \rho(\mathbf{r}) dV, \; z_C = \frac{1}{M} \int_V z \rho(\mathbf{r}) dV. \tag{7.55}$$

In this chapter we shall continue the study of material systems. For the sake of simplicity, we shall consider them discrete. The discussion of continuous systems is completely similar, just changing sums with integrals. The limitation to discrete systems does not subtract anything from the physics conclusions.

As examples, we shall now calculate the position of the center of mass in two examples of homogeneous bodies of simple geometrical shapes.

Example E 7.2 Figure 7.16 represents a thin sheet in the form of an isosceles triangle of height h and base b. It can be considered two-dimensional and the volume integral (7.54) becomes a surface integral. It is evident, for symmetry reasons, that the center of mass must be on the height of the triangle (the same quantity of mass must lay on the right and on the left). We need only to find its y coordinate. It is convenient to take as surface elements strips of height dy running from one side to the other. Indeed all points of such a strip have the same y and equally contribute to the integral. The length $l(y)$ of the strip at height y can be found considering the proportion $l(y){:}b = y{:}h$. Hence we have $l(y) = by/h$. The area

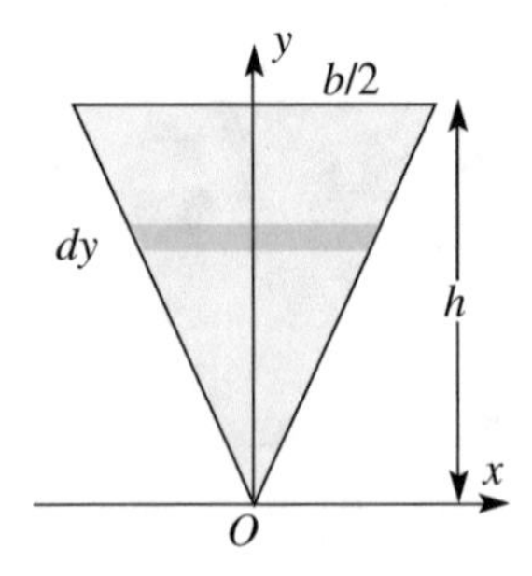

Fig. 7.16 Calculating the center of mass of a homogenous isosceles *triangle*

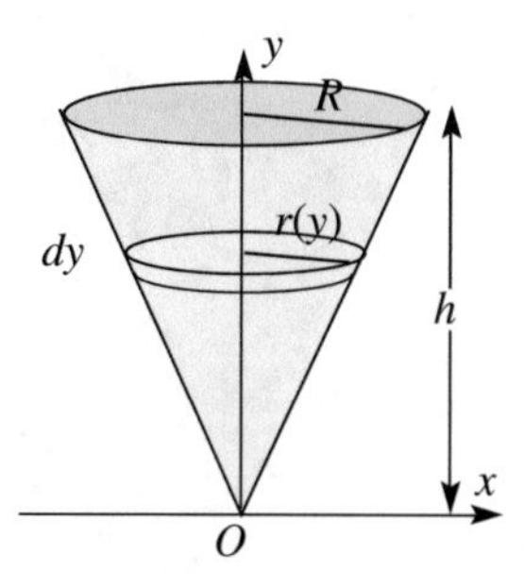

Fig. 7.17 Calculating the center of mass of a homogenous *cone*

of the strip is $dS(y) = (by/h)dy$ and, if σ is the surface density, namely the mass per unit area, its mass is $dm(y) = \sigma(by/h)dy$. We then calculate the integral (Fig. 7.17)

$$\int_0^h ydm = \sigma\frac{b}{h}\int_0^h y^2dy = \sigma\frac{bh^2}{3}.$$

The mass M of the body is σ times the area $hb/2$ and we have $y_C = (2/3)h$.

Example E 7.3 Figure 7.17 represents a homogeneous cone of height h and base radius R. As evident in this case too, the center of mass is on the axis. To calculate its height y, we take as volume elements thin sheets parallel to the base. All the points of a sheet have the same height y. The volume of the sheet at y is $dV = \pi R^2(y)\, dy$. But $r(y) = Ry/h$ and, if ρ is the density

$$\int_V y\rho(\mathbf{r})dV = \rho\frac{\pi R^2}{h^2}\int_0^h y^3dy = \rho\frac{\pi R^2h^2}{4},$$

which we must divide by the mass, that is $M = \pi R^2 h\rho/3$, obtaining $y_C = 3h/4$.

7.12 Angular Momentum

The fundamental equation Eq. (7.49), describes the evolution in time of the total linear momentum of a mechanical system. We shall now see how the total angular momentum varies in time. Figure 7.18 shows a mechanical system in a reference frame, which we choose to be inertial. We arbitrarily choose a geometric Ω to be the pole of the moments and angular momenta. The pole is not necessarily still, and we call $\mathbf{v}_\Omega$ its velocity.

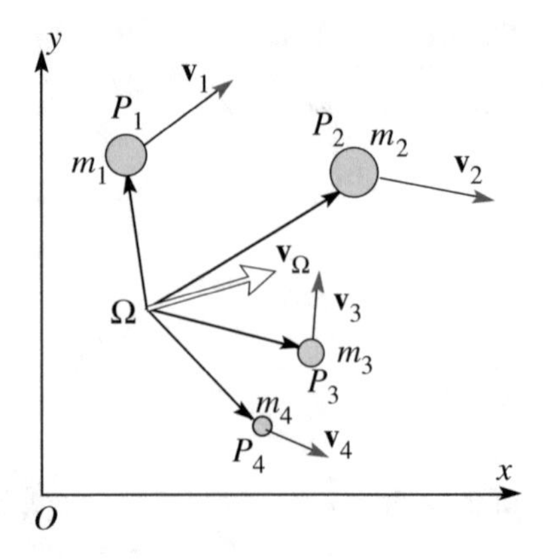

Fig. 7.18 The material system

The total angular momentum about the pole we have chosen is

$$\mathbf{L}_{\Omega} = \sum_{i=1}^{N} \mathbf{L}_{\Omega i} = \sum_{i=1}^{N} \boldsymbol{\Omega}\mathbf{P} \times \mathbf{p}_i. \tag{7.56}$$

We take the time derivative and obtain

$$\frac{d\mathbf{L}_{\Omega}}{dt} = \sum_{i=1}^{N} \frac{d\boldsymbol{\Omega}\mathbf{P}}{dt} \times \mathbf{p}_i + \sum_{i=1}^{N} \boldsymbol{\Omega}\mathbf{P}_i \times \frac{d\mathbf{p}_i}{dt}. \tag{7.57}$$

The vector $\boldsymbol{\Omega}\mathbf{P}_i$ is the difference between two vectors, $\boldsymbol{\Omega}\mathbf{P}_i = \mathbf{r}_i - \mathbf{r}_{\Omega}$, both of which vary in time. Consequently its time derivative is $d\boldsymbol{\Omega}\mathbf{P}_i/dt = \mathbf{v}_i - \mathbf{v}_{\Omega}$.

In the second term in the right-hand side we have the rates of change of the linear momenta of single points. As we are in an inertial frame, the rate of change of $\mathbf{p}_i$ is the resultant force, both internal and external, acting on the point P_i. We can write

$$\frac{d\mathbf{L}_{\Omega}}{dt} = \sum_{i=1}^{N} \mathbf{v}_i \times \mathbf{p}_i - \mathbf{v}_{\Omega} \times \sum_{i=1}^{N} \mathbf{p}_i + \sum_{i=1}^{N} \boldsymbol{\Omega}\mathbf{P}_i \times \mathbf{F}_i^{(e)} + \sum_{i=1}^{N} \boldsymbol{\Omega}\mathbf{P}_i \times \mathbf{F}_i^{(i)}.$$

The first term in the right-hand side is zero, being the sum of cross products of parallel vectors. The sum in the second term is the total linear momentum $\mathbf{P}$ of the system. The third term is the total moment of the external forces $\mathbf{M}^{(e)}$. The last term is the total internal moment, which is zero. In conclusion Eq. (7.57) becomes

$$\frac{d\mathbf{L}_{\Omega}}{dt} = \mathbf{M}_{\Omega}^{(e)} - \mathbf{v}_{\Omega} \times \mathbf{P}. \tag{7.58}$$

The expression becomes still simpler with two different choices of the pole.

If the pole is fixed in the (inertial) reference frame, $\mathbf{v}_\Omega = 0$ and

$$\frac{d\mathbf{L}_\Omega}{dt} = \mathbf{M}_\Omega^{(e)}. \tag{7.59}$$

This fundamental equation reads: *the rate of change of the total angular momentum of a mechanical system about a pole fixed in an inertial frame is equal to the moment of the external forces about the same pole.*

If the pole coincides with the center of mass, which generally moves, the second term in the right-hand side of Eq. (7.58) is again zero. It is the cross product of two parallel vectors, the velocity of the center of mass and the total linear momentum. We can write

$$\frac{d\mathbf{L}_{CM}}{dt} = \mathbf{M}_{CM}^{(e)}. \tag{7.60}$$

In words: *The rate of change of the angular momentum of a mechanical system about its center of mass as a pole is equal to the total external moment (about the same pole).*

7.13 Angular Momentum Conservation

The principle of conservation of angular momentum states that *in an isolated system the total angular momentum, about any pole fixed in an inertial frame is conserved.*

Indeed, in an isolated system the resultant external force and the total external moment are zero. If the pole stands still in an inertial frame Eq. (7.59) holds, we can state that the time derivative of the total angular momentum is zero.

Similarly for Eq. (7.60) we can state also that *in an inertial frame the total angular momentum of any isolated system about its center of mass is constant.*

Even if in the case of non-isolated systems, namely in the presence of external forces, it is sometimes possible to choose a fixed pole, such that the total external moment about it is zero. Then, the total angular momentum about that pole is conserved. We shall see some examples in the following.

Notice also that the total external moment may be zero and their resultant different from zero, or vice versa. Consequently, the linear momentum and the angular momentum conservations are in general independent issues. If the system is isolated however, both quantities are conserved.

Finally we observe the following. As we have seen, conservation of the total linear momentum is a consequence of one aspect of the action-reaction law: action and reaction are equal and opposite. The total angular momentum conservation is a consequence of the second aspect of the third law: action and reaction have the same application line. All the experimental evidence, without exceptions, is in favor of the angular momentum conservation. Consequently, also this second aspect of the third law must be considered experimentally proven.

The linear and angular momentum conservation laws are fundamental principles of physics, not only of mechanics. In advanced treatments we can show that they are consequences respectively of the homogeneity of the space (there are no privileged points in space) and of its isotropy (there are no privileged directions).

7.14 Energy of a Mechanical System

We continue to consider a material system of N material points P_i in an $\mathbf{r}_i$ the position vector of P_i, m_i its mass and $\mathbf{v}_i$ its velocity. The generic point P_i has the kinetic energy $U_{Ki} = m_i \upsilon_i^2/2$, and the total kinetic energy of the system is

$$U_K = \frac{1}{2}\sum_{i=1}^{N} m_i \upsilon_i^2. \tag{7.61}$$

During the motion of the system its kinetic energy will, in general, vary, because the single kinetic energies of the points vary under the action of the forces. Let $\mathbf{F}_i^{(e)}$ and $\mathbf{F}_i^{(i)}$ be the resultants of external and internal forces acting on P_i respectively. In the generic elementary time interval dt the displacement of the point is $d\mathbf{r}_i$. The corresponding elementary work of the forces is

$$dW_i = \mathbf{F}_i^{(e)} \cdot d\mathbf{r}_i + \mathbf{F}_i^{(i)} \cdot d\mathbf{r}_i = dW_i^{(e)} + dW_i^{(i)}.$$

Consider the point P_i moving on a certain trajectory from an initial position A in $\mathbf{r}_{iA}$ to a final position B in $\mathbf{r}_{iB}$. The variation of its kinetic energy is given by the kinetic energy theorem

$$U_K(B) - U_K(A) = \int_A^B \mathbf{F}_i^{(e)} \cdot d\mathbf{r}_i + \int_A^B \mathbf{F}_i^{(i)} \cdot d\mathbf{r}_i = W_{AB}^{(e)} + W_{AB}^{(i)}.$$

In words, the variation of the total kinetic energy of a system is equal to the works of both the external and internal forces. Differently from the cases of the total linear and angular momenta, the contribution of internal forces is not zero.

If all forces acting on the system are conservative, the work can also be expressed as a difference of potential energy. Calling U_P the total potential energy, which is the sum of the potential energies of all points of the system, we immediately find that

$$U_K(B) + U_P(B) = U_K(A) + U_P(A). \tag{7.62}$$

We define the *total energy* of the system U_{tot} as the sum of its potential and kinetic energy and we see that it has the same values in A and in B. Considering that these points are arbitrary, we conclude that the total energy is constant during movement of the system

$$U_{\text{tot}} = U_K + U_P = \text{constant}. \tag{7.63}$$

If the system is isolated, there are no external forces and only the internal ones make work. This does not imply that the total energy is conserved. For that to be the case all of the internal forces must be conservative. As an example consider a system made by a block and a trolley supporting it. The trolley can move on rails without appreciable friction, but there is friction between the plane of the trolley and the block. The block moves on that plane. The plane exerts a friction force on the block and so does the block on the plane. The two forces are equal and opposite with the same application line. During a motion, the total and angular momentum are conserved, but not kinetic energy.

7.15 Center of Mass Reference Frame

It is often useful to consider the motion of a mechanical system in the *center of mass frame*, CM for brevity, even if that frame is not usually inertial. We start from an inertial frame and we define as center of mass frame, the frame with origin in the center of mass of the system and with axes parallel to the axes of the inertial frame.

Figure 7.19 shows the two just mentioned reference frames and a generic point of the system P_i. The CM frame does not rotate relative to the inertial frame, it

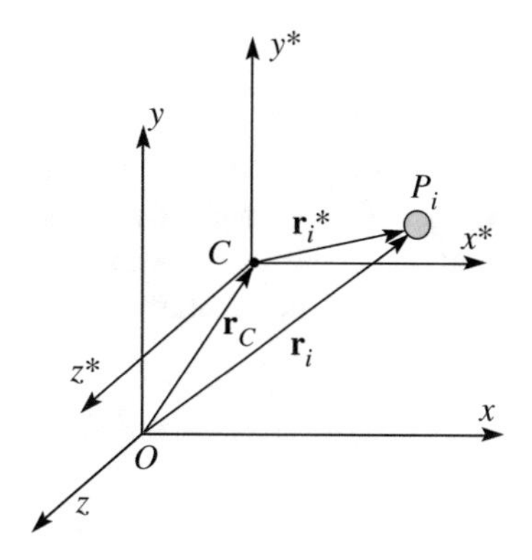

Fig. 7.19 The inertial frame xyz and the center of mass frame $x^*y^*z^*$

translates with the center of mass velocity. This may vary and, as a consequence, the CM frame is not in general inertial. It is so if the resultant external force is zero (even if the total external moment is not) because then the velocity of the center of mass is constant.

We shall indicate with an asterisk the quantities in the CM frame. The relation between the position vectors in the two frames is

$$\mathbf{r}_i^* = \mathbf{r}_i - \mathbf{r}_{CM} \tag{7.64}$$

and the relation between the velocities is

$$\mathbf{v}_i^* = \mathbf{v}_i - \mathbf{v}_{CM} \tag{7.65}$$

and similarly for the accelerations

$$\mathbf{a}_i^* = \mathbf{a}_i - \mathbf{a}_{CM}. \tag{7.66}$$

Obviously, the center of mass position vector and the velocity are null in its reference

$$\mathbf{r}_{CM}^* = 0, \quad \mathbf{v}_{CM}^* = 0, \quad \mathbf{a}_{CM}^* = 0. \tag{7.67}$$

In Sect. 7.9 we have found that in every reference frame, both inertial and not, the total linear momentum of a system is equal to the mass of the system times the velocity of the center of mass. But the latter is null in the CM frame and we have

$$\mathbf{P}^* = \sum_{i=1}^{N} m_i \mathbf{v}_i^* = 0. \tag{7.68}$$

The CM frame is also the frame in which the total linear momentum is zero. It is sometimes called the *center of momenta frame*.

We obtain another interesting property by expressing Eq. (7.42) in the CM frame. For the first of Eq. (7.67) this becomes

$$\sum_{i=1}^{N} m_i \mathbf{r}_i^* = 0. \tag{7.69}$$

We now consider the total angular momentum, which has an important role in mechanics. We might expect it to be different in the two frames of Fig. 7.19, the inertial and the center of mass frames. As a matter of fact they are equal. Indeed the total angular momentum in the inertial system is

$$\mathbf{L}_{CM} = \sum_{i=1}^{N} \mathbf{r}_i^* \times m_i \left(\mathbf{v}_i^* + \mathbf{v}_{CM} \right) = \sum_{i=1}^{N} \mathbf{r}_i^* \times m_i \mathbf{v}_i^* + \left(\sum_{i=1}^{N} m_i \mathbf{r}_i^* \right) \times \mathbf{v}_{CM}.$$

The first term in the last side is the angular momentum about the center of mass, as a pole, in the CM frame, while the second is zero for Eq. (7.69). Hence

$$\mathbf{L}^*_{CM} = \mathbf{L}_{CM}. \tag{7.70}$$

We conclude that the total angular momentum of a material system about its center of mass is an intrinsic characteristic of the system, independent of the reference frame.

7.16 The König Theorems

The two König theorems that we shall discuss in this section give two relations, one between the kinetic energy in the inertial and CM frames U_K and U^*_K respectively, and one between the angular momenta $\mathbf{L}$ and $\mathbf{L}^*$. In both cases the quantity in the inertial frame is equal to the sum of two terms, one of the system as a whole, the other corresponding to its motion relative to the center of mass. The theorems are named after Johann Samuel König (1712–1757)

König kinetic energy theorem.
The kinetic energy of the system in the inertial frame is

$$U_K = \frac{1}{2}\sum_{i=1}^{N} m_i \upsilon_i^2.$$

Using Eq. (7.65) it becomes

$$U_K = \frac{1}{2}\sum_{i=1}^{N} m_i\left(\mathbf{v}^*_i + \mathbf{v}_{CM}\right)^2 = \frac{1}{2}\sum_{i=1}^{N} m_i \mathbf{v}_i^{*2} + \frac{1}{2}M\mathbf{v}^2_{CM} + \left(\sum_{i=1}^{N} m_i \mathbf{v}^*_i\right)\cdot \mathbf{v}_{CM}.$$

The expression in parenthesis in the last term of the last side is the total momentum in the CM frame, hence is null. And we obtain

$$U_K = \frac{1}{2}\sum_{i=1}^{N} m_i \mathbf{v}_i^{*2} + \frac{1}{2}M\mathbf{v}^2_{CM} = U^*_K + \frac{1}{2}M\mathbf{v}^2_{CM}. \tag{7.71}$$

We read this expression as: the kinetic energy in the inertial frame is the sum of two terms. One term is the kinetic energy “of the center of mass”, if we think of it as being a material point with all the mass of the system. The second term is the kinetic energy in the center of mass system, namely relative to the motion of the parts of the system about the center of mass.

Example E 7.4 A child is sitting on a wheelchair near to a wall with his feet resting on it with folded legs. The child, in stretching his legs, pushes on the wall and accelerates backward. After his feet detach from the wall he continues to move at constant velocity (neglecting frictions). What forces have caused the acceleration? Which force is the variation of kinetic energy?

Our system is the child and the chair. We cannot consider it as point-like, because the stretching of the legs changes the shape of the system. The resultant external force is the normal reaction of the wall, **N**. This is the force causing the acceleration. If m is the mass and $\mathbf{a}_{CM}$ the center of mass acceleration, we have $\mathbf{N} = m\mathbf{a}_{CM}$.

The work of the external force **N** is, on the other hand, zero, because its application point does not move. Which is the cause of the kinetic energy variation?

In the analysis of this type of problem, the following mistake is often made. It consists in application of the kinetic energy theorem to the center of mass, in a form valid for the material point. Indeed, the center of mass behaves as a material point from several points of view, but not from this one. Let us look at that. We can write Eq. (7.50), which is valid for the center of mass, as

$$\mathbf{F}^{(e)} = m\frac{d\mathbf{v}_{CM}}{dt},$$

which is formally identical to the law and is valid for the material point. We try now to go ahead as we did in Sect. 2.10 to show the kinetic energy theorem for a material point. We indicate with $d\mathbf{s}_{CM}$ the elementary displacement of the center of mass in dt, in order to have $\mathbf{v}_{CM} = d\mathbf{s}_{CM}/dt$. We take the dot product of the above equation and $d\mathbf{s}_{CM}$ obtaining

$$\mathbf{F}(e) \cdot d\mathbf{s}_{CM} = m\frac{d\mathbf{v}_{CM}}{dt} \cdot d\mathbf{s}_{CM} = m\mathbf{v}_{CM} \cdot d\mathbf{v}_{CM}.$$

We indicate with Γ the trajectory of the center of mass and we consider two positions, A and B on Γ. As we did for the material point we integrate the above expression on Γ from A to B obtaining

$$\int_{\Gamma A}^{B} \mathbf{F}(e) \cdot d\mathbf{s}_{CM} = \frac{1}{2}mv_{CM}^2(B) - \frac{1}{2}mv_{CM}^2(A), \tag{7.72}$$

which has the same form as (2.36). Its meaning is however fundamentally different. While the right-hand side of Eq. (7.72) is indeed the difference of center of mass kinetic energy, the left-hand side is <u>not</u> the work of the resultant external force. This is because $d\mathbf{s}_{CM}$ is the displacement of the center of mass, not of the application point of the resultant. The latter may not even have been defined. It is defined only

Fig. 7.20 Two locks on a horizontal plane of negligible friction

if all the forces are applied to the same point. Consequently, Eq. (7.72) is not very useful in practice.

We can conclude that the work of the resultant external force has nothing to do with the variation of kinetic energy. The latter is due to an internal force, the one due to the muscles of the legs of the child.

Similarly, when a car accelerates, the force producing acceleration is the friction of the road on the tires. The work of this force is null. The kinetic energy variation is equal to the work of the internal forces due to the engine.

Example E 7.5 Figure 7.20 shows two blocks of masses m_1 and m_2 supported by a horizontal plane with negligible friction. A spring, in its natural length, is fixed to the left-hand side of the block on the right. Its elastic constant is k. The two blocks move with velocities $\mathbf{v}_1$ and $\mathbf{v}_2$ in the same direction and with $\upsilon_1 > \upsilon_2$. Block 1 reaches block 2 and hits it, compressing the spring.

We then calculate the maximum spring compression.

We shall solve the problem in two ways, using a trivial reasoning first, then using the König kinetic energy theorem. Let υ_1' and υ_2' be the velocities after the collision. The linear momentum (one dimension) and energy conservation give us two equations

$$m_1\upsilon_1 + m_2\upsilon_2 = m_1\upsilon_1' + m_2\upsilon_2' = P$$
$$\frac{1}{2}m_1\upsilon_1^2 + \frac{1}{2}m_2\upsilon_2^2 = \frac{1}{2}m_1\upsilon_1'^2 + \frac{1}{2}m_2\upsilon_2'^2 = U_{\text{tot}}.$$

From the first equation we express υ_2' as a function of υ_1'. Then, with the second equation, we express the spring energy as a function of υ_1'. We denote by x the compression of the spring

$$\upsilon_2' = \frac{m_1\upsilon_1 + m_2\upsilon_2}{m_2} - \frac{m_1}{m_2}\upsilon_1' = \frac{P}{m_2} - \frac{m_1}{m_2}\upsilon_1',$$
$$\frac{1}{2}kx^2 = U_{\text{tot}} - \frac{1}{2}m_1\upsilon_1'^2 - \frac{1}{2}m_2\left(\frac{P}{m_2} - \frac{m_1}{m_2}\upsilon_1'\right)^2$$
$$= U_{\text{tot}} - \frac{1}{2}m_1\upsilon_1'^2 - \frac{1}{2}\frac{P^2}{m_2} - \frac{1}{2}\frac{m_1^2}{m_2}\upsilon_1'^2 + \frac{m_1}{m_2}P\upsilon_1'.$$

The maximum spring deformation corresponds to the maximum elastic energy. We obtain the latter by taking the derivative of the last side, putting it equal to zero and solving for υ_1':

$$\upsilon_1' = \frac{P}{m_1 + m_2} = \frac{m_1\upsilon_1 + m_2\upsilon_2}{m_1 + m_2}.$$

By substituting this in the expression of the energy just found, we see that the velocity corresponding to the maximum compression is the center of mass velocity. Considering the symmetry of the problem, we expect υ_2' to be equal. This is immediately found from the above equation, as the reader can verify.

The second approach to solve the problem is much quicker. Moreover, it immediately shows the reason for both velocities being equal to the center of mass velocity. We write the energy in the form given by the König theorem.

$$U_{\text{tot}} = \frac{1}{2}kx^2 + \frac{1}{2}m_1\upsilon_1^2 + \frac{1}{2}m_2\upsilon_2^2 = \frac{1}{2}kx^2 + \frac{1}{2}(m_1 + m_2)\upsilon_{CM}^2 + \frac{1}{2}m_1\upsilon_1^{*2} + \frac{1}{2}m_2\upsilon_2^{*2}$$

where υ_{CM} is the center of mass velocity and $\upsilon_1^* = \upsilon_1 - \upsilon_{CM}$ and $\upsilon_2^* = \upsilon_2 - \upsilon_{CM}$ are the velocities relative to the center of mass. We then have

$$\frac{1}{2}kx^2 = U_{\text{tot}} - \frac{1}{2}(m_1 + m_2)\upsilon_{CM}^2 - \frac{1}{2}m_1\upsilon_1^{*2} - \frac{1}{2}m_2\upsilon_2^{*2}.$$

The first two terms in the right-hand side do not vary due to the energy and linear momentum conservation respectively. The elastic energy is then a maximum when the two last terms, namely the kinetic energies, and consequently the velocities, relative to the center of mass are zero.

The angular momentum König theorem.
With reference to the inertial frame of Fig. 7.19, we choose the pole in the origin O. The angular momentum is

$$\mathbf{L}_O = \sum_{i=1}^{N} \mathbf{r}_i \times m_i\mathbf{v}_i,$$

which, using Eqs. (7.64) and (7.65) becomes

$$\begin{aligned}\mathbf{L}_O &= \sum_{i=1}^{N} \left(\mathbf{r}_i^* + \mathbf{r}_{CM}\right) \times m_i\left(\mathbf{v}_i^* + \mathbf{v}_{CM}\right) \\ &= \sum_{i=1}^{N} \mathbf{r}_i^* \times m_i\mathbf{v}_i^* + \mathbf{r}_{CM} \times \sum_{i=1}^{N} m_i\mathbf{v}_i^* + \left(\sum_{i=1}^{N} m_i\mathbf{r}_i^*\right) \times \mathbf{v}_{CM} + \mathbf{r}_{CM} \times M\mathbf{v}_{CM}.\end{aligned}$$

The last side contains four terms. The first term is the total angular momentum in the center of mass about the center of mass as a pole, say $\mathbf{L}_{CM}^*$. The second term is the total linear momentum in the CM frame and is null. The third term is zero for Eq. (7.69). The fourth term is the cross product of the position vector of the center of mass and the total linear momentum in the inertial system, say $\mathbf{P} = M\mathbf{v}_{CM}$. We can write

$$\mathbf{L}_O = \mathbf{r}_{CM} \times \mathbf{P} + \mathbf{L}_{CM}. \tag{7.73}$$

We can state that the total angular momentum in the inertial frame is equal to the sum of two terms. One term is the angular momentum "of the center of mass", which is the angular momentum that the center of mass would have if it were a material point with the total mass of the system. The second term is the angular momentum relative to the center of mass.

7.17 Elastic Collisions

In this chapter we have already discussed collision experiments. From the observation of the linear momentum conservation we have deduced the validity of the action and reaction law. We shall now take the opposite point of view. Assuming the mechanics laws to be valid, we shall discuss in some detail collision phenomena. We shall limit the discussion to material points. This is an idealization. However, we can consider the real bodies as points coinciding with their centers of mass, in which all their mass is concentrated, as long as in the collision the kinetic energies of each body do not vary. If we are dealing, for example, with the collision of two rigid balls, their motions should be translations, with no rotation.

We specify that when talking of collision of two bodies we do not necessarily imply that the two bodies come into contact. Considering for example the Newton experiments on the collisions between two pendulums we might substitute the balls with two bar magnets, with their north poles facing each other. If we take the pendulums out of equilibrium and let them go, the two magnets will approach each other subject to the repulsive force between the magnets. This force will slow them down till they stop and bunch back, without touching each other. As another example, consider two ions of the same charge moving one towards the other. When they are far from one another they feel practically no force and move with constant velocities. But when they become close enough, the repulsive electric force will cause both trajectories to deflect. The two ions will move on curved trajectories, approaching to a minimum distance and then separating again. When they are far enough apart, the ions will again move practically with constant velocities. The final velocities are in general different in magnitude and direction from the initial ones.

In a collision process we can distinguish three phases. In the initial phase the two bodies are distant and do not interact, namely the force exerted by one on the other is negligible. The second phase is the phase of the proper collision, which has a limited duration, say Δt. During this time the two bodies *inter*act. The interaction forces are internal, an action and reaction pair. The internal forces are much larger than the external forces, which can consequently be neglected. As a matter of fact we talk of a collision when this condition is met. In the third phase the bodies no longer interact but move away from each other. Notice that during the collision the two bodies may change their shape, the internal energy can decrease or increase, one or both of them can break into a number of pieces, or they may join in a single body, etc. In other words, the bodies in the final state may be different, also in number, from the two initial ones.

As during the collision only internal forces are present, we can state in complete generality that the total linear momenta before and after the collision are equal. Assume for simplicity to have two bodies both in the initial and final state. We indicate with the subscripts i and f the quantities in the initial and final states respectively and write

$$\mathbf{p}_{i1} + \mathbf{p}_{i2} = \mathbf{p}_{f1} + \mathbf{p}_{f2}. \tag{7.74}$$

The collision is said to be *elastic* if each of the bodies after the collision is the same as before the collision, its internal energy included, and if the total energies after and before the collision are equal. As the external forces are negligible and as the internal forces are zero in the initial and final states, the initial and final total kinetic energies are equal as well. If m_1 and m_2 are the masses of the two bodies, we have

$$\frac{1}{2}m_1 \upsilon_{1i}^2 + \frac{1}{2}m_2 \upsilon_{2i}^2 = \frac{1}{2}m_1 \upsilon_{1f}^2 + \frac{1}{2}m_2 \upsilon_{2f}^2 \tag{7.75}$$

and Eq. (7.75) can be written as

$$m_1 \mathbf{v}_{i1} + m_2 \mathbf{v}_{i2} = m_1 \mathbf{v}_{f1} + m_2 \mathbf{v}_{f2}. \tag{7.76}$$

Equation (7.75) is one relation, Eq. (7.76) are three relations, in total four, between the initial and final states. We shall now consider a few important cases.

Often one of the particles is at rest. If it is not so, we can always change the reference frame by choosing a frame moving with one particle (think of an observer sitting on the particle). The frame in which one particle stands still is called a *laboratory frame*. The particle that is still, say particle 2, is called the *target particle*. In the laboratory frame Eqs. (7.75) and (7.76) become

$$m_1 \upsilon_{1i}^2 = m_1 \upsilon_{1f}^2 + m_2 \upsilon_{2f}^2 \tag{7.77}$$

$$m_1 \mathbf{v}_{i1} = m_1 \mathbf{v}_{f1} + m_2 \mathbf{v}_{f2}. \tag{7.78}$$

The velocity of the target particle after the collision is given by Eq. (7.78),

$$\mathbf{v}_{f2} = \frac{m_1}{m_2}\left(\mathbf{v}_{i1} - \mathbf{v}_{f1}\right). \tag{7.79}$$

Consider the case in which the mass of the target is very large, namely $m_1/m_2 \gg 1$. We see that the final velocity of the target particle is very small. Its final kinetic energy, namely the energy gained in the collision, is also very small. In the limit of infinite target mass, the final velocity and kinetic energy of the target are zero. For example, a standing railcar hit by a ping-pong ball does not move, neither does a billiard table when a ball hits one of its sides. As a consequence, the kinetic energies of a light particle hitting a very massive target particle before and after collision are equal.

We now consider the case of two equal mass particles, which are at rest in the laboratory frame. The masses being equal, we can eliminate it from Eqs. (7.77) and (7.78) and write

$$\mathbf{v}_{i1} = \mathbf{v}_{f1} + \mathbf{v}_{f2}, \; \upsilon_{1i}^2 = \upsilon_{1f}^2 + \upsilon_{2f}^2. \tag{7.80}$$

The first of these equations tells us that the three velocity vectors can be thought of as the sides of a triangle, as shown in Fig. 7.21. For the second equation we have a right triangle, the hypotenuse of which is $\mathbf{v}_{i1}$. The final velocities of two particles of equal masses in the laboratory frame are always at 90° from one another. This can be observed, for example in a billiard game.

Consider now Fig. 7.22, which represents the initial state of the collision between two spherical bodies. One is initially at rest. The distance between the line on which the center of the moving body travels and the center of the target is called *impact parameter*. It is b in the figure. Clearly, the final state depends on b. Suppose, for example, that the two bodies are rigid spheres. When they touch, they interact with a force in the direction of the normal to the contact surface, which depends on b. This is the direction also of the variation of the momenta.

The simplest case is when the impact parameter is zero. The collision is then said to be *central*. The incoming particle travels on a line passing through the center of the target. When the particles collide, the action and reaction forces are directed on that line, and so are consequently the final momenta. After the collision both

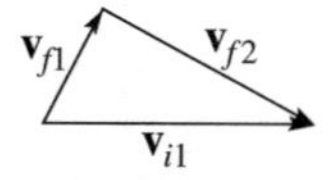

Fig. 7.21 Initial and final velocities in a collision of two equal mass particles in the CM frame

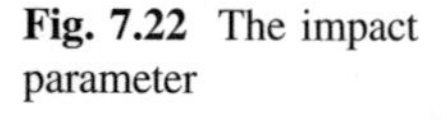

Fig. 7.22 The impact parameter

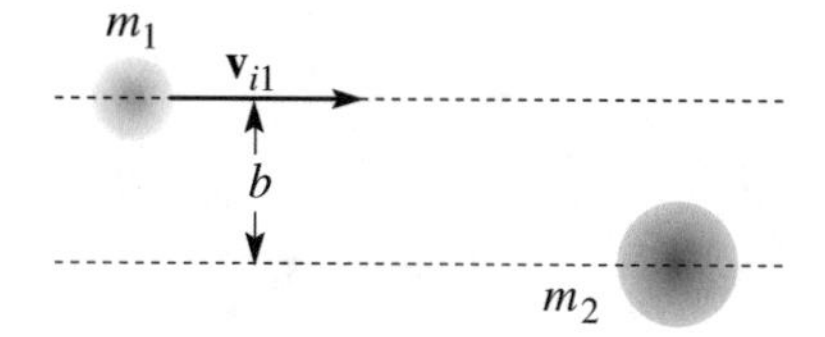

particles will travel on this line. The momentum conservation law Eq. (7.78) becomes a simple relation between magnitudes

$$m_2 v_{f2} = m_1 \left(v_{i1} - v_{f1} \right). \tag{7.81}$$

The energy conservation equation Eq. (7.77) becomes

$$m_2 v_{2f}^2 = m_1 \left(v_{1i}^2 - v_{1f}^2 \right). \tag{7.82}$$

We seek two final velocities as functions of the initial one v_{1i}. As Eq. (7.82) can be written as $v_{1i}^2 - v_{1f}^2 = m_1 \left(v_{1i} - v_{1f} \right) \left(v_{1i} + v_{1f} \right)$, we can usefully divide it by Eq. (7.81) obtaining $v_{f2} = v_{i1} + v_{f1}$. And finally

$$v_{f1} = \frac{m_1 - m_2}{m_1 + m_2} v_{i1}, \quad v_{f2} = \frac{2m_1}{m_1 + m_2} v_{i1}. \tag{7.83}$$

Let us discuss the first equation. If the mass of the incoming particle is smaller than the mass of the target ($m_1 < m_2$) its final velocity is negative, meaning that after the collision it bounces back. On the contrary, if its mass is larger than the mass of the target ($m_1 > m_2$), after the collision it continues to move forward, even if with a smaller velocity. An interesting case is when the two masses are equal. After the collision the velocities are $v_{f1} = 0$ and $v_{f2} = v_{i1}$. The two balls exchange their velocities. The phenomenon is easily seen hitting two pendulums of equal mass.

Finally, if $m_2 \gg m_1$, then $v_{f1} = -v_{i1}$ and $v_{f2} = 0$. This is the case of an elastic collision of a ball, for example a tennis one, against a wall, shown in Fig. 7.23. Here we suppose the wall to be smooth. In this case the force of the wall on the ball is normal to the surface. We decompose the quantity of motion of the ball in components normal and parallel to the wall. The latter is not changed by the collision. To the normal component we can apply the results we found for the central collisions. Particle 1 is the ball, particle 2 is the wall, hence $m_2 \gg m_1$. After the collision the wall is still at rest while the normal component of the ball velocity has changed its sign.

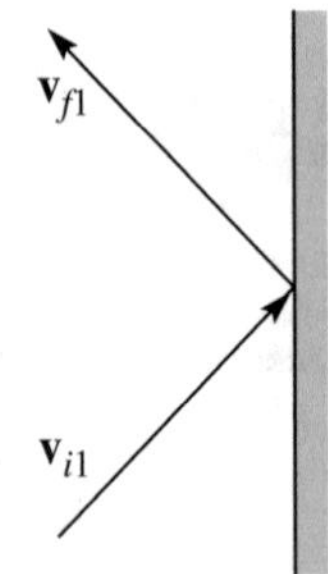

Fig. 7.23 Elastic collision on a wall

We now analyze the general case of the elastic collision between two particles. As during the collision the system is isolated, the center of mass velocity is constant and the CM frame is inertial. Recalling that $\upsilon_{i2} = 0$, the CM velocity in the laboratory frame is

$$\mathbf{v}_{CM} = \frac{m_1 \mathbf{v}_{1i} + m_2 \mathbf{v}_{2i}}{m_1 + m_2} = \frac{m_1 \mathbf{v}_{1i}}{m_1 + m_2}. \tag{7.84}$$

We obtain the velocities of the particles in the CM frame, which we indicate with an asterisk, by subtracting the CM velocity from their velocities in the laboratory frame

$$\mathbf{v}_{1i}^* = \mathbf{v}_{1i} - \mathbf{v}_{CM},\ \mathbf{v}_{2i}^* = 0 - \mathbf{v}_{CM},\ \mathbf{v}_{1f}^* = \mathbf{v}_{1f} - \mathbf{v}_{CM},\ \mathbf{v}_{2f}^* = \mathbf{v}_{2f} - \mathbf{v}_{CM}. \tag{7.85}$$

In the CM frame the total linear momentum is zero both before and after the collision. This means that the momenta of the two particles are equal and opposite before the collision and similarly after it. These quantities are called *center of mass momentum* before and after the collision respectively. If $\mathbf{p}_i^*$ is the momentum of particle 1 before the collision, the momentum of particle two is $-\mathbf{p}_i^*$. Similarly, after the collision the momenta are, say, $\mathbf{p}_f^*$ and $-\mathbf{p}_f^*$. We write the kinetic energy conservation as

$$\frac{\mathbf{p}_f^{*2}}{2m_1} + \frac{\mathbf{p}_f^{*2}}{2m_2} = \frac{\mathbf{p}_i^{*2}}{2m_1} + \frac{\mathbf{p}_i^{*2}}{2m_2}$$

and also

$$\mathbf{p}_f^{*2} = \mathbf{p}_i^{*2}. \tag{7.86}$$

In words, in an elastic collision in the CM frame, the magnitude of the linear momentum of each particle is equal after and before the collision. The only effect of the collision is to change the common direction of the momenta by an angle, say, θ, as shown in Fig. 7.24.

The angle θ is called a *scattering angle*. It cannot be found only on the basis of the conservation laws. First of all, it depends on the impact parameter b, which in the CM frame is the distance between the lines on which the center of mass of the two bodies travel in the initial state.

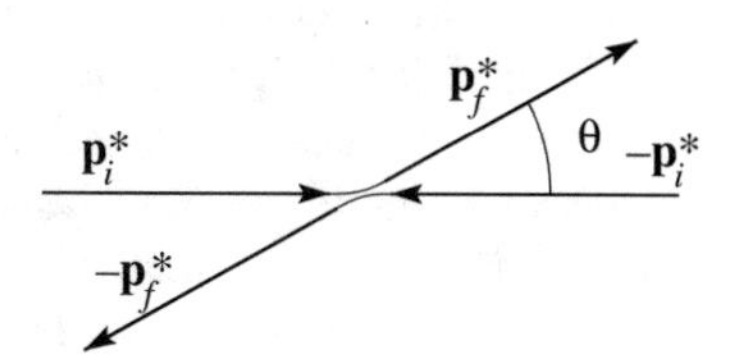

Fig. 7.24 Elastic collision in the CM frame

The dependence of the scattering angle on the impact parameter, given by the function $\theta(b)$, depends on the structure of the colliding bodies. Suppose, for example, that one of them, the incoming one in the laboratory frame, is point-like, while the target body has a structure. We can think of the first as an electron, the second an atom. We imagine the atom as a spherical cloud of negative electric charge with the positively charged nucleus at the center. This is very small and hard. If the impact parameter is larger than the atomic radius, the electron is not deflected in its motion, namely the scattering angle is $\theta = 0$. If the impact parameter is smaller than the atomic radius, the electron penetrates in the charged cloud, is deflected by the electric force and exits in a direction different from the incident one. The scattering angle is now $\theta \neq 0$, which is increasing with a decreasing impact parameter. In practice however it is never very large. When the impact parameter is smaller than the nuclear radius, the collision is with the nucleus, and is violent. The scattering angle is large. It can even reach 180°, namely the direction of motion can invert if the collision is central, $b = 0$, because the mass of the nucleus is much larger than that of the electron.

This example shows how the measurement of the function $\theta(b)$ in a scattering experiment (a it is called) can be extremely useful to understand the structure of the objects that, like atoms, are too small to be visible. As a matter of fact, the example we have just made is quite similar to the experiment performed in 1911 by Hans Wilhelm Geiger (1882–1945) and Ernest Marsden (1889–1970) that led Lord Ernest Rutherford (1871–1937) to discover the atomic nucleus. Geiger and Marsden used energetic α particles (rather than the electron in the example) sending them on a thin gold sheet and measuring how many of them were scattered at different angles. They found, in particular, that sometimes they were deflected backwards. If the atoms were soft clouds of charges, as in the current model, this could not happen. Rutherford concluded that a small hard nucleus had to be present inside the atom. In the same way the internal structure of the atomic nuclei was studied and, in 1967, the presence of the quarks in protons and neutrons was discovered.

7.18 Inelastic Collisions

As we have stated, linear momentum is always conserved in a collision. This is not the case for energy. When the final energy is different from the initial one the collision is said to be *inelastic*. Rigorously speaking, in real collisions between every day size objects, at least a small fraction of the mechanical energy is lost. For example, if we drop a steel ball on a rigid floor, it will bunch back but will not reach exactly the initial height. If we do the same experiment with a wax ball we see that it sticks on the floor. The real collisions are never perfectly elastic, but have a smaller or larger degree of inelasticity. We shall give a quantitative definition of this concept. Before doing that, let us consider the case of the *completely inelastic* collision (the case of the wax ball in the example).

Consider two spherical bodies of masses m_1 and m_2 and initial velocities $\mathbf{v}_{i1}$ and $\mathbf{v}_{i2}$. The collision is completely inelastic if the two bodies stick together, namely if their velocities after the collision are equal $\mathbf{v}_{f1} = \mathbf{v}_{f2}$. We can indicate simply with $\mathbf{v}_f$ the final velocity and write the momentum conservation as

$$\mathbf{P} = m_1\mathbf{v}_{1i} + m_2\mathbf{v}_{2i} = (m_1 + m_2)\mathbf{v}_f. \tag{7.87}$$

The final velocity is then

$$\mathbf{v}_f = \frac{m_1\mathbf{v}_{1i} + m_2\mathbf{v}_{2i}}{m_1 + m_2}, \tag{7.88}$$

which is the same as the center of mass velocity (that does not vary in the collision) as expected, considering that in the final state there is only one body. We write down the initial kinetic energy, using the König theorem

$$U_{K,i} = \frac{1}{2}m_1 v_{1i}^2 + \frac{1}{2}m_2 v_{2i}^2 = U_{K,i}^* + \frac{1}{2}(m_1 + m_2)v_{CM}^2,$$

where $U_{K,i}^*$ is the kinetic energy in the CM reference. The final kinetic energy is

$$U_{K,f} = \frac{1}{2}(m_1 + m_2)v_{CM}^2.$$

We see that in the completely inelastic collision all the kinetic energy relative to the center of mass $U_{K,i}^*$ is lost in the collision. If we want to look at the collision in the CM frame we can take over all the conclusions of the last section, with the exception of equality of the magnitudes of the initial and final momenta. If the collision is inelastic, the final center of mass momentum is smaller than the initial one, null if it is completely inelastic. Figure 7.25 shows the situation. In the completely inelastic collision all the momentum in the CM reference and all the kinetic energy relative to the center of mass are lost. In the laboratory frame not all the kinetic energy gets lost, because the velocity of the center of mass must be the same after and before the collision, due to the momentum conservation. Consequently, the kinetic energy "of the center of mass" cannot be lost. In the completely inelastic collision all the energy that can be lost is lost, but this is not all the energy.

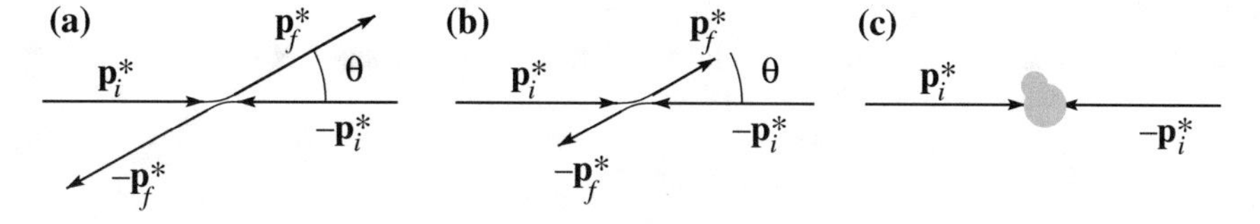

Fig. 7.25 Collisions in the CM frame. **a** Elastic, **b** inelastic, **c** completely inelastic

Fig. 7.26 Ballistic pendulum

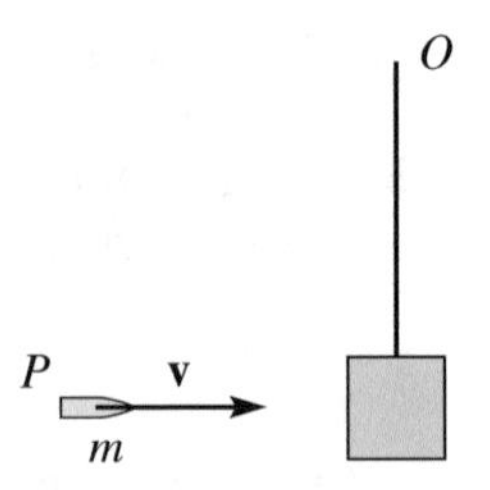

In the above example of the wax ball falling on the floor, the ball loses all its energy. However, in this case the mass of the target is enormous, practically infinite and consequently the velocity of the center of mass is zero.

An application of a completely inelastic collision is the *ballistic pendulum* used to measure the velocities of bullets. Figure 7.26 shows the device, which is made of a sand bag of mass M suspended with a bar to the pivot O. The bullet P of mass m and velocity υ to be measured hits the pendulum, penetrates the bag and sticks. By measuring the resulting oscillation amplitude we determine the velocity of the bag υ_f after the collision. Equation (7.81), considering that $\mathbf{v}_{2I} = 0$, and that $M \gg m$, becomes $m\upsilon = (m+M)\upsilon_f \simeq M\upsilon_f$, or $\upsilon = (m/M)\upsilon_f$. This gives the bullet velocity, the two masses being known.

Looking at Fig. 7.25 we understand immediately that all the intermediate cases between elastic and completely inelastic collisions are possible. The parameter characterizing the degree of elasticity is called the *coefficient of restitution* and is defined as the ratio between the center of mass momentum after and before the collision

$$e = p_f^*/p_i^*. \tag{7.89}$$

By definition, the coefficient is a non-negative number. It is equal to one in the elastic collision. Notice that it can be larger than one. Suppose for example that the target body contains a spring, which is compressed and blocked by a nail. In the collision the nail is broken, the spring expands and gives energy to the colliding bodies. The final center of mass momentum is larger than the initial ones. As another example, energy can be gained in collisions between two molecules. Such are the exothermic chemical reactions.

Let us write down the kinetic energy in the CM reference frame

$$U_{K,i}^* = \frac{p_i^{*2}}{2m_1} + \frac{p_i^{*2}}{2m_2} = \frac{1}{2}\frac{p_i^{*2}(m_1+m_2)}{m_1 m_2} = \frac{1}{2}\frac{p_i^{*2}}{\mu}$$

where μ is the reduced mass. Similarly the final kinetic energy is

$$U_{K,f}^* = \frac{1}{2}\frac{p_f^{*2}}{\mu}$$

and finally, for Eq. (7.89)

$$U^*_{Kf}/U^*_{Ki} = e^2, \tag{7.90}$$

namely the ratio of the final and initial kinetic energies relative to the center of mass is equal to the square of the coefficient of restitution.

Problems

7.1. What is the total momentum **P** of a system of particles in the CM frame?

7.2. A system of interacting bodies moves in the neighborhood of the earth's surface. Neglect air resistance. How does the center of mass move?

7.3. Two railcars move one against the other on a rail. The first one has a mass of 1000 kg and moves at the speed of 2 m/s. The second one has twice the mass. After the collision the two cars are at rest. What was the initial velocity of the second car? Did the kinetic energy change?

7.4. A railcar of 5 t mass and speed 10 m/s is stopped by bumpers in 0.5 s. Find the impulse and the average value of the force.

7.5. Two pendulums collide elastically. Initially, one of the two, of mass m_2 stands still in the equilibrium position, the other one, of mass m_1 is abandoned at a certain height above that. After the collision the two velocities are equal and opposite. (a) What is the ratio of their masses? (b) What is the ratio between the center of mass velocity and the velocity of pendulum 1 before the collision?

7.6. In Problem 7.5, knowing the kinetic energy $U_{Ki}(1)$ of pendulum 1 immediately before the collision, find: (a) the total kinetic energy in the CM reference, (b) the kinetic energy $U_{Kf}(1)$ of the first pendulum immediately after the collision.

7.7. In a first approximation, the moon revolves around the center of the earth. More precisely, earth and moon revolve around their common center of mass. Knowing that the mass of the earth is about 81 times that of the moon and that the distance between the two centers is about 60 earth radii, R_E, calculate the position of the center of mass (in R_E units).

7.8. A planet of mass M has a satellite of mass $m = M/10$. The distance between their centers is R. (a) Express the revolution period as a function of R and M. (b) Find the ratio between the (revolution) kinetic energies of the two bodies.

7.9. We have measured the period of T earth years of a binary system and the distance between the two stars in R astronomic units. Find the sum of the two masses in solar mass (M_S) units.

7.10. Two point-like bodies have a completely inelastic collision. The first body has a mass $m_1 = 2$ kg and the velocity before collision $\mathbf{v}_{1i} = (3, 2, -1)$ m/s. The second body has a mass $m_2 = 3$ kg and the velocity before collision

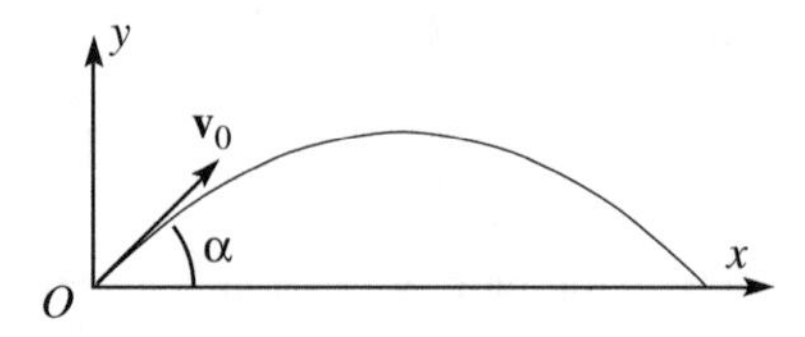

Fig. 7.27 The trajectory of problem 7.17

$\mathbf{v}_{2i} = (-2, 2, 4)$ m/s. (a) Find the velocity $\mathbf{V}$ of the composite body after the collision. (b) Find the total energy and the energy relative to the center of mass before the collision and compare with the kinetic energy after the collision.

7.11. A material point of mass m, moving with velocity $\mathbf{v}_{1i}$ collides with a second point, of mass $2m$, that is standing. We measure the velocity after the collision of the particle of mass m finding its direction at 45° with the incident one and its magnitude one half of the initial value. (a) Find the magnitude and direction of the velocity of a particle of mass $2m$. (b) Was the collision elastic?

7.12. The force $\mathbf{F} = (3, 4, 0)$ N is applied on the point P having coordinates (8, 6, 0) m. Find (a) its moment about the origin, (b) the lever arm b of the force, namely the distance of its application line from the pole. (b) the component F_n of the force perpendicular to the position vector $\mathbf{r}$.

7.13. A ball falls on the floor from 5 m. What are the heights it reaches when bouncing back the first, the second and the third times if the coefficient of restitution is 0.8? What are the corresponding energies? Neglect air resistance.

7.14. An air guide is a rail with a series of small holes through which compressed air is blown. A sledge can run on the guide practically without friction. We put two such sledges on the rail. The first one, of mass $m_1 = 2$ kg is still. On its right side lies a spring of elastic constant $k = 300$ N/m and 1 m long, in its natural length. The second sledge, of mass $m_2 = 3$ kg is launched towards the first with velocity 5 m/s. It hits the first sledge putting it and the spring in motion. What is the maximum deformation Δx of the spring?

7.15. A material system is made of a particle of mass $m_1 = 0.1$ kg in the point of coordinates (1, 2, 3) m, a particle of mass $m_2 = 0.2$ kg at the coordinates (2, 3, 1) m and a particle of mass $m_{13} = 0.3$ kg at the coordinates (3, 1, 2) m. Find the coordinates of the center of mass.

7.16. A body of mass $m = 2$ kg is shot vertically upwards with initial velocity $\upsilon_0 = 10$ m/s from a point with coordinates (0, 20, 0) m. The z-axis is vertical upwards. Find the difference $\Delta\mathbf{L}_O$ of the angular momentum of the body about the origin between the instant when it is back in the initial position and the initial instant.

7.17. A particle of mass m is launched with initial velocity $\mathbf{v}_0$ at an angle α with the horizontal. In the reference frame of Fig. 7.27, neglecting air resistance, find the time dependence of (a) the moment of the force about the origin O, (b) the angular momentum $\mathbf{L}_O$ of the body about the same pole.

Chapter 8
Rigid Bodies

In this chapter, we shall discuss the mechanics of an important class of extended systems, the rigid bodies. In a perfectly rigid body, the distance between any pair of its points does not vary for any acting forces or any motion. Clearly, this is an idealization, but, in practice, the solid objects are rigid in a good approximation.

The motion of the rigid bodies is governed by two differential equations. The known members are the ns of the external forces and of the external moments. The solution to these may involve advanced calculus. We shall limit the discussion to the simplest situations.

In Sect. 8.1, we shall define the rigid body and its motions, and then do the same for the properties of the systems of applied forces in Sect. 8.2. In Sect. 8.3, we shall consider the equilibrium conditions.

We then consider the rotations about a fixed axis, which are the simplest motions. We shall find the expressions of the kinetic energy and the angular momentum. We shall introduce a new kinematic quantity, the moment of inertia, and see how the dynamics of the rotations about a fixed axis have some similarities with the dynamics of the material point. In Sects. 8.9 and 8.10, we discuss two important examples, the torsion balance and the compound pendulum.

We shall then move on to more complex motions, those of a rigid body about a fixed point. We shall first find the expression of the angular momentum about that point, and of the kinetic energy in Sects. 8.12 and 8.13. We shall see that, in general, the directions of the angular momentum and the angular velocity are different. One consequence of that is that the forces develop on the supporting, as demonstrated in Sect. 8.14. In Sects. 8.15 and 8.16, we study the pure rolling motion of cylindrical and spherical rigid bodies on a plane.

In Sect. 8.17, we consider the gyroscopes, which are rigid bodies moving about a fixed point, a top being a good example.

Finally, in Sect. 8.18, we shall study the collisions between rigid bodies.

A. Bettini, *A Course in Classical Physics 1—Mechanics*,
Undergraduate Lecture Notes in Physics, DOI 10.1007/978-3-319-29257-1_8

8.1 Rigid Bodies and Their Movements

The solid bodies are approximately rigid. Upon first approximation, their shape does not change if we stretch, compress or torque them. Clearly, this is never rigorously true, because small deformations always take place. However, several dynamical properties of these bodies can be studied considering them as *rigid*. We define them as rigid if the distance between any pair of their points does not vary.

The space location of a rigid body is called its *configuration*, which is the set of the positions of its points. To define the configuration of a generic system of N points, we need 3 N coordinates, but only six for a rigid body. Let us see why.

We start by defining the position of one point, for example, A in Fig. 8.1, namely its three coordinates in the reference frame we have chosen. In a certain instant, it is in $A_1 = (x_{A1}, y_{A1}, z_{A1})$. We then shall give the coordinates of a second point, like B, which is in $B_1 = (x_{B1}, y_{B1}, z_{B1})$. But, wait a moment: we cannot do that arbitrarily. We can choose only two coordinates, because the distance between A and B is fixed, independent of the configuration, namely

$$(x_{B1} - x_{A1})^2 + (y_{B1} - y_{A1})^2 + (z_{B1} - z_{A1})^2 = \text{const.}$$

With this, we do not yet know the configuration. We need the position of a third point, like C, which is in C_1 in the considered instant. Of its three coordinates, we can arbitrarily choose only one, the other two being defined by the two conditions that the distances C_1A_1 and C_1B_1 are fixed. Now, the positions of all the points are defined, hence the configuration of the body too.

In total, the configuration of a rigid body is defined by six coordinates. We say that the system has six *degrees of freedom*.

Consider now two configurations. The transport from one configuration to another can always be obtained with a *translation* followed by a *rotation* around an axis. This is geometrical and not necessarily fixed. In Fig. 8.1, the translation brings point A from A_1 to A_2 and the body goes to the dotted configuration. To bring the other points to their final positions, we need a rotation about an axis through A_2 (this

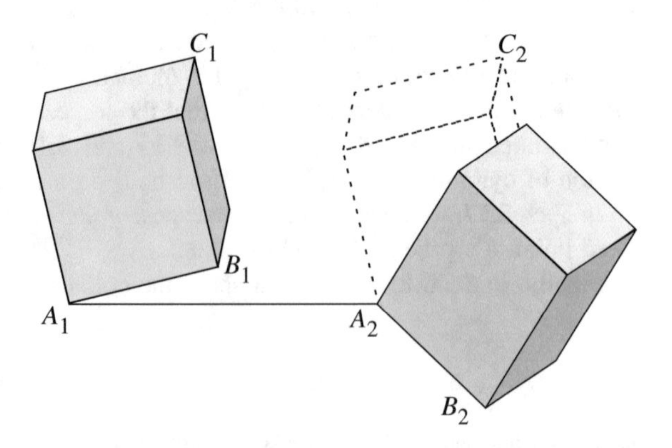

Fig. 8.1 Two configurations of a rigid body

point is already in position, and should not move any more). The axis should have the right direction and the rotation angle the right value.

In the above argument, the choice of the point A was arbitrary, but it determines the translation. If we had chosen, for example, B, the translation would have been different. Consequently, there are infinite translation-rotation pairs that produce the given displacement. It can be demonstrated, however, that once given the initial and final configurations, the rotation direction and the angle are determined.

Obviously, the same result can be obtained by performing the rotation first and then the translation, or a sequence of rotation and translation pairs. As a matter of fact, the motion of the rigid body can be thought of as a continuous series of infinitesimal *roto-translations*. The rotation axis, in general, varies continuously during the motion and we subsequently talk of an *instantaneous rotation axis*.

While the choice of the translating point is, as we said, arbitrary, it is in practice convenient to choose the center of mass, considering its privileged role in dynamics. We recall here the dynamical equations governing every mechanical system in an inertial frame that we found in Chap. 7. We choose a point Ω fixed in the inertial frame as the pole. Let $\mathbf{M}_\Omega$ be the total external torque about Ω and $\mathbf{L}_\Omega$ the total angular momentum about the same pole, $\mathbf{F}$ the external resultant force and $\mathbf{P}$ the total linear momentum. The two dynamical equations are

$$\mathbf{F} = \frac{d\mathbf{P}}{dt}, \tag{8.1}$$

$$\mathbf{M}_\Omega = \frac{d\mathbf{L}_\Omega}{dt}. \tag{8.2}$$

We also recall that the second equation is similarly valid when we choose a particular point, even if it is moving in the inertial frame, namely the center of mass of the system

$$\mathbf{M}_{CM} = \frac{d\mathbf{L}_{CM}}{dt}. \tag{8.3}$$

The two vector equations give six independent conditions. For any mechanical system, these are necessary conditions, but in general, they are not sufficient. They are, however, sufficient for a rigid body, which has six degrees of freedom, as many as the conditions. In other words, if we know the external resultant force and the total external torque (or moment) and the initial conditions, we can know the motion of the body solving the above differential equations.

We notice that Eq. (8.1) rules the motion of the center of mass of the body. Remembering that $\mathbf{P} = m\mathbf{v}_{CM}$, where m is the mass of the body and $\mathbf{v}_{CM}$ the velocity of its center of mass, we can write Eq. (8.1) in the equivalent form

$$\mathbf{F} = m\mathbf{a}_{CM} \tag{8.4}$$

where $\mathbf{a}_{CM}$ is the center of mass acceleration. The motion of the center of mass is exactly in the same way as the motion of a material point.

Equation (8.3) allows us to find the motion of the body about its center of mass. This is general around an axis through the center of mass but of varying direction and with varying angular velocity. The solution is, in general, quite complicated. We shall consider the simplest cases here.

We immediately notice an important property of the rigid motions: the work of the internal forces is always zero. Indeed, the internal forces come in couples acting on pairs of points in the direction of the line joining the points. The work done by one of the two for a given displacement of the body is equal to the force times the projection of the displacement of the point on which it acts on the direction of the force. The latter is the line joining the two points. The work done by the couple of forces is then equal to the magnitude of the force times the difference between the projections of the two displacements on the joining line. But this difference is the change in the distance between the two points, and this is zero, if the body is rigid.

8.2 Applied Forces

Suppose that several external forces are acting on a rigid body. As we have just seen, the motion of the body is determined by their resultant and total torque. Clearly, there is an infinite number of systems of forces having the same resultant and the same torque. All these force systems applied to the same rigid body produce the same motion, when starting from the same initial conditions. Consequently, from the observation of the motion, we can know the resultant force (from the center of mass acceleration) and the total torque (from the angular acceleration), but not the single acting external forces. We define as *equivalent* any system of applied forces with the same resultant and torque. Notice such force systems are equivalent for the motion of a rigid body, but they do not have the same effects if applied to a non-rigid body. Consider the very simple example of a couple on the same line. The resultant force and torque are zero. Acting on a rigid body, they tend to approach or separate the two points, namely to change their distance. This distance, the body being rigid, cannot vary. But if the body is a rubber band, the distance varies and both forces do work.

We now show a few simple properties of the force systems that will be useful in the following.

(1) A force system has resultant $\mathbf{F}$ and total torque about the fixed point Ω, $\mathbf{M}_\Omega$. We show that the torque about any other fixed pole Ω' is

$$\mathbf{M}_{\Omega'} = \mathbf{M}_\Omega + \Omega\Omega' \times \mathbf{F}. \tag{8.5}$$

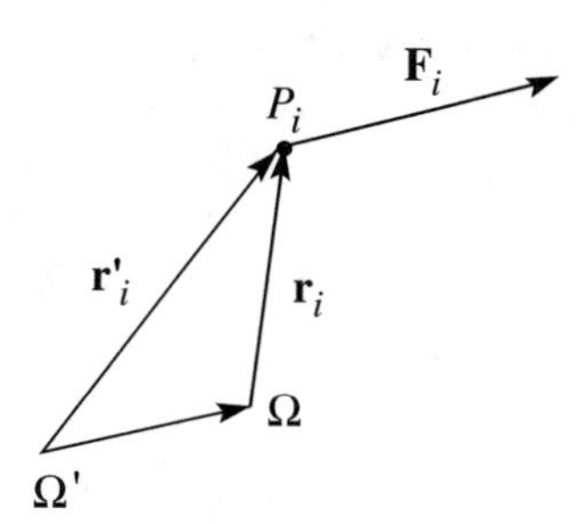

Fig. 8.2 A force and two different poles

With reference to Fig. 8.2, we can easily see that the relation between the torques about the two poles of the generic force $\mathbf{F}_i$ is

$$\boldsymbol{\tau}_{i\Omega'} = \mathbf{r}'_i \times \mathbf{F}_i = (\Omega\Omega' + \mathbf{r}_i) \times \mathbf{F}_i = \Omega\Omega' \times \mathbf{F}_i + \boldsymbol{\tau}_{i\Omega},$$

which, summed on all the forces, gives Eq. (8.5)

Corollary 1 *The torque of a force system of zero resultant is independent of the pole, namely* $\mathbf{M}_\Omega = \mathbf{M}_\Omega'$.

Corollary 2 *If two force systems have the same resultant and the same torque about the same pole, they have the same torque about any pole.*

(2) Consider a generic force system of resultant $\mathbf{F}$ and torque about Ω, $\mathbf{M}_\Omega$. The system is equivalent to a force system of a force $\mathbf{F}$ applied to the pole Ω plus a torque couple $\mathbf{M}_\Omega$. The demonstration is immediate. The two systems have the same resultant and the same torque, as the torque of $\mathbf{F}$ about the pole is null.

(3) A system of mutually *parallel forces* $\mathbf{F}_i$ applied to different points P_i of position vectors $\mathbf{r}_i$ is equivalent to their resultant $\mathbf{F}$ applied to the point C, having the position vector

$$\mathbf{r}_C = \sum_{i=1}^{N} F_i\mathbf{r}_i \Big/ \sum_{i=1}^{N} F_i. \tag{8.6}$$

The point C is called the *center of the force system*. The demonstration of the theorem is easy. First of all, the two systems obviously have the same resultant. As for the torque, let us take the origin O as the pole, as in Fig. 8.3. The forces being parallel, we can call $\mathbf{u}$ their common unit vector and write $\mathbf{F}_i = F_i\mathbf{u}$. The torque about O is

$$\mathbf{M}_O = \sum_{i=1}^{N} \mathbf{r}_i \times \mathbf{F}_i = \sum_{i=1}^{N} \mathbf{r}_i \times F_i\mathbf{u} = \left(\sum_{i=1}^{N} \mathbf{r}_iF_i\right) \times \mathbf{u},$$

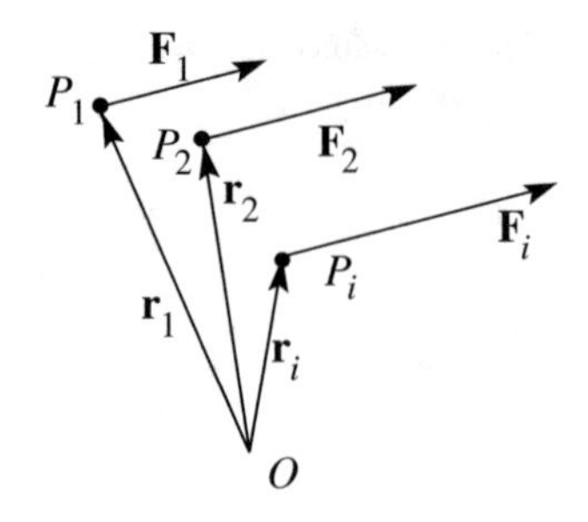

Fig. 8.3 A system of parallel forces

which, by definition of the center of the forces, becomes

$$\mathbf{M}_O = \left(\sum_{i=1}^{N} \mathbf{r}_i F_i\right) \times \mathbf{u} = \mathbf{r}_C \left(\sum_{i=1}^{N} F_i\right) \times \mathbf{u} = \mathbf{r}_C \times \mathbf{u} \sum_{i=1}^{N} F_i = \mathbf{r}_C \times \mathbf{F},$$

which proves the theorem.

The weight forces are a relevant example of parallel forces. Consider a system of n material points (the argument is also valid for a continuous system) P_i of position vectors $\mathbf{r}_i$ and masses m_i. The weights $m_i\mathbf{g}$ are parallel forces applied to the points P_i. The position vector of the center of the forces is

$$\mathbf{r}_C = \frac{\sum_{i=1}^{N} g m_i \mathbf{r}_i}{\sum_{i=1}^{N} g m_i} = \frac{\sum_{i=1}^{N} m_i \mathbf{r}_i}{\sum_{i=1}^{N} m_i}. \tag{8.7}$$

We see that the center of the weight forces, called the *barycenter*, is simply the center of mass of the system. The motion of a rigid body under the action of the weights of all its parts can be described as if a single force was acting, its total weight applied to the center of mass. This property, which we have already used, substantially simplifies several problems.

Notice, to be precise, that the coincidence between center of mass and center of the weight forces exists for bodies that are not too large, such that the weights of all their parts can be considered to be parallel. This is almost always true in practice.

8.3 Equilibrium of the Rigid Bodies

A configuration of a rigid body is said to be of equilibrium if, leaving the body at rest in that configuration, it keeps it indefinitely. The necessary and sufficient condition for the equilibrium, in an inertial frame, is that the external resultant force and the external moment are zero. Indeed, if the body is in equilibrium, the acceleration of its center of mass is zero; hence, the resultant force is zero. In addition, the angular momentum that is initially zero must remain as such. Hence,

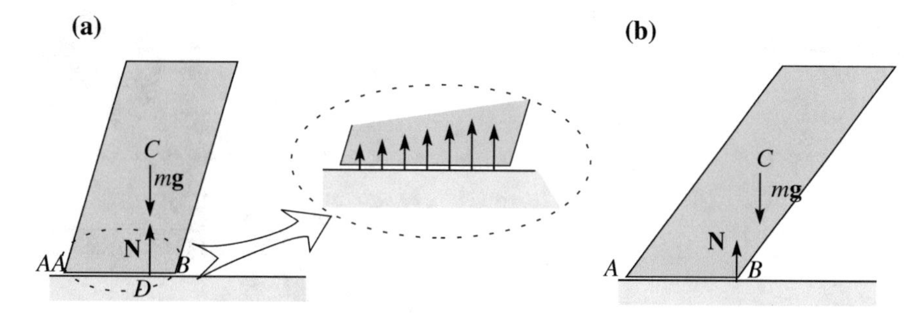

Fig. 8.4 **a** An equilibrium position, **b** a non-equilibrium position

the total moment is zero. On the other hand, if the resultant is zero, the center of mass does not change its velocity, which is initially zero, and, if the moment is zero, the angular momentum is constant and remains zero, if it is so initially.

Notice that the two conditions are independent from one another. For example, a couple of forces have zero resultant and non-zero moment, while a force applied to the pole has zero moment and nonzero resultant.

Example E 8.1 Consider a rigid body on a horizontal plane under the action of its weight. The position is of an equilibrium position if the vertical through the center of mass of the body intersects its support base. Indeed, the external forces are the weights of its elements and the constraint forces. The former are equivalent to the total weight applied to the center of mass, the latter are normal to the base and consequently are a system of parallel forces too. Consequently, they are equivalent, with their resultant $\mathbf{N}$ applied to their center of forces D, as shown in Fig. 8.4a. The constraint automatically adjusts its reaction in such a way that the magnitude of $\mathbf{N}$ and the center D guarantee the equilibrium, in other words, that $m\mathbf{g}$ and $\mathbf{N}$ are a couple with the same line of application. This implies that $\mathbf{N} = -m\mathbf{g}$ and that D should be on the vertical from C. This is possible if the foot of this vertical is between A and B, namely inside the base. The insert in the figure shows a possible configuration of the constraint forces. They are applied between A and B. Consequently, their center must be a point of AB.

In the configuration of Fig. 8.4b, the equilibrium is not possible. Even if the constraint normal reaction $\mathbf{N}$ is concentrated in the extreme point B of the basis, this is not enough to produce a couple of zero moments. The body overturns. During the fall, the normal reaction is less than the weight, because the center of mass is accelerating downwards. The difference $mg - N$ is equal to the acceleration of the center of mass times the mass of the body.

The center of the constraint forces can, however, be brought outside the segment AB, and the equilibrium is also guaranteed in the conditions of Fig. 8.5b, if part of the constraint forces is directed upwards. We can, for example, drive a nail in A, as in Fig. 8.5, or attach a hook. If $\mathbf{R}$ is the reaction of the nail, or of the hook, and $\mathbf{N}$ the reaction of the plane, the equilibrium is when the resultant force and moment are zero, namely $\mathbf{N} = \mathbf{R} + m\mathbf{g} = 0, R\overrightarrow{AB} = mg\overrightarrow{BD}$.

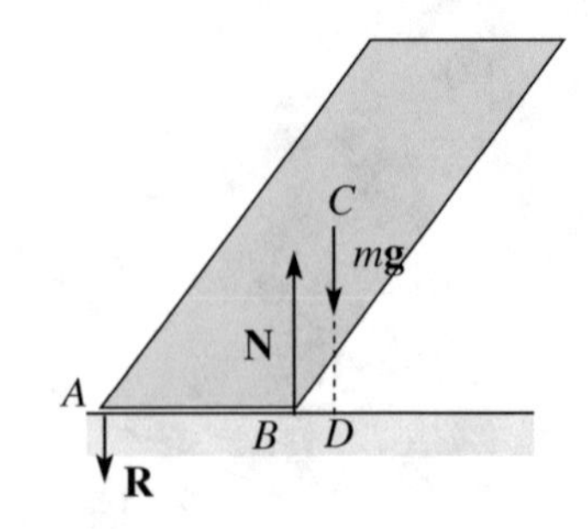

Fig. 8.5 The constraint **R** guarantees equilibrium

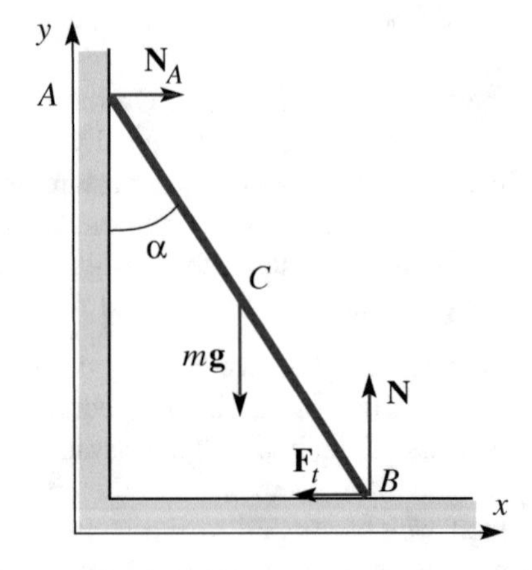

Fig. 8.6 The forces acting on a ladder

Example E 8.2 The ladder shown in Fig. 8.6 of length l is supported by a vertical wall, at an angle of α. Suppose the friction on the wall to be negligible, while the coefficient of static friction on the horizontal plane is μ_S. Let us discuss the equilibrium conditions.

In Fig. 8.5, C is the center of mass, and A and B are the footholds. We take the reference frame with the x-axis horizontal in the plane of the figure, the z-axis horizontal directed out of the figure and the y-axis vertical upwards. The external forces are: the weight $m\mathbf{g}$, applied to the center of mass, the constraint reaction applied in B, which we consider decomposed in a vertical component, $\mathbf{N}$, and a horizontal component, $\mathbf{F}_t$, and finally, the constraint reaction applied in A, $\mathbf{N}_A$ that is horizontal (no friction here). At equilibrium, their resultant is zero:

$$\mathbf{N} + \mathbf{F}_t + \mathbf{N}_A + m\mathbf{g} = 0.$$

This equation gives two independent relations, its x and y components, the z component being identically zero. The two relations are $N = -mg$, which gives the unknown N, and $F_t = -N_A$, which links the other two unknowns. We now state that the external moment should be zero too, namely

$$M_z = mg(l/2)\sin\alpha - N_A l\cos\alpha = 0.$$

We have written the signs in this equation taking into account that $\mathbf{N}_A$ must be in the positive x direction, because the wall can only push. Consequently, $\mathbf{N}_A$ tends to rotate the ladder clockwise and the z component of its moment is negative. On the other hand, for the above written equation, for the equilibrium of the horizontal forces, $\mathbf{F}_t$ must be in the opposite x direction. The z component of its moment is consequently positive. Solving the two equations for F_t and N_A, we immediately have $F_t = -N_A = -(mg/2)\tan\alpha$.

The friction force cannot be too large, namely $F_t \le \mu_s N$. On the other hand, $|F_t|/N = (\tan\alpha)/2$. Consequently, to be in equilibrium, the leaning angle should not be too large, namely $\tan\alpha \le 2\mu_s$. For larger angles, the ladder slides down.

We have assumed the vertical wall to be smooth and its reaction to be normal. If there is friction, as there always is in practice, there is a vertical component to the wall reaction too. We would have one more unknown, with the same number of equations. Under these conditions, the problem is undetermined. Indeed, there is an infinite number of pairs of the two tangential reactions that lead to equilibrium. Another example of an undetermined problem is the problem of finding the constraint reactions on the four wheels of a car, or the four legs of a table, on a plane. These problems have a solution if more information is available, such as the nature of the elastic forces of the tires on the car or the lengths of the legs of the table.

8.4 Rotation About a Fixed Axis

An important and relatively simple class of rigid movements is the class of movements about a fixed axis. Consider a rigid body of arbitrary shape, as represented in Fig. 8.7, which can move around the axis a, which is fixed in an inertial frame. Let $\mathbf{u}_a$ be the unit vector, arbitrarily chosen in one of the two directions of a. The configuration of the body is defined by the rotation angle, which we call ϕ, around the axis a, relative to a fixed plane, which we choose as the origin of the angles.

We now choose a point Ω on the axis as the pole of the moments and call $\mathbf{M}_\Omega$ the total external moment and $\mathbf{L}_\Omega$ the total angular momentum about Ω. The dynamic equation is

$$\mathbf{M}_\Omega = \frac{d\mathbf{L}_\Omega}{dt}. \tag{8.8}$$

We now take the dot product of the two members with the unitary vector of the rotation axis $\mathbf{u}_a$. We have

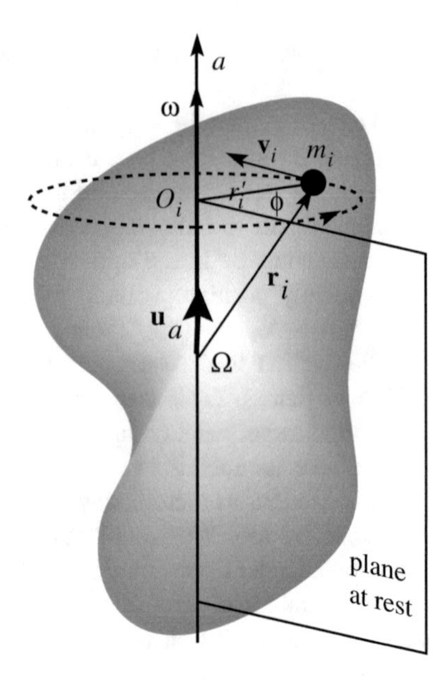

Fig. 8.7 A rigid body with a fixed axis

$$\mathbf{M}_\Omega \cdot \mathbf{u}_a = \frac{d(\mathbf{L}_\Omega \cdot \mathbf{u}_a)}{dt}. \tag{8.9}$$

In this equation, we have the projections on the a-axis of the external moment and of the angular momentum, namely

$$M_a \equiv \mathbf{M}_\Omega \cdot \mathbf{u}_a, \quad L_a \equiv \mathbf{L}_\Omega \cdot \mathbf{u}_a. \tag{8.10}$$

These quantities are called the *external moment or the torque about the axis* and the *angular momentum about the axis*. Both quantities are the components of a pseudo-vector. They can have both signs. It can be easily shown that they are independent of the choice of the pole Ω, provided it is on the rotation axis.

We can write Eq. (8.9) as

$$M_a = \frac{dL_a}{dt}, \tag{8.11}$$

which expresses the *theorem of the angular momentum about an axis*. In other words, the rate of change of the angular momentum about a fixed axis, in an inertial frame, is equal to the external moment about the same axis.

Let us find the expression of the angular momentum. The angular velocity, which we call $\boldsymbol{\omega}$, is parallel to the axis. Its magnitude and its sign relative to the axis can vary in time, but not its direction. We start by considering, for simplicity, the body consisting of particles of mass m_i, in the positions $\mathbf{r}_i$ relative to Ω, distance from the axis r'_i and velocity $\mathbf{v}_i$, as shown in Fig. 8.7. The trajectory of the generic particle is a circle normal to the axis of radius r'_i. Its velocity is tangent to this circle and has the magnitude $\upsilon_i = \omega r'_i$.

We profit by the fact that the angular momentum about the axis is independent of the pole on the axis and take it, for each particle, in the center O_i of its orbit. The angular momentum of the particle about this pole is

$$\mathbf{l}_i = \mathbf{r}'_i \times m_i \mathbf{v} = \mathbf{r}'_i \times m_i \left(\boldsymbol{\omega} \times \mathbf{r}'_i \right),$$

which, as in figure, has the direction of the axis. What we need is its component on the axis. Its sign is the same as the sign of the projection on the axis of the angular velocity, ω_a. We have $l_{ai} = r_i'^2 m_i \omega_a$. We now sum over all the particles and obtain the total angular momentum about the axis

$$L_a = \omega_a \sum_{i-1}^{N} m_i r_i'^2 = \omega_a I_a \tag{8.12}$$

where we have introduced the quantity

$$I_a = \sum_{i-1}^{N} m_i r_i'^2, \tag{8.13}$$

which is the *moment of inertia* of the body about the axis a.

We now consider the body as a continuous distribution of masses. Instead of point particles of mass m_i, we consider infinitesimal volume e dV, in the position $\mathbf{r}$ and having mass $dm = \rho(\mathbf{r})\, dV$, where ρ is the density (that can be different from point to point). Following the same arguments as for the discrete body, one finds the same result

$$L_a = \omega_a I_a, \tag{8.14}$$

but now with an integral in place of the sum, namely

$$I_a = \int r_i'^2 \rho(\mathbf{r}) dV. \tag{8.15}$$

In Sect. 8.7, we shall calculate the moments of inertia of several bodies of simple geometry. We observe here that the moment of inertia depends on the axis, not only on the body. What matters is how the masses are distributed about the axis. The equation of motion Eq. (8.11) can be written in equivalent forms.

$$M_a = \frac{dL_a}{dt} = I_a \frac{d\omega_a}{dt} = I_a \frac{d^2\phi}{dt^2} \tag{8.16}$$

and also

$$M_a = I_a \alpha \tag{8.17}$$

where

$$\alpha = \frac{d^2\phi}{dt^2} \tag{8.18}$$

is the angular acceleration.

The last expression looks very similar to the dynamical equation for a point moving along a straight line. If x is its coordinate, m the mass and F_x the component of the acting force, the equation of motion is, as we know,

$$F_x = ma_x = m\frac{d^2x}{dt^2}.$$

Equation (8.17) is the same differential equation with ϕ in place of x (and consequently, angular velocity and acceleration in place of the linear ones), the external moment about the axis in place of the force and the moment of inertia in place of the mass. Consequently, the solutions to Eq. (8.17) are the same as those for the linear motion of a point.

The simplest case is when the external moment about the axis M_a is constant. Then, the angular acceleration $\alpha = M_a/I_a$ is constant too and, analogous to the uniformly accelerated rectilinear motion, we have

$$\phi(t) = \phi_0 + \omega_0 t + \frac{1}{2}\alpha t^2 \tag{8.19}$$

where ϕ_0 and ω_0 are the angle and the angular velocity, respectively, at $t = 0$.

Example E 8.3 Figure 8.8 shows a rigid disk, say a pulley, that can rotate around a horizontal axis a passing through it center of mass. A wire, to which a mass m is attached, is wrapped around the pivot. The radius of the pivot is r. The external moment about the axis is clearly constant, $M_a = mgr$. Suppose the disk to be initially at rest and choose the origin of the angles such that $\phi_0 = 0$. The motion is then $\phi(t) = \frac{1}{2}\frac{mgr}{I_a}t^2$. Namely, the angle through which the system has turned is proportional to the square of the time.

8.5 Conservation Angular Momentum About an Axis

We still consider a rigid body with a fixed (in an inertial frame) rotation axis a. If the external moment about the axis is zero, the angular momentum about the axis is constant, namely

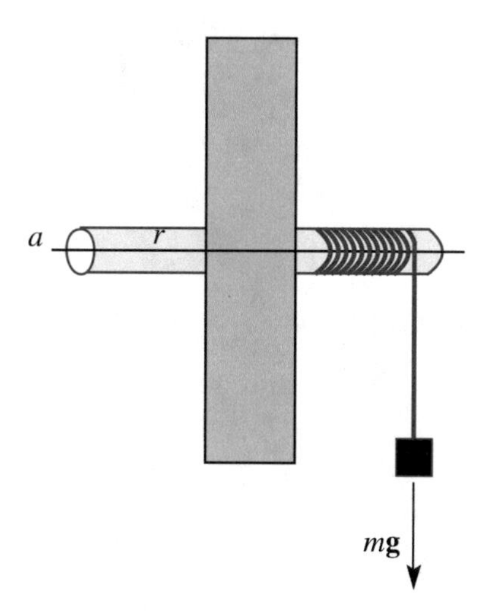

Fig. 8.8 A pulley and a weight

$$L_a = \text{constant.} \tag{8.20}$$

The external moment about the axis is zero, a part of the trivial case of absence of forces, in two important cases: (1) the directions of all the external forces are parallel to the axis, and (2) the application lines of all the external forces meet the axis. In these cases, for Eq. (8.14), as the moment of inertia is constant, the angular velocity is constant too

$$\omega_a = \text{constant.} \tag{8.21}$$

Notice that, for zero external moment about the axis, Eq. (8.20) is also valid for non-rigid bodies, while Eq. (8.21) is not. A simple experiment follows. A person sits on a turntable stool holding in his hands two heavy objects with arms horizontally outstretched. A second person pushes the first in rotation. The first brings hands and heavy objects near his chest. His angular velocity increases substantially. The initial moment of inertia of the body was, say, I_1 and was quite large because heavy masses were far from the axis, while the final one, I_2, is much smaller because the masses are close to the axis. We can say that the external moment about the axis is zero, if we neglect frictions, because the external forces, the weights, are parallel to the axis. The angular momentum is conserved, and, if ω_1 and ω_2 are the initial and final velocities, we have $I_1\omega_1 = I_2\omega_2$ and consequently, as $I_2 \ll I_1$, $\omega_2 \gg \omega_1$. This trick is used by skaters in their figures.

Example E 8.4 As an example, consider the system in Fig. 8.9, which shows an electrical motor fixed on a support that can rotate about a vertical axis, coinciding with the axis of the motor. The motor has two parts: the external one (stator) is fixed to the platform, while the internal one (rotor) is free to rotate and has a flywheel (V in the figure). The two parts are coaxial rigid bodies with moments of inertia, I_1 being the internal and I_2 the external.

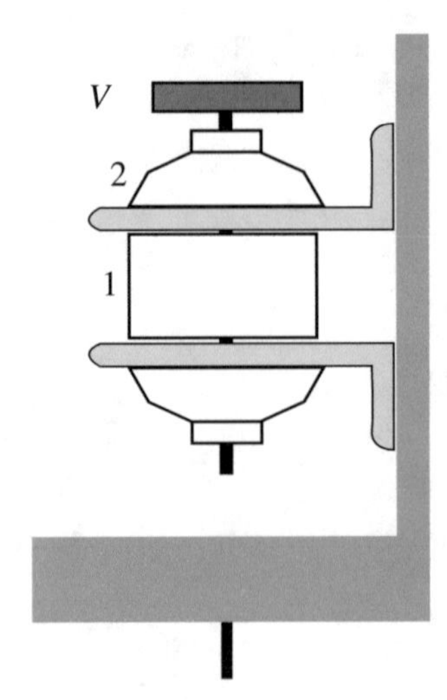

Fig. 8.9 An electrical motor

Suppose that, starting from rest, we switch on the motor for some time and then switch it off. We neglect frictions. We observe that the two parts rotate at angular velocities ω_1 and ω_2, respectively.

The initial angular momentum is zero. The final one is zero as well, because during the action of the motor, the forces are only internal. Hence, again, $L_a = I_1\omega_{a1} + I_2\omega_{a2} = 0$ or $\omega_{a2} = -\omega_{a1} I_1/I_2$. We can measure the initial and final angular velocities, repeat the experience with different flywheels, and verify if the prediction is correct.

8.6 Work and Kinetic Energy

We continue our study of the rigid body rotating about the fixed axis a, represented in Fig. 8.7. Its generic particle of mass m_i, as shown in Fig. 8.10, moves in a circle. We call O_i its center and $\mathbf{r}'_i$ the position vector of the particle from it.

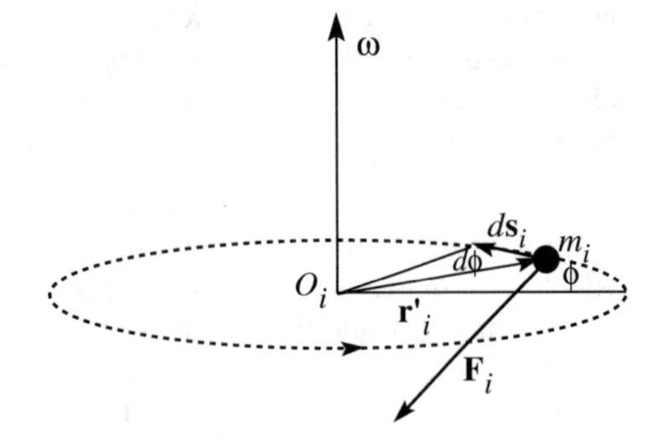

Fig. 8.10 The motion of a particle of a rigid body rotating about an axis

We now calculate the total moment about the axis of the external forces $\mathbf{F}_i$ acting on the particle. We start from the moment $\boldsymbol{\tau}_i$ about any pole on the axis. Once more, we take the center O_i of the trajectory of m_i as the pole. The force $\mathbf{F}_i$ can be thought of as the sum of three components, one parallel to the axis, one to $\mathbf{r}'_{i,}$ and one tangent to the trajectory. The contribution of the first is normal to the axis and has no axial component. The contribution of the second is zero, because it is parallel to the arm. The only contribution is the third.

We call $\mathbf{u}_t$ the unit vector tangent to the trajectory with positive direction in accordance with the direction of increasing angles (which is not necessarily the direction of motion). Let F_{ti} be the component of the external force on $\mathbf{u}_t$. The component of $\boldsymbol{\tau}_i$ on the axis is then, in magnitude and sign, $\tau_{ai} = r'_i F_{ti}$.

Consider now the infinitesimal rotation of the body along the angle $d\phi$, and calculate the corresponding total work of the forces. As we know, the body being rigid, the total work of the internal forces is zero. As for the work of the external forces, we start with the work on one particle. The displacement of the particle is $ds_i = r'_i d\phi$ and the elementary work $dW_i = F_{ti} ds_i = F_{ti} r'_i d\phi = \tau_{ai} d\phi$. To find the total work of the external forces, we have now only to add up all the particles. Taking into account that $d\phi$ is the same for all and calling $M_a = \sum \tau_{ai}$, we have

$$dW = M_a d\phi. \tag{8.22}$$

This important relation tells us that the elementary work of the external forces for an infinitesimal rotation is equal to the external moment about the axis times the rotation angle. Again, we have found an analogy with the elementary work of the force on a point $F_x dx$.

The work for a finite rotation, say from ϕ_1 to ϕ_2, is obtained by integration

$$W = \int_{\phi_1}^{\phi_2} M_a d\phi. \tag{8.23}$$

For the rotations about a fixed axis, the kinetic energy theorem has a simple expression. Recalling Eq. (8.16), we write

$$dW = M_a d\phi = I_a \frac{d\omega}{dt} d\phi = I_a d\omega \frac{d\phi}{dt} = I_a \omega d\omega.$$

For a finite rotation, the work is equal to the difference of the kinetic energies

$$W = I_a \int_1^2 \omega d\omega = \frac{1}{2} I_a \omega_2^2 - \frac{1}{2} I_a \omega_1^2. \tag{8.24}$$

We see that the kinetic energy of a rigid body rotating about a fixed axis is (once again similar to the material point)

$$U_K = \frac{1}{2} I_a \omega^2. \tag{8.25}$$

8.7 Calculating Inertia Moments

In this section, we shall calculate the moments of inertia of a few geometrically simple bodies. We shall consider all of them as homogeneous, namely having density independent on point. Consequently, their geometric centers coincide with their centers of mass.

Cylindrical bar. Figure 8.11a represents a bar of mass m and length L. We assume it to be thin, namely of transverse dimensions much smaller than the length. We assume the faces to be perpendicular to the geometrical axis. The shape of the faces is arbitrary. They can be circles, squares, anything. We calculate its moment of inertia about the axis c normal to the bar through its center C.

We take a coordinate x along the bar originating in its center. We cut the bar into infinitesimal slices between x e $x + dx$ of mass dm. As the diameter of the slice is very small, we can consider all the points of the slice at the same distance from the axis c. The mass of the slice is clearly $dm = (m/L)dx$. We notice that there are two slices at the same distance from c, on its two sides. Their contribution to the moment of inertia is $dI_c = 2x^2 dm = 2(m/L)x^2 dm$. We integrate it on half of the bar, namely from 0 to $L/2$, and obtain

$$I_c = \frac{2m}{L} \int_0^{L/2} x^2 dx = \frac{mL^2}{12}. \tag{8.26}$$

Ring. Figure 8.12 represents a thin ring of mass m and radius R. We assume the diameter of the section to be small compared to R. All the points of a section can be considered at the same distance R from the center.

We calculate the moment of inertia about the axis c normal to the plane of the ring through its center C. As all the mass sits at the same distance, we immediately have

$$I_c = mR^2. \tag{8.27}$$

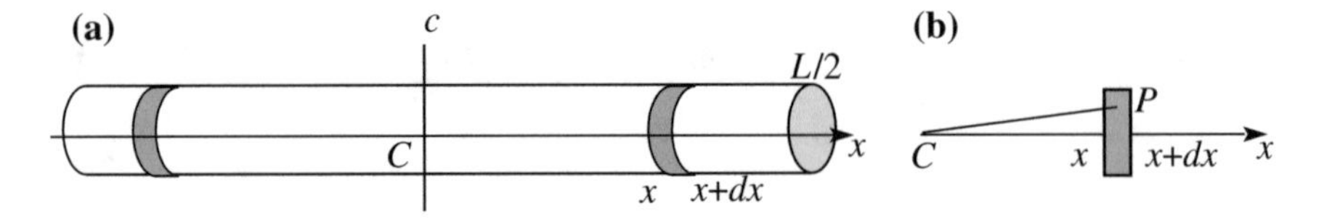

Fig. 8.11 Calculating the moment of inertia of a thin bar about a central transverse axis

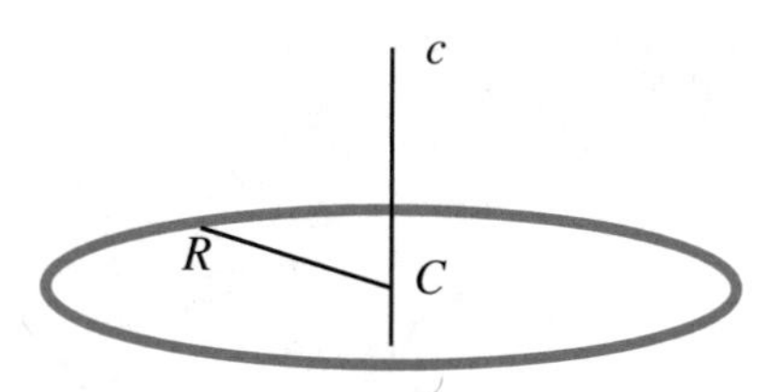

Fig. 8.12 Calculating the moment of inertia of a thin ring about the central axis

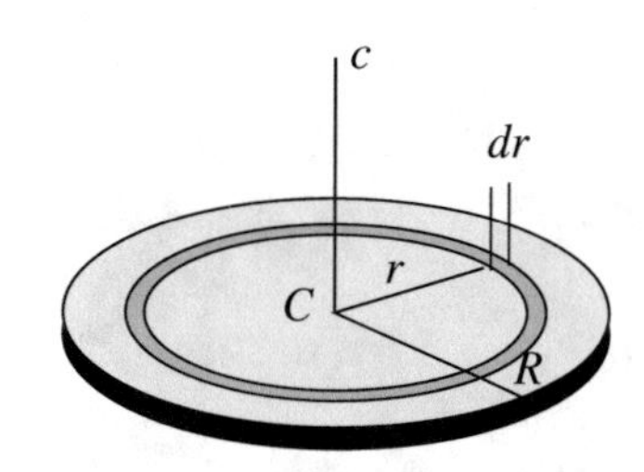

Fig. 8.13 Calculating the moment of inertia of a disk

Cylindrical surface. The moment of inertia of a cylindrical surface (namely of negligible thickness) about the geometrical axis is given by Eq. (8.27) as well, because all the masses in this case are also at the same distance R from the axis.

Homogenous disk. Figure 8.13 represents a disk of radius R and mass m. We calculate the moment of inertia about the geometric axis c shown in the figure. We divide the disk into infinitesimal rings of rays between r and $r + dr$. The area of a ring is $2\pi r\, dr$, to be compared with the area πR^2 of the entire disk. The mass of the ring is then $dm = m(2\pi r dr)/(\pi R^2) = (2m/R^2)rdr$. Its contribution to the moment of inertia is $dI_c = r^2 dm = (2m/R^2)r^3 dr$. Integrating, we obtain

$$I_c = \frac{2m}{R^2}\int_0^R r^3 dr = \frac{mR^2}{2}. \tag{8.28}$$

Homogenous cylinder. Figure 8.14 shows a homogenous cylinder. It can be thought of as a pile of disks. Hence, the moment of inertia about the symmetry axis is given by Eq. (8.28).

Homogeneous rectangular parallelepiped. Figure 8.15 represent a parallelepiped of uniform density ρ, mass m and side lengths a, b and c. We calculate the moment of inertia about the axes through the center C parallel to the sides. These we call x, y and z and take as reference axes.

As a matter of fact, it will be enough to calculate the moment of inertia about one axis, say z, and this will be analogous for all the axes. We have

$$I_z = \int_{-a/2}^{+a/2} dx \int_{-b/2}^{+b/2} dy \int_{-c/2}^{+c/2} dz\left(x^2+y^2\right)\rho = abc\rho\frac{a^2+b^2}{12} = m\frac{a^2+b^2}{12}.$$

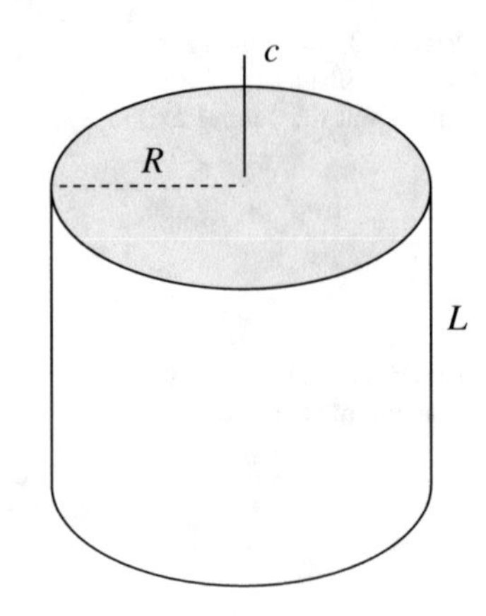

Fig. 8.14 Homogenous cylinder and its axis

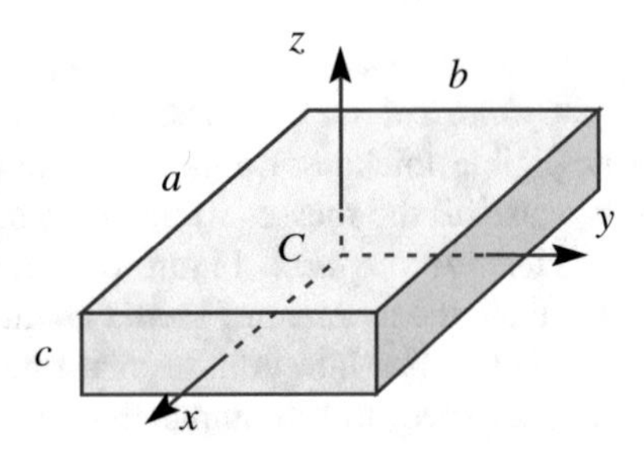

Fig. 8.15 Calculating the moment of inertia of a parallelepiped about three central axes

Analogous expressions holding for the other axes, we can conclude that

$$I_x = m\frac{b^2+c^2}{12}, \quad I_y = m\frac{a^2+c^2}{12}, \quad I_z = m\frac{a^2+b^2}{12}. \tag{8.29}$$

Homogeneous cube. The moment of inertia about one, of the three, symmetry axes is a particular case of what we have just found. If L is the length of the side, we have

$$I_a = m\frac{L^2}{6}. \tag{8.30}$$

Homogeneous sphere. We give only the result without developing the calculation. The moment of inertia about an axis through the center is

$$I_a = \frac{2}{5}mR^2. \tag{8.31}$$

8.8 Theorems on the Moments of Inertia

In this section, we shall show two theorems that will help in several cases of computing moments of inertia. The first one, the Steiner theorem, after Jacob Steiner (1796–1863), concerns rigid bodies of arbitrary shape, while the second one is for thin bodies, namely of negligible thickness.

Theorem of the parallel axes, or Steiner theorem. The theorem of the parallel axes states that the moment of inertia about an arbitrary axis is equal to sum of the moment of inertia about the parallel axis through the center of mass and the product of the mass of the system and the square of the distance between the two axes.

Figure 8.16 represents a rigid body of arbitrary shape. The c axis is through its center of mass C. The moment of inertia about c is I_c, the one relative to the parallel axis a, at distance h, is I_a. Let $\mathbf{h}$ be the vector from a to c in a plane normal to the axes. Consider an arbitrary element of the body, of mass dm and the plane normal to the axes through the element. In this plane, let $\mathbf{r}_c'$ and $\mathbf{r}_a'$ be the vectors to dm from c and a, respectively. Clearly, $\mathbf{r}_a' = \mathbf{h} + \mathbf{r}_c'$.

The contribution dI_a of the moment of inertia about a is

$$dI_a = r_a'^2 dm = \left(\mathbf{h} + \mathbf{r}_c'\right)^2 dm = h^2 dm + 2\mathbf{h} \cdot \mathbf{r}_c' dm + r_c'^2 dm.$$

Taking into account that the last term is dI_c and integrating on the body, we have

$$I_a = h^2 \int dm + 2\mathbf{h} \cdot \int \mathbf{r}_c' dm + I_c.$$

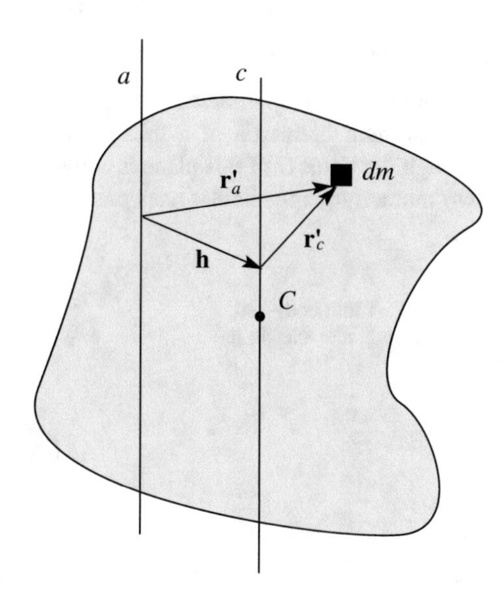

Fig. 8.16 A rigid body, an axis through the center of mass and a parallel axis

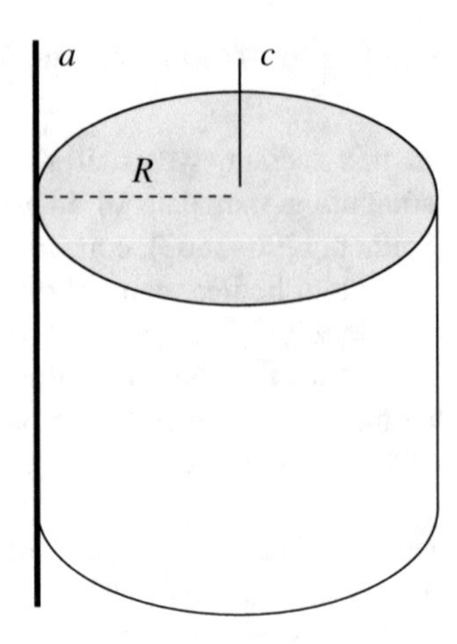

Fig. 8.17 Moment of inertia of a *cylinder* about a generator

The integral in the first term is the mass of the body, while the second term is the component on the considered plane of the position vector of the center of mass from the center of mass, and is zero. We have

$$I_a = mh^2 + I_c, \tag{8.32}$$

which is the parallel axes theorem. Notice that mh^2 is a positive definite quantity. For all the axes of a given direction, the moment inertia is minimum for the axis through the center of mass.

Example E 8.5 Consider the right cylinder in Fig. 8.17, of mass m and radius R, its central axis c and its generator a.

The moment of inertia relative to c is given by Eq. (8.28). Hence, for the parallel axes theorem, $I_a = m\frac{R^2}{2} + mR^2 = \frac{3}{2}mR^2$.

Theorem of the perpendicular axes.
The moment of inertia of a thin body about an axis perpendicular to its plane through the point O of this plane is equal to the sum of its moments of inertia about two mutually perpendicular axes passing through O.

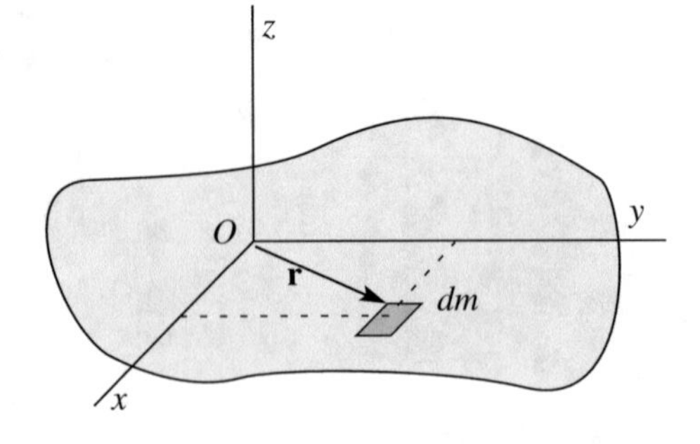

Fig. 8.18 A thin body and two perpendicular axes in its plane

Fig. 8.19 A rectangular plate and the considered axes

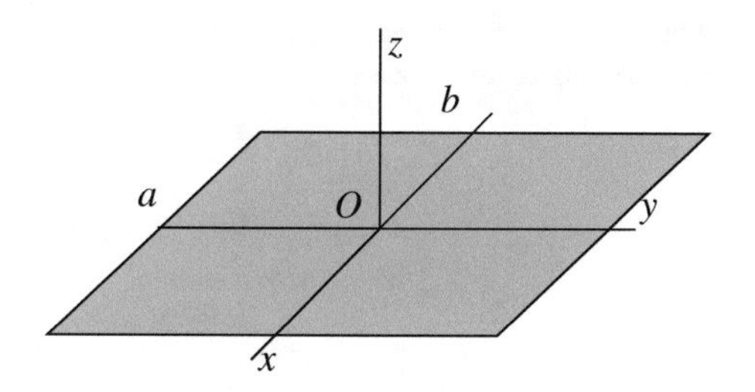

Consider the body represented in Fig. 8.18. O is an arbitrary point of the body that we take as the origin of the coordinates axes, z normal to the plane, and x and y in the plane. The moment of inertia about z is $I_z = \int r^2 dm$, where r is the distance of the element dm from z. As $r^2 = x^2 + y^2$, we have, $I_z = \int x^2 dm + \int y^2 dm$ namely

$$I_z = I_x + I_y, \tag{8.33}$$

which is the theorem of the perpendicular axes.

Example E 8.6 Calculate the moment of inertia of a rectangular plate of sides a and b about the perpendicular axis through its center, as in Fig. 8.19.

Equation (8.29), with $c = 0$, gives $I_x = mb^2/12$ and $I_y = ma^2/12I_y$ and, for the theorem of the perpendicular axes

$$I_z = m\left(a^2 + b^2\right)/12, \tag{8.34}$$

which is the third of Eq. (8.29)

Example E 8.7 Calculate the moment of inertia of a circular plate of radius R about a diameter, as in Fig. 8.20.

The moment of inertia about the central perpendicular axis I_z is given by Eq. (8.24). On the other hand, obviously, $I_x = I_y$ and, for the theorem of the perpendicular axes, $I_z = I_x + I_y = 2I_x = mR^2/2$, thus giving us

$$I_x = mR^2/4. \tag{8.35}$$

Example E 8.8 Find the moment of inertia of a circular disk about an axis tangent to its rim, as in Fig. 8.21.

Fig. 8.20 Circular plate and axes

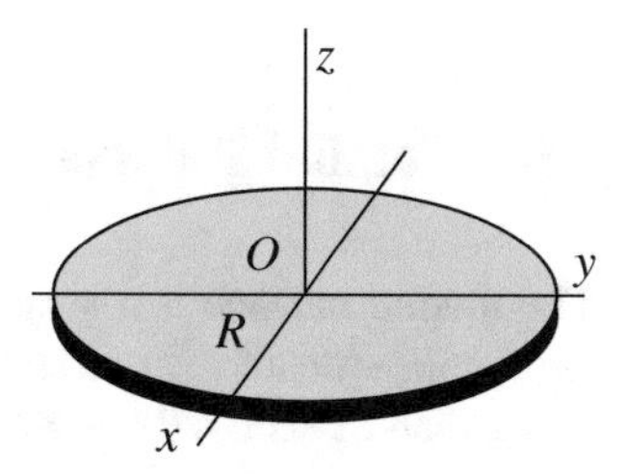

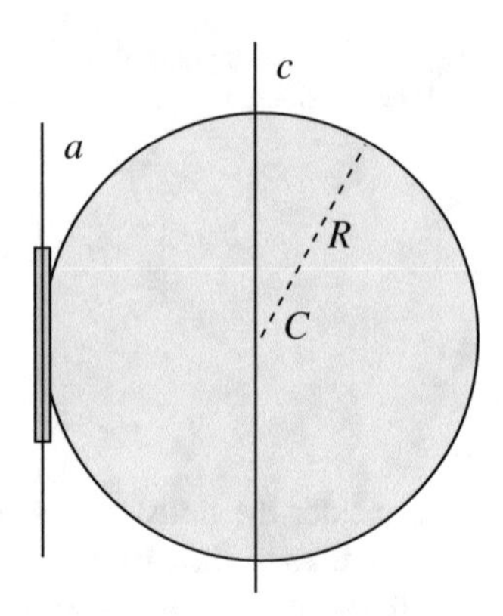

Fig. 8.21 A circular disk and an axis tangent to its rim

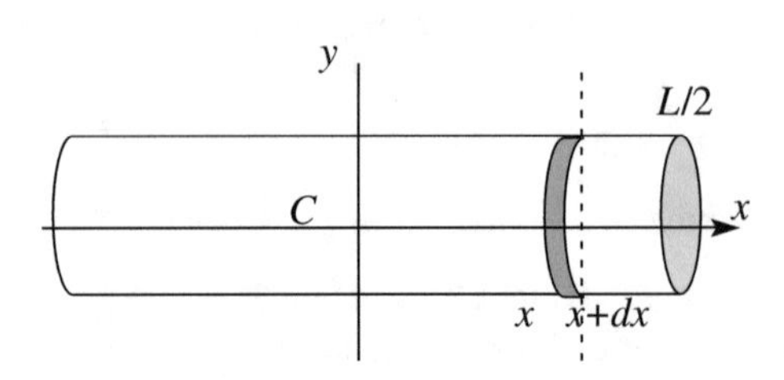

Fig. 8.22 The cylinder and its longitudinal and perpendicular central axes

We just have to apply the theorem of the parallel axes to the result we just found to have $I_a = mR^2/4 + mR^2 = 5mR^2/4$.

Moment of inertia of a cylinder about the normal axis through the center. Consider the (homogeneous) cylinder of radius R and length L represented in Fig. 8.22.

We want the moment of inertia about the axis y in the figure. This is the same situation as we discussed in the previous section, but here, we do not assume the section of the cylinder to be negligible. We call λ the linear density, namely the mass per unit length of the cylinder. Consider an infinitesimal slice between x and $x + dx$. Its mass is $dm = \lambda\, dx$. We can use Eq. (8.35) to find the moment of inertia of the slice about the axis through it parallel to y (dotted in the figure). For the theorem of parallel axes, we have dI_y by adding to it x^2dm, namely $dI_y = R^2dm/4 + x^2dm$. Integrating along the entire length, namely in x from $-L/2$ to $L/2$, we have

$$I_y = mL^2/12 + mR^2/4. \tag{8.36}$$

8.9 Torsion Balance

The torsion balance is a very sensitive instrument used to measure small moments and, consequently, small forces. We have already seen how it was used by Cavendish in Sect. 4.7 and by Eötvös in Sect. 5.8. We shall discuss it in more detail

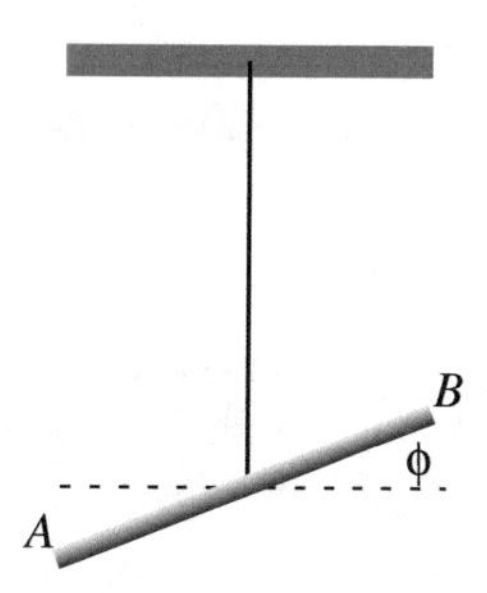

Fig. 8.23 Schematics of the torsion balance

now. Figure 8.23 shows the scheme of the device. A rigid bar *AB* is suspended from a vertical wire through its center. The equilibrium position of the bar is determined by the configuration of the wire at rest.

When we apply a moment τ, the bar rotates about its center. The rotation gives origin to an elastic moment τ_e in the wire in the opposite direction, proportional to the rotation angle ϕ

$$\tau_e = -k\phi \tag{8.37}$$

where the minus sign indicates that the elastic moment tends to bring the bar back into its original position. The elastic constant k depends on the length and the section of the wire and on its material. We can choose this constant when we design the balance, depending on the torques we have to measure. For example, thin quartz wires can be used for sensitivities down to several femtonewton.

The new equilibrium is reached when the rotation angle is such that the elastic moment is equal to the applied one, $\tau = \tau_e$. Hence, we can measure τ by measuring ϕ, and knowing k.

The most accurate measurement of k is done using a dynamical method. We rotate the bar at an angle ϕ_0 and let it go. It is the motion of a rigid body about a fixed axis under the action of the external torque τ_e. If I is the moment of inertia, the equation of motion, Eq. (8.16), is

$$\tau_e = -k\phi = I\frac{d^2\phi}{dt^2} \tag{8.38}$$

or

$$\frac{d^2\phi}{dt^2} + \omega_0^2\phi = 0 \tag{8.39}$$

with

$$\omega_0^2 = k/I. \tag{8.40}$$

We recognize the differential equation of the oscillator. Its solution is an harmonic motion in the angular coordinate ϕ with period

$$T = 2\pi/\omega_0. \tag{8.41}$$

The period can be measured with high accuracy, because we can measure it over many oscillations and count them. Once we know the period and the moment of inertia by construction, we know the elastic constant.

8.10 Composite Pendulum

The composite pendulum is a rigid body, of mass m, which can rotate around a fixed horizontal axis, an axis which should not be through the center of mass. In Fig. 8.24, O is the trace of the axis, C is the center of mass and ϕ is the angle to the vertical, taken to be positive counter-clockwise. The distance of the center of mass from the axis is h.

We take the pole for the moments to be the fixed point O. Two forces act on the pendulum, the weight, which we can think of as being applied to the center of mass, and the constraint reaction, applied to the axis of rotation. This is a cylinder of radius r, as shown in the insert of the figure. The constraint reaction is applied to the point P of its lateral surface. In the presence of friction, the force has a direction different from the direction of the segment OP and its moment about O is different from zero. If, however, the friction is negligible, as we shall assume, the direction of the force is OP and its moment is zero. The external moment on the system is,

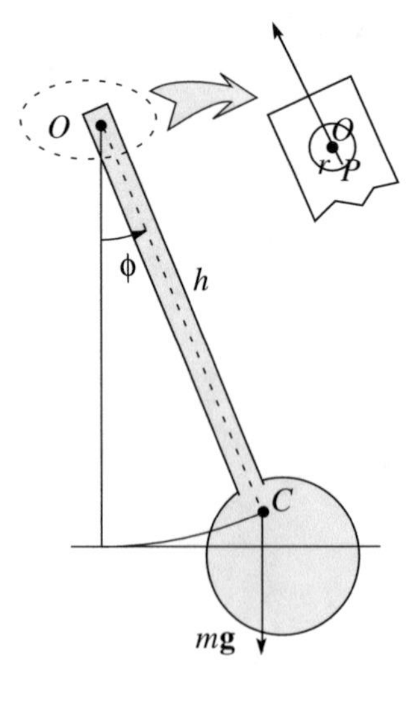

Fig. 8.24 The composite pendulum

under these conditions, the moment of the weight, which, at the angle ϕ, is $-mgh \sin \phi$. The equation of motion is

$$-mgh \sin \phi = I \frac{d^2\phi}{dt^2} \tag{8.42}$$

where I is the moment of inertia about the axis. For small angles, we can approximate the sine with the angle, obtaining

$$\frac{d^2\phi}{dt^2} + \omega_0^2 \phi = 0 \tag{8.43}$$

with

$$\omega_0^2 = mgh/I. \tag{8.44}$$

Equation (8.43) is equal to that of the simple pendulum. Hence, the motion of the composite pendulum is a harmonic motion in ϕ. Its period is

$$T = \frac{2\pi}{\omega_0} = 2\pi\sqrt{\frac{I}{mgh}}. \tag{8.45}$$

The device is used, in particular, to measure g, knowing from construction the other quantities in Eq. (8.45).

The period of the composite pendulum is equal to the period of the simple one of length

$$l = \frac{I}{mh}, \tag{8.46}$$

which is then called the *reduced length* of the composite pendulum

8.11 Dumbbell

We have discussed several examples of rotations of rigid bodies around a *fixed* axis. However, the axis will move if we do not provide the proper supports to keep it fixed. In general, the axis is supported by a massive body at rest, on which the axis rotates through a number of ball bearings to reduce the frictions as much as possible. The relevant kinematic quantities are the angular velocity and the angular momentum. Both are vector quantities. The former is by definition parallel to the axis, the latter not necessarily so. Up to now, we have used only the component on the axis of the angular momentum. In general, there are also components perpendicular to the axis, which, in addition, vary in time. Consequently, an external

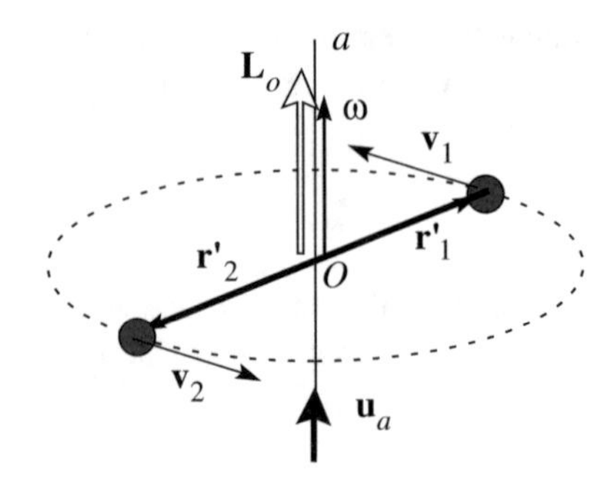

Fig. 8.25 A dumbbell rotating about a symmetry central axis

moment must be present. This is the action of the supports. We shall now turn our attention to this action.

We shall start from the particularly simple case of the dumbbell in Fig. 8.25. It is made of two equal spheres of mass m at the extreme ends of a rigid bar of length $2d$ of negligible mass.

The axis, a in the figure, is vertical through the center of the system O perpendicular to the bar, namely a symmetry axis of the body. The frictions in the rotation are negligible. The system is very similar to those we have already discussed.

We indicate with $\mathbf{r}_1'$ and $\mathbf{r}_2'$ the position vectors of the two masses from O and with $\mathbf{u}_a$ the unit vector of the axis. The angular velocity has the direction of the axis $\boldsymbol{\omega} = \omega\, \mathbf{u}_a$.

The angular momentum about the fixed point O, $\mathbf{L}_O = \mathbf{r}_1' \times m\mathbf{v}_1 + \mathbf{r}_2' \times m\mathbf{v}_2$. Considering that $|\mathbf{r}_1'| = |\mathbf{r}_2'| = d$, we can write

$$\mathbf{L}_O = 2dm\upsilon\mathbf{u}_a = 2md^2\omega\mathbf{u}_a = I_a\boldsymbol{\omega} \tag{8.47}$$

where I_a is the moment of inertia about a. In this case, the angular momentum is parallel to the rotation axis. The external moment is zero. Indeed, the moments of the weights of the two masses are equal and opposite and we are neglecting the frictions. Under these conditions, angular momentum and angular velocity are constant in time. If initially the system rotates at a certain angular velocity, it will continue to do so forever. The ball bearings that keep the axis must support the total weight, but do not exert any moment.

We now suppose the fixed rotation axis to be still through the center, but not perpendicular to the bar, at the angle, say $\pi/2 - \theta$, with it, as in Fig. 8.26. The angular velocity still has the direction of the axis, $\boldsymbol{\omega} = \omega\, \mathbf{u}_a$. If $\mathbf{r}_1$ and $\mathbf{r}_2$ are the position vectors of the two masses, the angular momentum about O is

$$\mathbf{L}_O = \mathbf{r}_1 \times m\mathbf{v}_1 + \mathbf{r}_2 \times m\mathbf{v}_2. \tag{8.48}$$

Looking at the figure, we see that the two terms are equal both in magnitude and in direction. The latter is the direction perpendicular to the bar in the plane of the

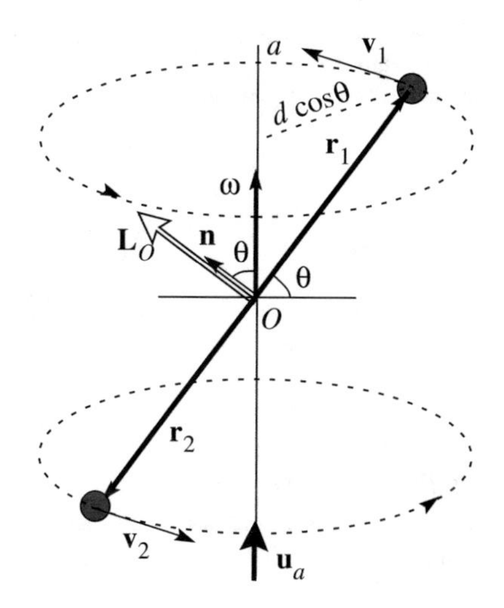

Fig. 8.26 A dumbbell rotating about a central, non-symmetry, axis

bar and the axis. We call **n** its unit vector. The plane rotates with angular velocity **ω**. The magnitude of the angular momentum is

$$L_O = 2dmv = 2md^2\omega\cos\theta. \tag{8.49}$$

The moment of inertia relative to the axis is now $I_a = 2m(d\cos\theta)^2$. Multiplying Eq. (8.49) by cos θ, and indicating with L_a the angular momentum about the axis, we have

$$L_a = L_O\cos\theta = I_a\omega. \tag{8.50}$$

We already knew this result. The axial angular momentum is equal to the moment of inertia times the angular velocity. However, the axial angular momentum is only one of the components of the angular momentum vector. Equation (8.49) gives its magnitude, while its direction is **n**. We have

$$\mathbf{L}_O = 2md^2\omega\cos\theta\mathbf{n}. \tag{8.51}$$

Even if the angular velocity is constant, the angular momentum vector is not. It rotates at constant velocity on the cone of semi-vertex angle θ around the fixed axis. Consequently, $d\mathbf{L}_O/dt \neq 0$ and the external moment is not zero. It is due to the supporting ball bearings.

Let us look more closely at the situation. We decompose the momentum of Eq. (7.59) in its components parallel and perpendicular, transverse, to the axis

$$\mathbf{M}_P^e = \frac{d\mathbf{L}_P}{dt}, \quad \mathbf{M}_T^e = \frac{d\mathbf{L}_T}{dt},$$

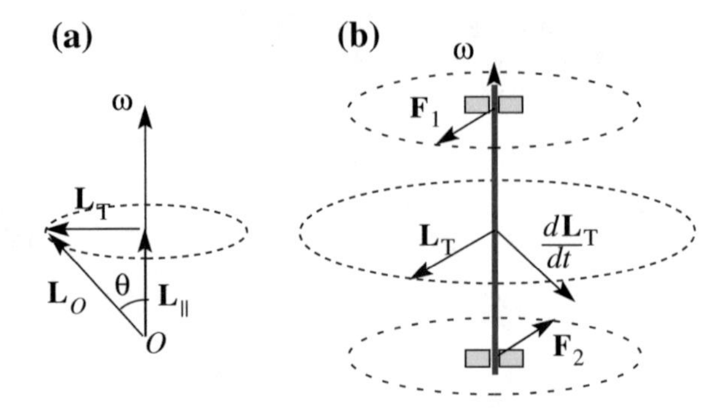

Fig. 8.27 **a** The angular momentum and its components, **b** the external torque

as shown in Fig. 8.27a for the angular momentum. Its component parallel to the axis $\mathbf{L}_P$ is constant, and consequently, $\mathbf{M}_P = 0$. $\mathbf{L}_T$ is constant in magnitude and rotates around the axis at a constant angular velocity. Its derivative is

$$\frac{d\mathbf{L}_T}{dt} = \boldsymbol{\omega} \times \mathbf{L}_T.$$

As seen in Fig. 8.27b, this derivative is also a vector rotating at constant angular velocity $\boldsymbol{\omega}$ in a plane perpendicular to the axis. It is at 90° with $\mathbf{L}_T$. This derivative is just the external moment, which is exerted by the ball bearings. These act with two forces, $\mathbf{F}_1$ and $\mathbf{F}_2$ in the figure, of constant magnitude and rotating direction.

The situation is similar, for example, when the rotation axis of the reel of a car is not exactly the symmetry axis. The periodic stress on the ball bearings would induce vibrations in the vehicle.

8.12 Angular Momentum About a Fixed Pole

Let us summarize what we have established up to this point on the motions of rigid bodies. The simplest is the rotation about a fixed axis. In this case, the configuration of the body is defined by a single (angular) coordinate. Its rate of change is the angular velocity. The axial angular momentum is the component on the axis of the angular momentum about any point of the axis (and is independent of its choice). The axial angular momentum is equal to the angular velocity times the moment of inertia about the axis. The rate of change of the axial angular momentum is equal to the component of the moment of the external forces on the axis. This is the differential equation ruling the dynamics of the system. In the last section, we saw the consequences of the angular momentum components perpendicular to the axis on the constraints that guarantee the stability of the axis. We shall now further study the relation between angular velocity and angular momentum and the motion of a rigid body about a *fixed point*, which is at rest in an inertial reference frame. We call it the *pole*, *O*.

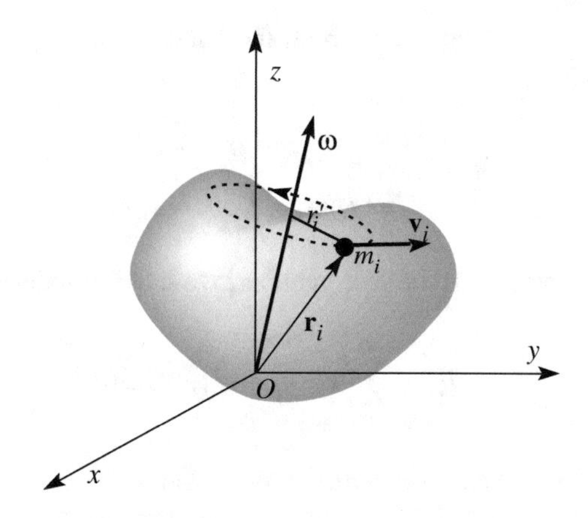

Fig. 8.28 Motion of a rigid body about a fixed point O

The rigid motion about a fixed pole is still a rotation with an angular velocity $\boldsymbol{\omega}$, which now may vary both in magnitude and in direction. In other words, in every instant, the body rotates about an "instantaneous rotation axis" that passes through O and has the direction of $\boldsymbol{\omega}$, which continuously changes. We choose the inertial reference frame as shown in Fig. 8.28, with origin O. We shall also take O as the pole.

We consider a system of material points. A continuous system can be treated through the same arguments with integrations in place of sums. Let m_i be the mass of the generic point and $\mathbf{r}_i$ its position vector. Its velocity is

$$\mathbf{v}_i = \boldsymbol{\omega} \times \mathbf{r}_i, \tag{8.52}$$

which is obviously the same for all the points. The angular momentum of the point about O is

$$\mathbf{l}_{Oi} = \mathbf{r}_i \times m_i \mathbf{v}_i = m_i \mathbf{r}_i \times (\boldsymbol{\omega} \times \mathbf{r}_i). \tag{8.53}$$

We now use Eq. (1.29) to express the double cross product in the last member, obtaining

$$\mathbf{l}_{Oi} = m_i r_i^2 \boldsymbol{\omega} - m_i \mathbf{r}_i (\boldsymbol{\omega} \cdot \mathbf{r}_i). \tag{8.54}$$

The sum of these quantities is the total angular momentum we want to find. In doing that, we would find a set of quantities analogous to the moment of inertia about an axis. These are the nine elements of a 3 × 3 matrix. We shall work on the Cartesian components. We start with the x component of the just found equation. After simplification, we have

$$l_{Oi,x} = m_i \left(y_i^2 + z_i^2\right) \omega_x - m_i x_i y_i \omega_y - m_i x_i z_i \omega_z.$$

We now add up all the points, obtaining

$$\begin{aligned} L_{O,x} &= \left[\sum_{i=1}^{N} m_i\left(y_i^2+z_i^2\right)\right]\omega_x - \left[\sum_{i=1}^{N} m_i x_i y_i\right]\omega_y - \left[\sum_{i=1}^{N} m_i x_i z_i\right]\omega_z \\ &= I_{xx}\omega_x - I_{xy}\omega_y - I_{xz}\omega_z \end{aligned}$$

where, in the last member, we have introduced the quantities

$$I_{xx} = \sum_{i=1}^{N} m_i\left(y_i^2+z_i^2\right), \quad I_{xy} = \sum_{i=1}^{N} m_i x_i y_i, \quad I_{xz} = \sum_{i=1}^{N} m_i x_i z_i. \tag{8.55}$$

The first quantity is immediately recognized as the moment of inertia about the x-axis, while the second and third ones are the *products of inertia*. The same argument for the other two components of the angular momentum lead to analogous expressions. The final result can be expressed in a compact form with the matrix formalism as

$$\begin{pmatrix} L_{Ox} \\ L_{Oy} \\ L_{Oz} \end{pmatrix} = \begin{pmatrix} I_{xx} & -I_{xy} & -I_{xz} \\ -I_{yx} & I_{yy} & -I_{yz} \\ -I_{zx} & -I_{zy} & I_{zz} \end{pmatrix} \begin{pmatrix} \omega_x \\ \omega_y \\ \omega_z \end{pmatrix}. \tag{8.56}$$

The 3×3 matrix is mathematically a tensor and is called the *tensor of inertia*. Its elements in the first line are given by Eq. (8.55), and those of the other two by analogous expressions. We shall not need to know its mathematical properties. We only notice that the matrix is symmetric, namely the elements in symmetric positions about the diagonal are equal, $I_{xy} = I_{yx}$, etc.

The situation looks quite complicated, but we can make it simpler with an appropriate choice of the directions of the coordinate axes. This is because the matrix of inertia is square and symmetric. Indeed, mathematics shows that this type of matrix can always be put in diagonal form by a rotation of the axes. We still refer to x, y, z as such axes, pose, for simplicity, $I_x = I_{xx}$, $I_y = I_{yy}$, $I_z = I_{zz}$, and write

$$\begin{pmatrix} L_{Ox} \\ L_{Oy} \\ L_{Oz} \end{pmatrix} = \begin{pmatrix} I_x & 0 & 0 \\ 0 & I_y & 0 \\ 0 & 0 & I_z \end{pmatrix} \begin{pmatrix} \omega_x \\ \omega_y \\ \omega_z \end{pmatrix}. \tag{8.57}$$

Another form of Eq. (8.57) that we shall use is

$$\mathbf{L}_O = I_x\omega_x\mathbf{i} + I_y\omega_y\mathbf{j} + I_z\omega_z\mathbf{k}. \tag{8.58}$$

We have found that the angular momentum of the rigid body about the pole O is the sum of three vectors. Each of them is directed as one of the axes, with the magnitude equal to the product of the component of the angular velocity on that axis and the moment of inertia about that axis. This is true only for the particular choice of axes that makes the matrix of inertia diagonal. These are called the *principal axes of inertia relative to O*. Their position is fixed relative to the body and they move with it. Consequently, the reference $Oxyz$ is NOT generally an inertial one. If the pole O is the center of mass, the principal axes are called *central axes of inertia*.

We shall now state without demonstration a few important properties of the principal axes of inertia.

Firstly, as intuition suggests, if the body has symmetry axes relative to O, these are also the principal axes.

For example, the principal axes of a homogeneous rectangular parallelepiped relative to its center O are the axes parallel to its sides through O. If two sides of the parallelepiped are equal, two of its moments of inertia are equal, say $I_x = I_y$. Consider now an axis through O in an arbitrary direction in the plane xy (namely defined by the two equal moments of inertia). It can be shown that the moment of inertia about it is $I = I_x = I_y$, even if that axis is not a symmetry axis. We notice that the symmetry of the moments of inertia is larger than the symmetry of the distribution of the masses.

Consider as a second example a right homogenous cylinder. Its geometric axis is both a symmetry axis and a central axis of inertia. Any axis in the plane perpendicular to it through the center is a central axis too. Hence, again, there are infinite central axes.

There are also cases in which all three moments about the central axes are equal. Consider, for example, the symmetry axes of a homogenous cube parallel to its sides. These are clearly central axes of inertia, with equal moments. However, any other axis through the center is also a central axis of inertia with the same moment. Again, we see that the symmetry of the moments of inertia is larger than the symmetry of the distribution of the masses. The former is for a cube, a spherical symmetry. Obviously, all the axes through the center of a homogeneous sphere are central axes of inertia.

Consider now again a homogeneous cylinder, with height h and base radius R. We put the origin of the reference in its center and the z on its axis. The other two (central) axes are on the normal section. We already know the expression of the moments of inertia, Eqs. (8.28) and (8.36), which give

$$I_x = I_y = mh^2/12 + mR^2/4, \quad I_z = mR^2/2.$$

We notice that all of them are equal if $h = \sqrt{3}R$. In these particular cases, all the axes through the center are central axes of inertia. All the moments of inertia about them are equal. Again, *the symmetry of the moments of inertia is larger than the symmetry of the masses*. In other words, if there are symmetry axes, these are principal axes of inertia, but a principal axis of inertia may not be a symmetry axis. Indeed, any rigid body of whatever shape, with no symmetry at all, like an irregular stone, has three principal axes of inertia about any point at rest with it, even outside the body.

We state without proof that the principal axes of inertia about a point O and those about another point O' are not parallel, in general.

We shall now discuss a few important aspects of Eq. (8.58). First, it tells us that angular velocity and angular momentum are not, in general, parallel vectors. However, they are so if the rotation is around a principal axis, namely $\boldsymbol{\omega}$ is parallel to a principal axis. Consequently, the principal axes are also called *permanent rotation axes* or *spontaneous rotation axes*. Consider a rotation about a fixed point in an inertial frame. Its generic motion is a rotation about an instantaneous axis through the fixed point, whose direction varies continuously in time. As a consequence, the angular momentum about the point varies too. This implies the existence of a non-zero external moment.

Consider now a rigid body with a fixed point which is otherwise free. The external moment is zero. Consequently, its angular momentum about the fixed point is constant. If, at a certain instant, the body rotates about a principal axis with angular velocity $\boldsymbol{\omega}$, it is simply $\mathbf{L} = I\boldsymbol{\omega}$. $\mathbf{L}$ being constant, $\boldsymbol{\omega}$ is constant too, in magnitude and direction. If, on the contrary, the body rotates around a non-principal axis, $\mathbf{L}$ is constant, but $\boldsymbol{\omega}$ is not necessarily so.

The same arguments are valid for the motion of a rigid body without any constraint, provided the center of mass is chosen as the pole, for Eq. (7.60).

8.13 Kinetic Energy

In this section, we shall discuss the kinetic energy of a rigid body moving about a fixed point O, which is not necessarily in an inertial frame. Figure 8.29 shows the situation at a certain instant. The vector $\boldsymbol{\omega}$ is the instantaneous angular velocity, which, in general, varies both in magnitude and direction.

As usual, we think of the body as being made of material points of mass m_i. The kinetic energy of the generic point is $U_{K,i} = (1/2)m_i v_i^2 = (1/2)m_i r_i'^2 \omega^2$, where v_i is the magnitude of the velocity of the point and r'_I is its distance from the

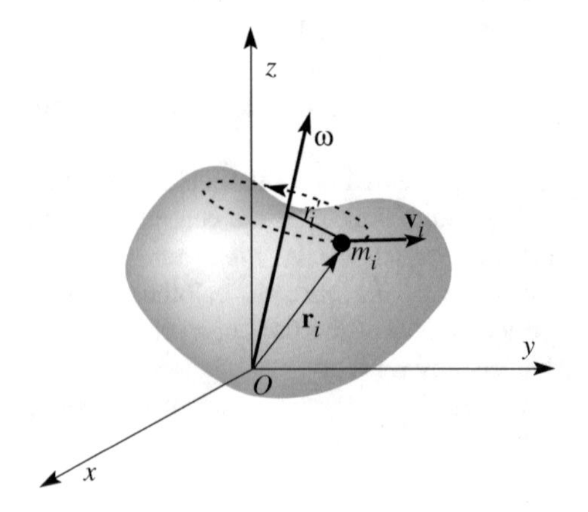

Fig. 8.29 A rigid body moving about a fixed point

instantaneous rotation axis. We obtain the kinetic energy of the body adding up all the points. If I_ω is the moment of inertia about the instantaneous rotation axis, we have

$$U_K = \frac{1}{2} I_\omega \omega^2. \tag{8.59}$$

We had already found this expression, Eq. (8.25), in the case of rotation about a fixed axis.

If the reference is an inertial one and if the body is not subject to external forces, the kinetic energy is constant in time, but the direction of the angular velocity relative to the body does, in general, vary. Also in general, both ω and I_ω vary, while the product of the square of the former and the latter are constant. In practice, Eq. (8.59) is not very useful. Let us find a more useful expression proceeding in a way similar to what we did in Sect. 8.12 for the angular momentum. We work in the reference frame of Fig. 8.29, with origin in the fixed point O. The velocity of the generic point P_i at the position vector $\mathbf{r}_i$ is

$$\mathbf{v}_i = \boldsymbol{\omega} \times \mathbf{r}_i. \tag{8.60}$$

The kinetic energy of the point is

$$\begin{aligned} U_{K,i} &= \frac{1}{2} m_i v_i^2 = \frac{1}{2} m_i (\boldsymbol{\omega} \times \mathbf{r}_i)^2 = \frac{1}{2} m_i \left[\left(\omega_y z_i - \omega_z y_i \right)^2 + \left(\omega_z x_i - \omega_x z_i \right)^2 + \left(\omega_x y_i - \omega_y x_i \right)^2 \right] \\ &= \frac{1}{2} m_i \left(\omega_y^2 z_i^2 + \omega_z^2 y_i^2 - 2\omega_y \omega_z y_i z_i + \omega_z^2 x_i^2 + \omega_x^2 z_i^2 - 2\omega_z \omega_x z_i x_i + \omega_x^2 y_i^2 + \omega_y^2 x_i^2 - 2\omega_x \omega_y x_i y_i \right) \end{aligned}$$

We should now add up all the points. In the above expression, we have, for example, the term $m_i \omega_z^2 \left(x_i^2 + y_i^2 \right)/2$. Adding up the points, this gives $I_z \omega_z^2/2$, and is analogous for the other axes. The sums of the terms with the products of two coordinates give terms propositional to the products of inertia. It is then convenient to choose the coordinates on the principal axes relative to O, because the products of inertia are zero. With this choice, we have

$$U_K = \frac{1}{2} \left(I_x \omega_x^2 + I_y \omega_y^2 + I_z \omega_z^2 \right). \tag{8.61}$$

which we can write, recalling Eq. (8.58), as

$$U_K = \frac{1}{2} \mathbf{L}_O \cdot \boldsymbol{\omega}. \tag{8.62}$$

In this expression, the components on the axes no longer appear. Consequently, it is valid independent of the reference frame. We also notice that, in absence of external forces, both kinetic energy and angular momentum are conserved. Consequently, the component of the angular velocity on $\mathbf{L}_O$ is constant too.

8.14 Rotation About a Fixed Axis. Forces on the Supports

We often deal, in practice, with symmetric rigid bodies that rotate about a fixed axis at high angular velocities. This is the case with the rotating parts of electric and internal combustion engines, with the reels of cars and bikes, turbines, helices, etc. There are two vector quantities in the game: the angular velocity, which is, by construction, parallel to the axis, and the angular momentum, which can have a direction. We have seen an example of this situation in Sect. 8.11. In Sect. 8.12, we have seen that angular velocity and angular momentum are parallel only if the rotation axis is a principal axis of inertia. If this is not the case, an external moment must be applied to maintain the rotation axis as fixed. This is done through the mechanical structure that supports the axis, in general, through a ball bearing to reduce frictions.

To be concrete, consider the example in Fig. 8.30. The rotation axis is through the center of mass C of the body, but is not the symmetry axis. The axis is kept in position by two ball bearings, represented in the figure. The central axes of inertia are the symmetry axis of the disk, that we shall take as coordinate z, and any two mutually perpendicular directions in the plane through C normal to z, which we take as coordinates x and y. The figure is a shot of the movement when the x axis goes through the plane of the figure.

The total force exerted by the supports is just equal to the weight of the body, both if it rotates and if it is at rest. It will not enter into our arguments.

We shall take as the pole of the moments of the forces and of the angular momentum the center of mass C, which is also a fixed point in this case. The symmetry axis of the body forms an angle α with the rotation axis. Consequently, angular momentum and angular velocity are not parallel. We shall soon find the direction of the former.

We observe that the angular momentum can be usefully decomposed in one component parallel and one perpendicular to the axis. The direction of the latter rotates around the axis with angular velocity ω.

The component of the angular momentum on the axis is, with obvious meaning of the symbols,

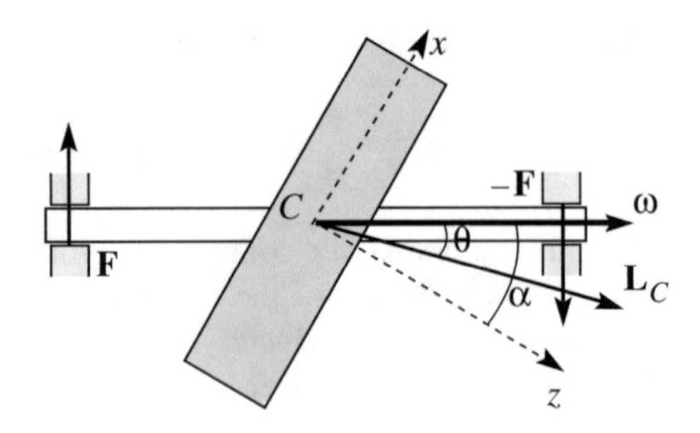

Fig. 8.30 Rotation of a rigid body about a central non-principal axis

$$L_\omega = I_\omega \omega. \tag{8.63}$$

To vary the magnitude of the angular velocity, we must apply a moment parallel to the axis. This is what engines do, when they accelerate or decelerate. As a matter of fact, the ball bearings are used to decrease the friction, which, however, cannot be completely eliminated. The friction moment opposes the motion. If we abandon the body in rotation, we observe its angular velocity gradually decreasing due to the moment of the frictions.

We now study the rotation of the components normal to the axis of the angular momentum and of the moment exerted by the support. We assume the frictions to be negligible and the moment of the forces to be perpendicular to the axis. Consequently, both the magnitude of the angular velocity and the axial component of the angular momentum are constant.

Equation (8.58) becomes, in the case under consideration,

$$\mathbf{L}_C = I_x \omega_x \mathbf{i} + I_z \omega_z \mathbf{k}. \tag{8.64}$$

If θ is the angle between the angular momentum and the rotation axis, as seen in Fig. 8.30, we have

$$\tan(\alpha - \theta) = \frac{L_{C,x}}{L_{C,z}} = \frac{I_x \omega \sin\alpha}{I_z \omega \cos\alpha} = \frac{I_x}{I_z} \tan\alpha. \tag{8.65}$$

Both the ratio I_x/I_z and the relation between α and θ depend on the shape of the body. If the body is a disk, as we saw in Sect. 8.8, $I_x/I_z = 1/2$, and Eq. (8.65) gives $\tan(\alpha - \theta) = \tan(\alpha)/2$. If, as is often the case, the angles are small and we can approximate the tangent with its argument, it is $\theta \approx \alpha/2$. Hence, the angle between angular momentum and rotation axis is constant in time. In addition, as we have already observed, the component of the angular momentum on the axis is also constant and, as a consequence, the magnitude of the angular momentum is constant. In conclusion, the normal component of the angular momentum is constant in magnitude and rotates around the axis with angular velocity $\boldsymbol{\omega}$. The dynamical equation is

$$\mathbf{M}_C = \frac{d\mathbf{L}_C}{dt} = \boldsymbol{\omega} \times \mathbf{L}_C \tag{8.66}$$

where $\mathbf{M}_C$ is the external moment exerted by the ball bearings. The couple of forces is shown in the figure. In the considered instant, the plane of the couple is the plane of the figure. The magnitude of the moment is $M_C = \omega L_C \sin\theta$. And also, writing Eq. (8.63) as $L_C \cos\theta = I_\omega \omega$,

$$M_C = I_\omega \omega^2 \tan\theta. \tag{8.67}$$

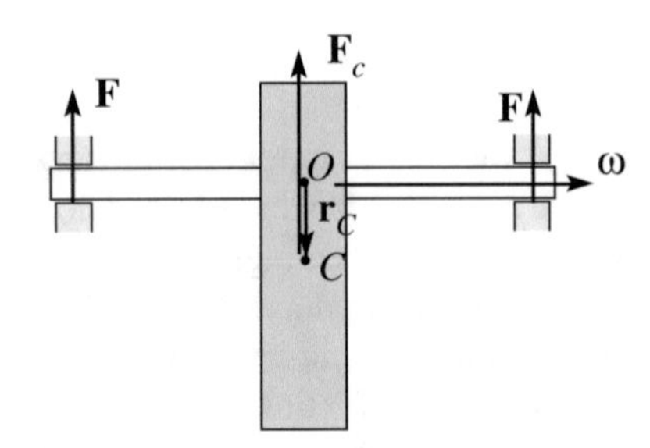

Fig. 8.31 Rotation about a principal non-central axis

In conclusion, the stress on the support is periodic, with period $2\pi/\omega$, and proportional to the square of the angular velocity. If the latter increases, for example, by a factor of ten, the moment increases by one hundred.

We now consider a rotation at constant angular velocity around a fixed axis, which is principal of inertia, but not through the center of mass, as in Fig. 8.31. In this case, the angular momentum is parallel to the axis and, consequently, is constant in time. The moment exerted by the ball bearings is zero. The force they exert, however, must be equal to the centripetal force that is necessary to maintain the center of mass in its circular motion, namely

$$\mathbf{F}_C = m\mathbf{a}_C = -m\frac{\omega^2}{r_C}\mathbf{u}_C \tag{8.68}$$

where $\mathbf{r}_C$ is the position vector of the center of mass relative to the point O on the axis (see figure) and $\mathbf{u}_C$ is its unit vector. The force is exerted by the ball bearings. Its direction rotates at angular velocity ω, its magnitude is constant, proportional to the square of the angular velocity.

In conclusion, the ball bearings during the rotation must develop forces that periodically vary in direction, having resultant $\mathbf{F}_C$ and total moment $\mathbf{M}_C$. The former is zero if the center of mass is on the axis; the latter is zero if the rotation axis is a principal axis of inertia. Both are zero if the axis is central of inertia. Clearly, this is the configuration engineers try to realize, especially if the velocities are high. Under such conditions the system is said to be *dynamically balanced*. Dynamic balance is obtained, for example, for car wheels, by inserting small lead counterweights where necessary along the tire rim.

8.15 Rolling Motion

The wheels of a bike or of a car moving down the street normally roll without slipping. If the friction between wheel and street is lower due to rain or snow, slipping can set in, a situation that should obviously be avoided. The wheel can be considered a disk. The hub is a central axis of inertia.

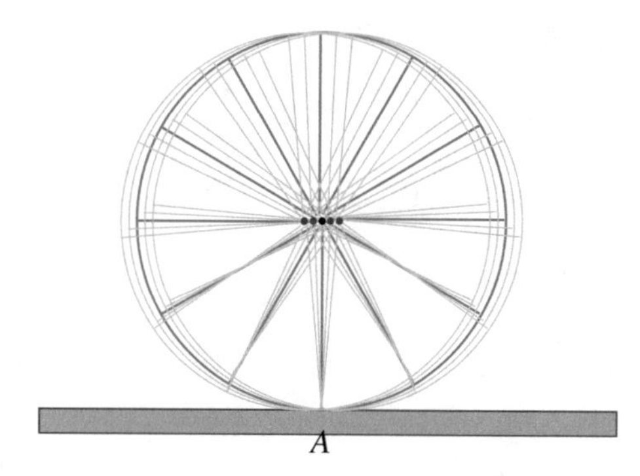

Fig. 8.32 A bike wheel moving down the road

Consider, for example, a bike wheel. If we lift the bike and begin rotating the wheel, which does not touch the ground, it rotates around its axis. If we put it down and ride it, the motion of the wheel is the sum of a translation, with the velocity of its center, and of a rotation.

We shall consider rolling without slipping here. If this is true, in every instant, the contact point of the wheel with the ground is still. Figure 8.32 represents the wheel at a certain instant in full color and in four near instants, two before and two after that in pale color. As one can see, the extreme of the spoke near the ground is almost at rest, while all the other points move, to the degree that they are farther from the contact point.

As a matter of fact, there are two equivalent ways to describe the rolling motion, shown in Fig. 8.33.

1. a translation with the velocity of the center of mass with a superposed rotation around the symmetry axis with angular velocity ω
2. a rotation, again with angular velocity ω, around the instantaneous rotation axis, which is the axis parallel to the symmetry one in contact with the ground in the considered instant.

The type of motion we are discussing, rolling without slipping, can take place for cylindrical and spherical shapes. To be concrete, we shall continue considering a cylinder, of radius R, rolling on a plane, with reference to Fig. 8.34.

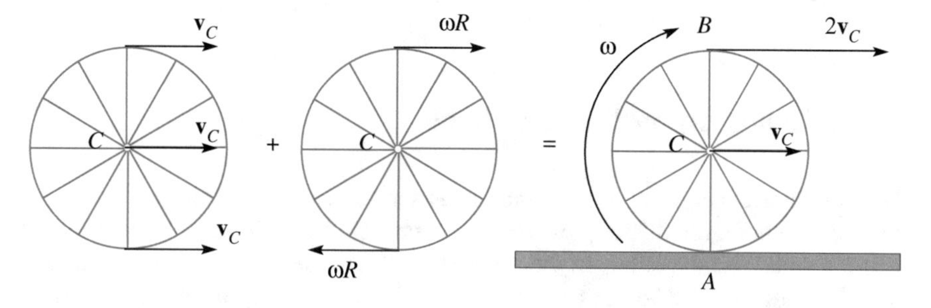

Fig. 8.33 Two possible representations of rolling without slipping

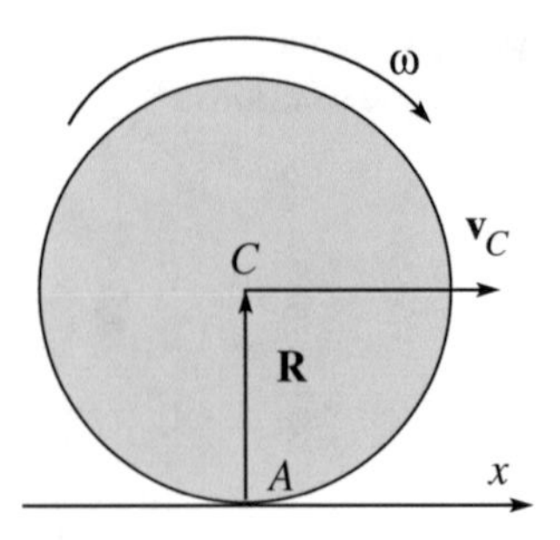

Fig. 8.34 Cylinder rolling on a plane

We take the x axis on the ground in the direction of the motion. If there is no slipping, the magnitude υ_C of the velocity $\mathbf{v}_C$ of the center of mass and the angular velocity ω are linked by the relation

$$\omega R = \upsilon_C. \tag{8.69}$$

The direction of the angular velocity vector $\boldsymbol{\omega}$ is normal to the plane drawn towards the inside. If $\mathbf{R}$ is the position vector of the center C relative to the contact point A, we can write

$$\mathbf{R} \times \boldsymbol{\omega} = \mathbf{v}_C. \tag{8.70}$$

We now find the expression of the kinetic energy of the body in both of the above-mentioned points of view and verify that the result is the same.

In the first point of view, the kinetic energy is the sum of the kinetic energy "of the center of mass", $m\upsilon_C^2/2$, where m is the mass of the cylinder, and that of the motion relative to the center of mass, $I_C\omega^2/2$, where I_C is the moment of inertia relative to the central axis

$$U_K = \frac{1}{2}m\upsilon_C^2 + \frac{1}{2}I_C\omega^2 = \frac{1}{2}\left(mR^2 + I_C\right)\omega^2. \tag{8.71}$$

In the second point of view, the motion is a pure rotation, with the same angular velocity. The moment of inertia is, for the theorem of parallel axes, $mR^2 + I_C$. Hence, the kinetic energy is given by the last member of Eq. (8.71).

8.16 Rolling on an Inclined Plane

An important example of rolling motion is the descent of a rigid sphere on an inclined plane. Figure 8.35 represents the system. The plane forms the angle θ with the horizontal, and the radius of the sphere is R. The forces acting on the sphere are its weight $m\mathbf{g}$, applied to the center of mass C, and the reaction of the constraint applied to the contact point A. The latter can be decomposed in two components,

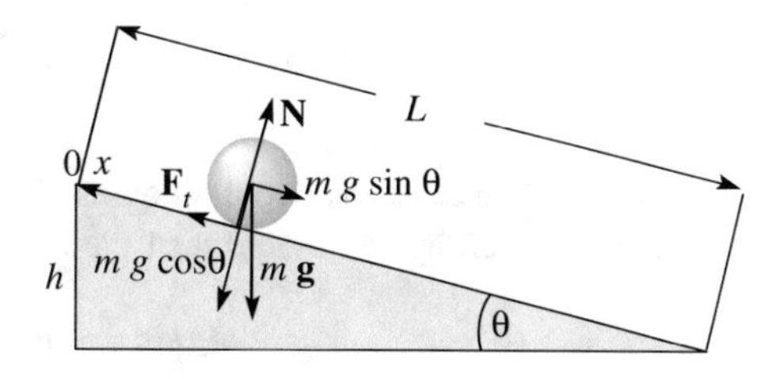

Fig. 8.35 A sphere rolling on an inclined plane without slipping

one normal, **N**, and one tangent, $\mathbf{F}_t$. Notice that, in the case of the latter, the friction force must be present in order to prevent slipping. As a matter of fact, the magnitude of $\mathbf{F}_t$ cannot be larger than $\mu_S N$, where μ_S is the coefficient of static friction. After that, slipping occurs. We shall assume that the condition of pure rolling is satisfied.

We shall deal with the problem through three different methods.

Method 1. We consider the moment of the forces, M_A, about the instantaneous rotation axis through the contact point A. If I_A is the moment of inertia relative to this axis, we can write the dynamical equation

$$M_A = I_A \frac{d\omega}{dt}. \tag{8.72}$$

The moment of the constraint reaction, which is applied in A, is zero. The moment of the weight is, in magnitude, $mgR\sin\theta$ and we have $mgR\sin\theta = I_A \frac{d\omega}{dt}$. The velocity of the center of mass is $\upsilon_C = \omega R$, because the motion does not include slipping, and its acceleration is $a_C = Rd\omega/dt$. Substituting in the above equation, we find

$$a_C = \frac{mgR^2 \sin\theta}{I_A},$$

but, for the theorem of parallel axes, $I_A = I_C + mR^2$, where I_C is the moment of inertia about the central symmetry axis. Consequently,

$$a_C = \frac{g\sin\theta}{1 + \frac{I_C}{mR^2}}. \tag{8.73}$$

Method 2. We consider the moments about the horizontal central axis (through C), M_C, and use the equation

$$M_C = I_C \frac{d\omega}{dt}. \tag{8.74}$$

The moment of the weight is zero because it is applied to C. The moment of the normal reaction **N** is also zero because the force is parallel to the arm. The magnitude of the tangent reaction of the constraint is $F_t R$. We can write

$$RF_t = I_C \frac{d\omega}{dt}. \tag{8.75}$$

This equation contains two unknowns, the angular acceleration and F_t. A second equation is given by the theorem of the center of mass motion

$$mg \sin \theta - F_t = ma_C. \tag{8.76}$$

Recalling that $a_C = Rd\omega/dt$, we find back for a_C Eq. (8.73) and for F_t

$$F_t = \frac{I_C}{I_C + mR^2} mg \sin \theta. \tag{8.77}$$

Method 3. In the process, we are considering that the mechanical energy is conserved. Indeed, even if a non-conservative force is present, such as the friction, its work is zero, because the contact point A, where it is applied, does not move. Suppose that the body starts from rest at the point O of the plane at the height h (see Fig. 8.35). We call x a coordinate along the inclined plane directed downwards with the origin in O. The velocity of the center of mass is $\upsilon_C = dx/dt$. We take the zero of the potential energy at $h = 0$. Initially, the energy of the body is only potential, and its value is mgh. When the body is at the generic coordinate x, its potential energy is $mg(h - x \sin \theta)$. Its kinetic energy is the sum of the kinetic energies of the center of mass, $m\upsilon_C^2/2$, and of the rotation about the center of mass, $I_C\omega^2/2$. The energy conservation equation is then

$$mgh = mg(h - x \sin \theta) + \frac{1}{2} m\upsilon_C^2 + \frac{1}{2} I_C \omega^2$$

or

$$mgx \sin \theta = \frac{1}{2} m\upsilon_C^2 + \frac{1}{2} I_C \omega^2 = \frac{1}{2} m\upsilon_C^2 + \frac{1}{2} \frac{I_C}{R^2} \upsilon_C^2, \tag{8.78}$$

from which we obtain the center of mass velocity at the generic x

$$\upsilon_C = \sqrt{\frac{2gx \sin \theta}{1 + I_C/(mR^2)}}. \tag{8.79}$$

At the end of the inclined plane, the center of mass velocity is then

$$\upsilon_{Cf} = \sqrt{\frac{2gh}{1 + I_C/(mR^2)}}. \tag{8.80}$$

The ratio I_C/m that appears in this expression has the physical dimensions of a length squared. This length, k, is called the *radius of gyration* of the body about the central axis, namely

$$k = \sqrt{I_C/m}. \tag{8.81}$$

Using this quantity, the final center of mass velocity is

$$\upsilon_{C,f} = \sqrt{\frac{2gh}{1 + (k/R)^2}}. \tag{8.82}$$

Using energy conservation, we have directly found the center of mass velocity. Taking its time derivative, we get back Eq. (8.73) written in terms of the gyration radius.

$$a_C = \frac{g \sin\theta}{1 + (k/R)^2}. \tag{8.83}$$

In the denominators of the expressions, we have found we have the ratio of two lengths, the gyration radius and the geometric radius of the body. This ratio depends on the distribution of the masses, as we shall now see in some examples. Notice that the acceleration and the final velocity from a given height are smaller for larger values of k/R. Indeed, as we have seen, part of the initial potential energy becomes kinetic energy of the translation, while part becomes kinetic energy of the rotation. The ratio between these two energies is

$$\frac{U_{K,\text{rot}}}{U_{K,\text{transl}}} = \frac{I_c\omega^2/2}{m\upsilon_C^2/2} = \left(\frac{k}{R}\right)^2. \tag{8.84}$$

For example, using the expressions for the moments of inertia we found in Sect. 8.7, we find for an empty cylinder $k^2 = R^2$ and $a_c = (g \sin\theta)/2$, for a full homogeneous cylinder $k^2 = R^2/2$ and $a_c = (2g \sin\theta)/3$, and for a full homogenous sphere $k^2 = 2R^2/5$ and $a_c = (5g \sin\theta)/7$. In general, the empty bodies descend slowly, followed by the full ones. This is because, for the same total mass, the former have larger moments of inertia, and consequently, the fraction of kinetic energy associated with the rotation is larger. To enhance the effect, we can build the device shown in Fig. 8.36a, which is a disk with a cylindrical axis. The radius R of the latter is much smaller than that of the disk. The axis lays on two parallel inclined rails. The ratio k/R can be made very small, obtaining a quite slow downward acceleration. Contrastingly, in the configuration of Fig. 8.36b, the instantaneous axis of rotation is close to the central axis and the larger fraction of the kinetic energy the energy of the center of mass.

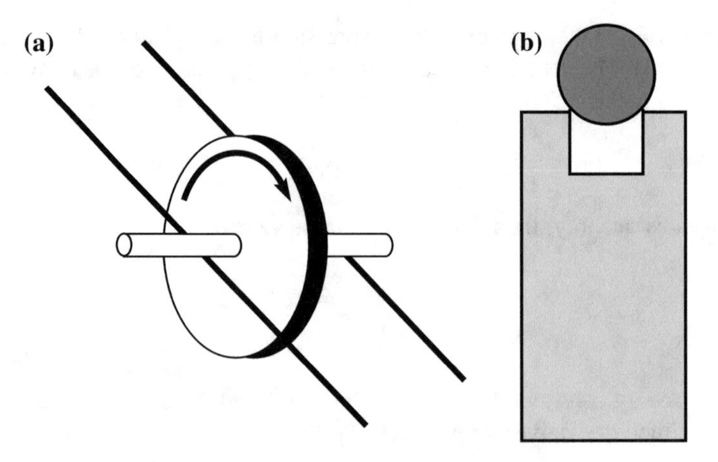

Fig. 8.36 The fraction of kinetic energy in rotation is **a** large, **b** small

We shall now analyze when the conditions of pure rolling are satisfied. We have already found the expression Eq. (8.77) for the tangential force that the constraint must provide. We now write it in the form

$$F_t = \frac{(k/R)^2}{1+(k/R)^2} mg \sin\theta. \tag{8.85}$$

The maximum tangential force the constraint can provide is $F_{t,\max} = \mu_s N$. The normal reaction should equilibrate the normal component of the weight, because there is no acceleration in that direction, namely $N = mg\cos\theta$. Hence, $F_{t,\max} = \mu_s mg\cos\theta$. The no-slipping condition is then

$$F_t = \frac{(k/R)^2}{1+(k/R)^2} mg \sin\theta \le \mu_s mg \cos\theta.$$

This is a condition on the slope angle θ, namely, simplifying,

$$\theta \le \arctan\left[\frac{1+(k/R)^2}{(k/R)^2}\mu_s\right]. \tag{8.86}$$

Suppose we study the motion of a sphere rolling on an inclined plane and we gradually increase its slope. When we reach slopes larger than the value of Eq. (8.86), we observe the contact point slipping on the inclined plane.

Let us briefly go back to what we saw in Sect. 2.12, as to how Galilei experimentally established that the velocity of a sphere at the end of an inclined plane is independent on its slope, depending only on the drop h. He did not know, that part of the kinetic energy is in the rotation motion. However, we can now show that this conclusion was independent of that. In the configuration of Fig. 8.35, the velocity of the sphere after a drop h is

$$\upsilon_{C,f} = \sqrt{2\frac{5}{7}gh} \tag{8.87}$$

to be compared to that of a material point

$$\upsilon_{C,f} = \sqrt{2gh}. \tag{8.88}$$

Consequently, the motion of the center of mass of the sphere is the same for a material point with 5/7 g in place of g. We notice, in addition, that he very likely was using a cross-section of the beam similar to Fig. 8.36b for which the factor in front of g is closer to 1. However, this factor is irrelevant, because the scope of his experiments was the study of the accelerated motion, not the measurement of the gravity acceleration.

8.17 Gyroscopes

A *gyroscope* is a rigid disc with a fixed point. Often, but not always, the fixed point is the center or mass or, at least, a point on the symmetry axis. The construction is such that the rotation axis is free to assume any orientation. If the fixed point is the center of mass, the external moment is zero, and consequently, the angular momentum is conserved when the disk rotates. The direction of the axis is unaffected by tilting or rotation of the mounting. For this property, gyroscopes of this type are useful for measuring or maintaining orientation. Another example of a gyroscope is the spinning top.

The gyroscope in Fig. 8.37 is the disk in the center. The mounting, called a *Cardan mounting*, after Girolamo Cardano (1501–1576), guarantees a complete freedom to rotate in any direction with the center of mass fixed. The support is made of three "gimbals" or rings. The outer gimbal is a half circular, or fully circular, ring fixed on the support basis. The second gimbal is mounted on the outer one. It is free to pivot about an axis in its own plane (a in the figure) that is always perpendicular to the pivot axis of the outer gimbal. The third gimbal is mounted on the second one and is free to pivot about an axis in its own plane perpendicular to the first axis (b in the figure). Finally, the axis of the disk is mounted on the third gimbal, free to pivot around an axis in its plane perpendicular to the second axis (c in the figure). This is a central axis of the disk and, as such, a permanent rotation axis.

All the pivots are joined through ball bearings to minimize the frictions. Notice that in the figure, the three axes are not only mutually perpendicular, but also that b is vertical, and a and c are horizontal. The latter condition is not necessary, however. Indeed, if one takes the basis in one's hand and rotates the external

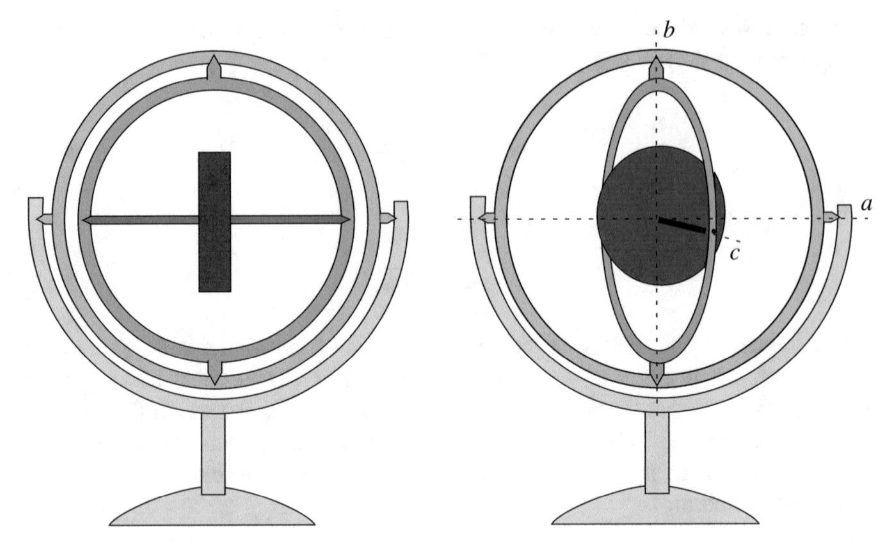

Fig. 8.37 A gyroscope with Cardan mounting

support, b will not be vertical and a and c will not be horizontal, but they remain mutually perpendicular.

If we take the disk in our hand, we feel how it can be rotated in any direction without any effort. Indeed, the disk is in an indifferent equilibrium configuration and, as we just said, the frictions are negligible. We can give a rapid spinning motion to the disk by wrapping many turns of wire around its axis and then drawing it quickly. The rotation can last a long time, because the frictions are very small.

The Cardan mounting is not necessary for gyroscopes having the fixed point on the symmetry axis, but not in the center of mass. The most well-known example is the spinning top. The top, the motion of which we shall study soon, is a body of approximately conic shape supported on a horizontal plane spinning about its axis. If the friction between the tip of the top and the plane is enough, the support point remains (approximately) at rest and the top is a gyroscope.

We anticipate that the motions of the gyroscopes, when we apply an external action onto them, look quite strange. Gyroscopes do not behave as our intuition would suggest to us. To understand them, we should fix our attention on the fact that the characteristic kinematic quantities of a rigid body in rotation are the angular velocity and the angular momentum. Both are vectors. Pay attention to the fact that, to modify the angular momentum, we need to apply a torque, or a couple of forces, rather than one force. The induced *change* of angular momentum (another vector) has the direction of the applied torque, which is perpendicular to the force. If we apply a torque parallel to the angular momentum, we modify its magnitude and not its direction, whereas if we apply the torque perpendicular to the angular momentum, we modify its direction and not its magnitude.

Let us now discuss a few simple examples.

In the first case, represented in Fig. 8.37, the fixed point is the center of mass and the axis is the symmetry axis, which is an axis of permanent rotation. The angular momentum $\mathbf{L}_C$ and the angular velocity $\boldsymbol{\omega}$ are parallel and

$$\mathbf{L}_C = I_C\boldsymbol{\omega}. \tag{8.89}$$

If the external moment is zero, the angular momentum is constant and the angular velocity as well:

$$\mathbf{L}_C = \text{constant} \quad \Rightarrow \quad \boldsymbol{\omega} = \text{constant}. \tag{8.90}$$

We observe that, if we take the support in one hand and change its orientation, the spinning direction relative to the ground, which is an inertial frame, does not change. The support gimbals change direction about the invariable direction of the rotation axis (c, in this case).

Torpedoes, for example, make use of this property. One mounts a gyroscope inside the torpedo and guarantees a continuous spinning with a motor. If the torpedo deviates from the straight trajectory, due to a submarine current or some other factor, the direction of the spinning axis changes relative to the torpedo. A servomechanism then enters into action to modify the route acing on the helm.

The second case is the same gyroscope in the presence of an applied torque. We can, for example, suspend a mass m to a point A of the c axis at a certain distance from the center, as in Fig. 8.38. The angular velocity and the angular momentum are still parallel and Eq. (8.89) is still valid. But now, the angular momentum varies, according to the equation

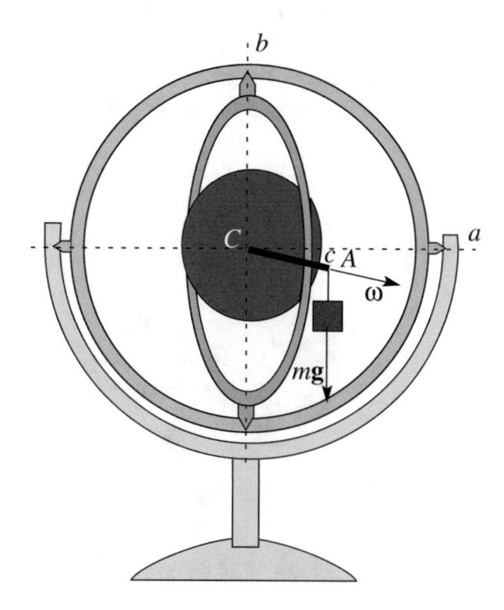

Fig. 8.38 A gyroscope with an external torque

$$\mathbf{M}_C = \frac{d\mathbf{L}_C}{dt}. \tag{8.91}$$

Our intuition suggests that we would see the point A lower under the action of the weight. But this is not what we observe. Point A does not lower, but, on the contrary, slowly moves in a horizontal circle. This motion of the rotation axis c is called *precession*.

To understand this, as we have already stated, we must think about the direction of the applied moment, not of the force. Let us start considering the instant in which the axis of the gyroscope is still at rest and we apply the weight. The vertical weight force exerts on the gyroscope a moment, or torque, the direction of which is horizontal and perpendicular to the c axis and, consequently, perpendicular to the angular momentum. In the time interval dt, the variation of angular momentum is, for Eq. (8.91),

$$d\mathbf{L}_C = \mathbf{M}_C dt,$$

which has a direction perpendicular to $\mathbf{L}_C$ in the horizontal plane, as shown in Fig. 8.39. Consequently, $\mathbf{L}_C$ varies in direction and not in magnitude. The c axis, which has the direction of the angular velocity and, consequently, of the angular momentum, also rotates in accord. The torque, which is perpendicular to the axis, rotates as well, because the weight is applied to the point A of the axes. The torque always remains perpendicular to the angular momentum. Consequently, the just described situation is such in every instant, not only in the initial one.

The angular velocity of the precession motion, which we call $\boldsymbol{\Omega}$, is directed vertically, in our case, upwards. To have it downwards, namely to have the gyroscope preceding in the opposite direction, we just have to attach the weight at the other extreme of the c axis, or have the disk spinning in the opposite direction and the weight still in A.

The motion of the gyroscope is a rotation with angular velocity equal to the (vector) sum of $\boldsymbol{\omega}$ and $\boldsymbol{\Omega}$. Rigorously speaking then, the rotation axis is not exactly

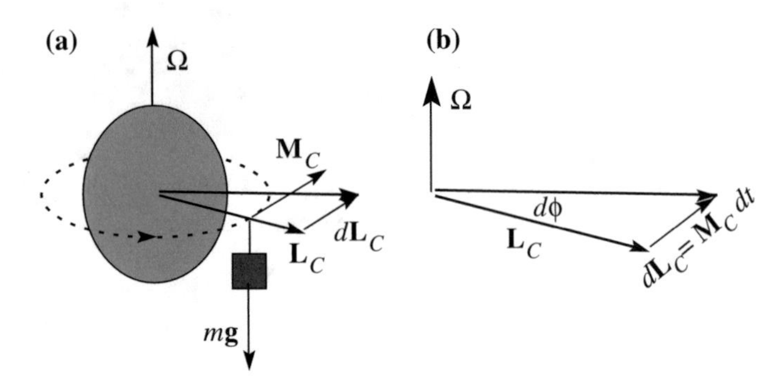

Fig. 8.39 The precession of the gyroscope

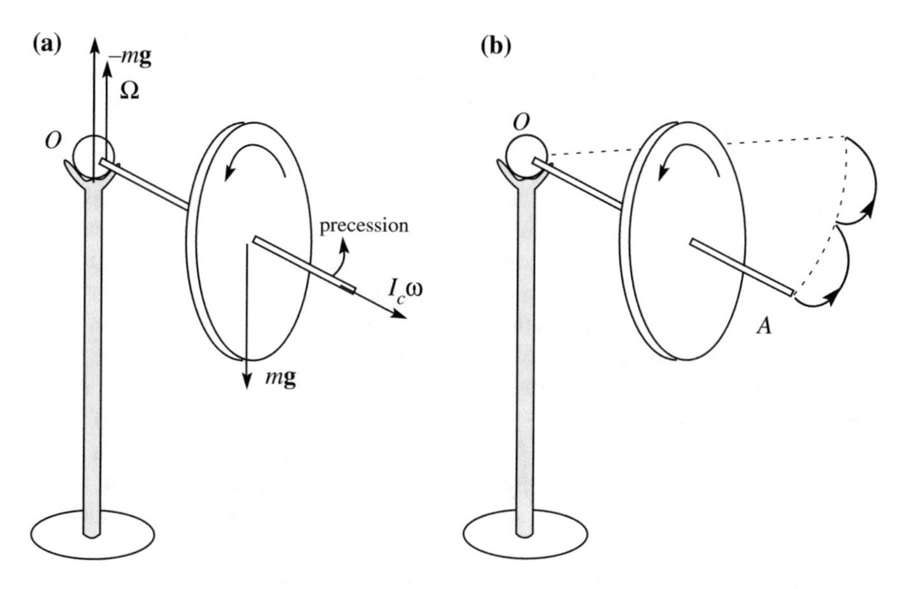

Fig. 8.40 **a** A gyroscope with suspension point on the symmetry axis, but not in the center. **b** precession and nutation

the symmetry axis. However, in practice, the spinning velocity is always much larger that of the precession. The rotation is in a good approximation about the symmetry axis and we can consider Eq. (8.89) valid. With reference to Fig. 8.39b, we can write $dL_C = L_C d\phi$. If l is the distance of point A from the center C, the magnitude of the moment is $M_C = mgl$ and the precession velocity is

$$\Omega = \frac{d\phi}{dt} = \frac{mgl}{I_C\omega}.$$

The third case we will consider is the following. The suspension point is not the center of mass, but is on the symmetry axis anyway. The external moment is not zero; its direction is always perpendicular to the rotation axis. Figure 8.40a shows how such conditions can be realized. The axis of the disk ends with a small sphere. The sphere lays on a concave support on top of a column, allowing the axis to spin and to change its direction freely. We now give to the gyroscope a rapid spin about its axis, keeping it horizontal with our hand. When we abandon the axis, it does not fall downwards, but rotates in a precession motion in the horizontal plane. The analysis of the motion is identical to the preceding case, with the only difference being that the weight is now the weight of the gyroscope itself.

Notice how the behavior of the system is completely different when the disk is spinning from when it is not. In the latter case, if we take the extreme of the axis in our hand and then abandon it, the axis falls, rotating in the vertical plane. If we do the same with the disk spinning, the axis rotates in the horizontal plane. The acting torque, the moment of the weight, is equal in both cases, and so is the change of the angular momentum in any time interval *dt*. This, however, in the case of the spinning disk, adds to a pre-existent angular momentum, modifying its direction, while, contrastingly, in the case of no spinning, the change is solely to the angular momentum, which consequently has the direction of the torque.

To be sure, the angular velocity is the sum of $\boldsymbol{\omega}$ and $\boldsymbol{\Omega}$, and, consequently, is not exactly parallel to a principal, permanent rotation axis. The just-made description is valid only in a first approximation. Let us look more carefully into the issue.

As a matter of fact, the gyroscope's strange immunity to its own weight is not completely true. If we set the gyroscope spinning with the point A in our hand, when we abandon it, it initially falls down vertically a bit. However, as soon as the precession starts, the extreme A rises again, reaching the horizontal plane, as shown in Fig. 8.40b. This is not all, however. The axis does not remain horizontal. The precession has slowed down somewhat due to the rise of the axis and is no longer fast enough to neutralize the weight. The extreme falls again to the height of the first descent, the precession velocity increases and the extreme rises again, and so on; the motion continues with a series of up and down oscillations, which are ideally all equal. The motion is similar to the motion of the head of somebody that nods, and is called *nutation*, which means 'nodding' in Latin.

We shall not further analyze this motion, which is quite complex. Rather, we shall make a few observations. When a gyroscope spins about its symmetry axis with angular velocity $\boldsymbol{\omega}$ and precedes at the same time with angular velocity $\boldsymbol{\Omega}$, its total angular velocity is not parallel to the symmetry axis. Consequently, angular velocity and angular momentum are not exactly parallel. The effects are generally small because $\omega \gg \Omega$, but it is the basis of the nutation phenomena.

Consider a gyroscope rotating about an axis a bit different from a symmetry axis in absence of external torque. In that case, the angular momentum is constant and the angular velocity rotates around it, describing a cone. In our case, however, an external torque exists. It is the moment of the weight that is directed horizontally, perpendicular to the axis. The vertical component of the angular momentum is constant, because the external torque is horizontal. The magnitude of the angular momentum is constant too, because the torque is perpendicular to its direction. Consequently, the angular momentum vector rotates uniformly in the horizontal plane. This is the precession. The angular velocity contemporarily describes a cone around the angular momentum. The extreme of the axis describe a cycloid curve, as shown in Fig. 8.40b. This is the nutation.

We can look at the phenomenon from another slightly different point of view. When we abandon the axis horizontal of the spinning gyroscope, the precession starts. This adds to the angular momentum the vector quantity $I_\Omega \boldsymbol{\Omega}$ where I_Ω is the moment of inertia about the vertical axis through O. The external torque being horizontal, the vertical component of the angular momentum is conserved. Consequently, the spinning axis *must* fall a bit, or even better, rotate downwards, in such a way that $I_C \boldsymbol{\omega}$ has a vertical component equal and opposite to $I_\Omega \boldsymbol{\Omega}$ (see Fig. 8.41). An oscillation starts in which $I_\Omega \boldsymbol{\Omega}$ increases and decreases alternatively, and so does the angle with the horizontal of $I_C \boldsymbol{\omega}$.

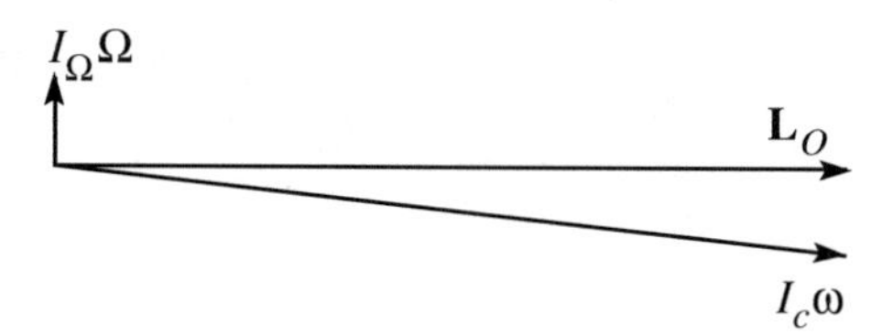

Fig. 8.41 The vectors playing roles in the nutation

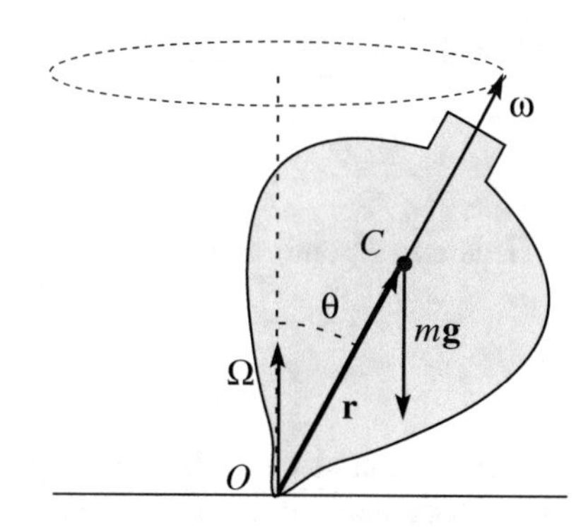

Fig. 8.42 A *top* and its precession

As a last example of precession, we consider the top. The top is a rigid body of approximately conical shape, ending with a tip. Initially, we give the top a rapid spin about its symmetry axis with angular velocity $\boldsymbol{\omega}$. The tip O lays on a horizontal floor, as in Fig. 8.42. We assume that the friction is enough to keep point O at rest. We observe that, beyond spinning, the top also has a precession motion, with angular velocity that we shall call $\boldsymbol{\Omega}$.

Let $\mathbf{r}$ be the position vector of the center of mass C relative to O and $m\mathbf{g}$ the weight of the top, applied, as usual, to the center of mass. The constraint forces are applied in O, which we choose as the pole of the moments. Consequently, the constraint forces do not contribute to the external moment. We have

$$\mathbf{M}_O = \mathbf{r} \times m\mathbf{g} = \frac{d\mathbf{L}_O}{dt}. \tag{8.92}$$

The moment $\mathbf{M}_O$ is horizontal, perpendicular to the spin axis. Consequently, the direction, but not the magnitude of the angular momentum, varies. More precisely, the angular momentum rotates with angular velocity $\boldsymbol{\Omega}$. Hence, for the Poisson formula

$$\frac{d\mathbf{L}_O}{dt} = \boldsymbol{\Omega} \times \mathbf{L}_O. \tag{8.93}$$

Considering that $\Omega \ll \omega$, we can assume that $\mathbf{L}_O \approx I\boldsymbol{\omega}$ and, for the above equations, that $\boldsymbol{\Omega} \times I\boldsymbol{\omega} = \mathbf{r} \times m\mathbf{g}$. Now, both $\boldsymbol{\Omega}$ and $\mathbf{g}$ are vertical, something we can express as $\mathbf{g} = g\boldsymbol{\Omega}/\Omega$. Substituting in the last expression, we have

$$\boldsymbol{\Omega} \times I\boldsymbol{\omega} = -\boldsymbol{\Omega} \times \frac{mg}{\Omega}\mathbf{r}$$

or

$$I\boldsymbol{\omega} = -\frac{mg}{\Omega}\mathbf{r}$$

and finally, for the magnitudes

$$\Omega = \frac{mgr}{I\omega}. \tag{8.94}$$

This corresponds to the period of the precession

$$T = 2\pi\frac{I\omega}{mgr}. \tag{8.95}$$

Let us look at the orders of magnitude. We approximate the top with a homogeneous cylinder of radius $R = 2$ cm. Let $r = 3$ cm be the distance from the center of mass to the tip. Suppose that the spinning angular velocity is $\omega = 120\ \text{s}^{-1}$ (that is, about 20 turns per second). Let us calculate the precession period. The moment of inertia is $I = mR^2/2$. We have

$$T = 2\pi\frac{I\omega}{mgr} = 2\pi\frac{mR^2}{2mgr}\omega = \frac{\pi R^2}{gr}\omega = 0.8\ \text{s}.$$

The corresponding precession angular velocity is $\Omega = 2\pi/T \simeq 8\ \text{s}^{-1}$, which is qiute small in comparison to the spinning velocity.

8.18 Collisions Between Material Systems

In Chap. 7, we studied the collision phenomena. In that discussion, we considered, for each colliding body, only the motion of its center of mass. We did not consider there the motion of each body relative to its center of mass. For example, when a football player kicks the ball, the momentum of the ball varies. However, the player may wish to give an angular momentum to the ball as well, to make it follow a curved trajectory. In general, in a collision, both the linear and the angular momentum of each body vary. We shall now discuss this aspect of collisions. We shall limit the discussion to the cases in which one of the bodies can be considered as point-like, while the other is extended and rigid.

There are two different possible situations: the target body may be free or may be subject to constraints. In the first case, but not in the second, we can neglect the external forces during the collision, as in the case of point-like objects.

Consequently, the total linear momentum and the angular momentum are conserved. The latter does not have any role in collisions between point-like objects. It adds nothing to the linear momentum conservation. The angular momentum conservation has, contrastingly, observable consequences in collisions between extended objects. Even better, we observe that the angular momentum conservation in an isolated system is a consequence of a particular aspect of the action-reaction law, namely that action and reaction have the same application line. This aspect cannot be experimentally controlled with collisions between point-like bodies. We need to look at the angular momentum conservation in collisions between extended bodies.

If constraints are present, there are external forces acting during the collision. Consider, for example, a ball hitting a bar pivoted on an axis initially at rest. The motion after the collision is bound to be a rotation about the axis. The forces exerted by the constraints must be such as to equilibrate some of the components of the impulsive forces that develop during the collision. Consequently, their intensity is great, and cannot be neglected. The collision with a constrained body is not, consequently, a process in an isolated system. Linear and angular momentum, in general, are not conserved.

We shall limit the discussion to two examples.

Example E 8.9 A homogeneous disk of mass M and radius R lays on a horizontal plane. It is initially at rest. A bullet of mass m and velocity $\mathbf{v}_{i1}$ hits the disk on its rim tangentially, as in Fig. 8.43, and sticks. Find the motion of the system after the collision.

There are no constraints. The system can be considered as isolated. Linear and angular momentum are conserved. As in Fig. 8.43, we take a reference frame at rest on the support plane, with the x axis in the direction of the initial velocity of the bullet and through the center A of the disk. The y is perpendicular to x in the support plane and the z axis is perpendicular to both. The angular momentum of the system, which we take to be about the center of mass C, is in the z direction. After the collision, we have one lone body moving at the velocity of the center of mass v_C. The angular momentum conservation gives the two equations

$$m\upsilon_{i1} = (m+M)\upsilon_{C,x}; \quad 0=(m+M)\upsilon_{C,y}.$$

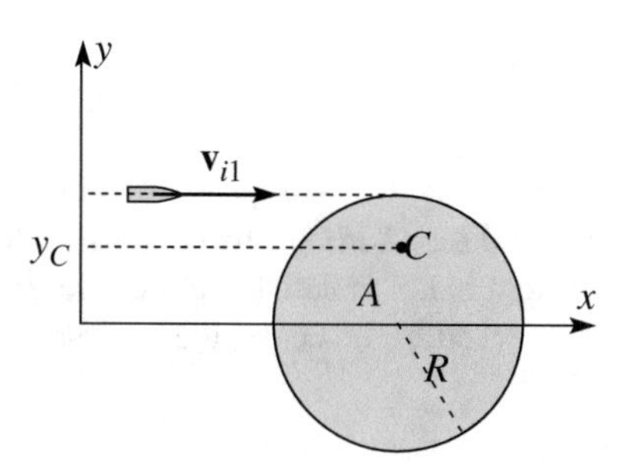

Fig. 8.43 A bullet hitting a disk

The second equation tells us that the y component of the velocity of the center of mass is zero, which is obvious. The first equation gives the velocity of the center of mass

$$\upsilon_C = \upsilon_{C,x}\frac{m\upsilon_{i1}}{m+M}. \quad (8.96)$$

We choose the center of mass as being the pole of the angular momentums. Its y coordinate does not vary during the motion and is given, by definition, by

$$y_C = \frac{Rm+0M}{m+M} = R\frac{m}{m+M}. \quad (8.97)$$

The initial angular momentum is that of the bullet, because the disk is not moving. Its direction is opposite to the z axis and its magnitude is

$$L_{C,i} = (R-y_C)m\upsilon_{i1} = \frac{mM}{m+M}R\upsilon_{i1}. \quad (8.98)$$

In the final state, the system disk plus the bullet rotates with angular velocity ω, which we must determine. Its angular momentum about the center of mass is $L_{C,f} = I_C\omega$. The angular momentum conservation then gives

$$I_C\omega = \frac{mM}{m+M}R\upsilon_{i1}, \quad (8.99)$$

which gives us ω once we know the moment of inertia I_C. This is the sum of the moments of the inertia of the bullet, I_b and of the disk, I_d. The former is $I_b = m(R-y_c)^2 = mR^2\left(\frac{M}{m+M}\right)^2$, and the latter can be found with the theorem of parallel axes, giving $I_d = \frac{1}{2}MR^2 + My_C^2 = \frac{1}{2}MR^2\frac{3m^2+M^2+2mM}{(m+M)^2}$. In conclusion,

$$I_C = I_b + I_d = MR^2\frac{M+3m}{2(m+M)}$$

and finally, from Eq. (8.99),

$$\omega = \frac{\upsilon_{i1}}{R}\frac{2m}{3m+M}.$$

Example E 8.10 Suppose now the disk in the previous example is constrained by a vertical axis through its center A, about which it can rotate freely. In this case, the system is not isolated. Neither the linear nor the angular momentum are necessarily conserved. In this case, the external forces are the constraint ones, which are applied to the point A. Their moment about A is zero. Consequently, the angular momentum about A is conserved, say $L_{A,f} = L_{A,i}$.

In the final state, the system is a rigid body, disk plus bullet, rotating with the angular velocity ω, still to be found. We can write $L_{A,i} = m\upsilon_{i1}R = L_{A,f} = I_A\omega$. Here, I_A is the moment of inertia of the system about A, namely $I_A = MR^2/2 + mR^2$. Hence, the final angular velocity is

$$\omega = \frac{2m}{M+2m}\frac{\upsilon_{i1}}{R}.$$

Notice the difference from the previous example.

8.19 The Virtual Works Principle

In this section, we shall discuss a method that often turns out to be useful for establishing the equilibrium conditions for mechanical systems, rigid or not. The method is based on the so-called *virtual works principle.*

A *virtual displacement* of a mechanical system is defined as any infinitesimal displacement compatible with the constraints to which the system is subject. For example, for a rigid body pivoted on a fixed axis, a rotation of an infinitesimal angle about the axis, or for a carriage on a rail, an infinitesimal translation in the direction of the rail, etc.

The work dW_i that would be done for that displacement by the ith force acting on the system is called the *virtual work* of that force. The virtual works principle states that a mechanical system is in equilibrium in a given configuration if the sum of the virtual works done by the forces acting on the system for any virtual displacement from that configuration is zero.

$$\sum_{i=1}^{N} dW_i = 0. \tag{8.100}$$

We shall discuss a few examples.

Example E 8.11 Figure 8.44 shows a rigid bar, pivoted in O. Two forces, $\mathbf{F}_1$ and $\mathbf{F}_2$, are applied to its extremes A_1 and A_2 perpendicular to the bar. The distances of the extremes from O are b_1 and b_2, respectively.

The virtual displacements $d\mathbf{s}_1$ and $d\mathbf{s}_2$ of the extremes are infinitesimal arcs of the circles of the center in O of radiuses A_1 and A_2. Indeed, the only degree of freedom is the rotation angle ϕ about the axis.

Let us start with $\mathbf{F}_1$. Its virtual work for the displacement $d\mathbf{s}_1$ is $dW_1 = \mathbf{F}_1 \cdot d\mathbf{s}_1 = F_1 b_1 d\phi = \tau_{1,z} d\phi$, where $\tau_{1,z}$ is the moment of the force about the z rotation axis, with positive direction pointing outside the page of the drawing. Once $d\mathbf{s}_1$ is chosen, the displacement $d\mathbf{s}_2$ of A_2 is fixed. The virtual work of the

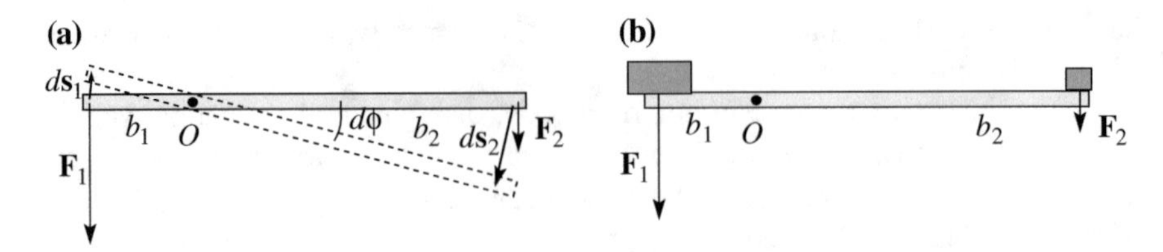

Fig. 8.44 **a** Finding the equilibrium condition for a lever, **b** the same with two weights

force $\mathbf{F}_2$ is $dW_2 = \mathbf{F}_2 \cdot d\mathbf{s}_2 = -F_2 b_2 d\phi = \tau_{2,z} d\phi$. According to the virtual work principle, the configuration is of equilibrium if $F_1 b_1 - F_2 b_2 = \tau_{1,z} + \tau_{2,z} = 0$. We recognize the known result that to have equilibrium, the total moment about the rotation axis must be zero.

As a matter of fact, the virtual works principle is a consequence of the energy conservation law. To see that, let us look at the present example from a slightly different point of view, as in Fig. 8.44b. The force $\mathbf{F}_1$ is the weight of a block placed in A_1. We want to establish the equilibrium by placing another block of weight $\mathbf{F}_2$ in A_2. Which is the value of F_2 requested for that? We imagine that when we place the second block, A_2 moves $d\mathbf{s}_2$ down and A_1 moves $d\mathbf{s}_1$ up. The corresponding variation of potential energy is $-F_1 ds_1 + F_2 ds_2$.

Being the total energy conserved, the variation of potential energy might be compensated by an opposite variation of kinetic energy. However, in the virtual change we are considering, the system is at rest both before and after the displacement and the kinetic energy is always zero. We conclude that the potential energy cannot vary, $F_1 ds_1 - F_2 ds_2 = 0$. This is what the virtual works principle states.

Example E 8.12 Figure 8.45a shows two blocks of masses m_1 and m_2 resting on two inclined planes tilted to the horizontal at the angles θ_1 and θ_2 and connected by a rope. Frictions are negligible. We want to know which is the ratio of the two masses to have equilibrium.

We think to move block 1 of ds upwards on the plane. The work done by the weight is $dW_1 = - m_1 g(\sin\theta_1)ds$. At the same time, block 2 moves on its plane of the same ds downwards, because we want the rope to remain invariant. The work of its weight is $dW_2 = +m_2 g(\sin\theta_2)ds$. The constraint forces are normal to the displacements and do no work. The virtual works principles then requires for equilibrium that $dW_1 + dW_2 = 0$. The ratio of the masses must be $m_1/m_2 = \sin\theta_2/\sin\theta_1$.

Historically, as we have already mentioned, Galilei established the law of the free fall with experiments on inclined planes of different slopes. He then extended the validity of the law to the vertical motion with the exact above argument, with one of the angles being equal to 90°.

His contemporary Simon Stevin (1548–1620) demonstrated the rule as drawn in Fig. 8.45b. The chain is in equilibrium. In the case of this particular right triangle,

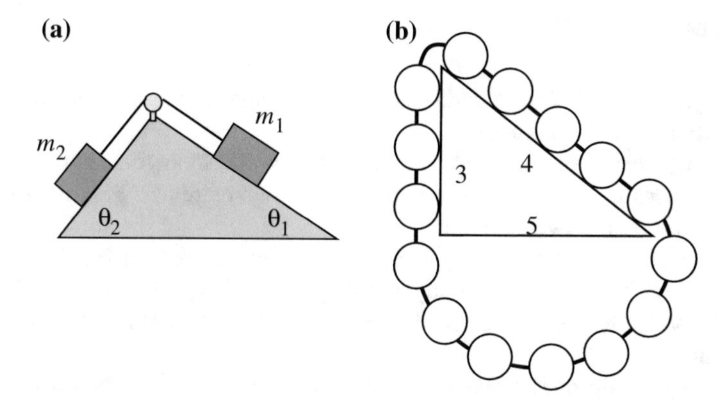

Fig. 8.45 **a** Two blocks in equilibrium on different slopes, **b** the basis of the Stevin argument

five rings balance three rings. In those times, the trigonometric functions were still not known.

Example E 8.13 Figure 8.46 shows a rigid bar of mass m pivoted at its lowest point and held by a rope. The bar holds, in turn, a block of mass M. Find the tension T of the rope. This problem also has one degree of freedom, the rotation angle θ. To respect the constraint, we can only diminish it, say by $-d\theta$. The acting forces are three: the weight of the bock Mg, the weight of the bar mg and the tension of the rope T.

The displacement corresponding to $-d\theta$ of the block is $d(b \cos \theta) = b \sin \theta \, d\theta$ upwards. The work of its weight is $dW_1 = -Mgb \sin \theta \, d\theta$. The weight of the bar is applied to its center of mass. Its work is $dW_2 = -mg(b/2) \sin \theta \, d\theta$. The application point of the rope has the displacement $d(a \sin \theta) = a \cos \theta \, d\theta$ and the work of the tension is $dW_3 = a \cos \theta \, d\theta$.

Imposing $dW_1 + dW_2 + dW_3 = 0$, we have $T = (M + m/2)g(b/a) \tan \theta$.

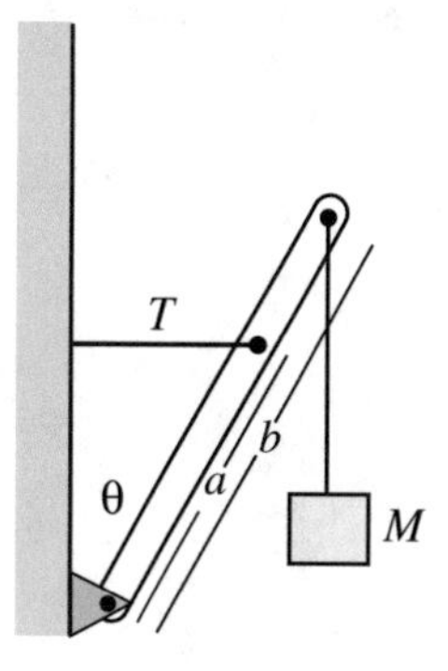

Fig. 8.46 Find the equilibrium configuration

Problems

8.1. Fig. 8.47 represents a rigid bar b, and $\mathbf{v}_1$ and $\mathbf{v}_2$ are the velocities of its extremes. Is it possible?

8.2. A rigid bar of length $L = 8$ m and mass $m = 100$ kg lays on two supports at distances $L_1 = 2$m and $L_2 = 1$ m from the two extremes. Find the forces F_1 and F_2 on the two supports.

8.3. On which of the following elements does the moment of inertia of a body depend? The mass of the body, the shape of the body, the angular velocity of the body, the position of the axis relative to the body, or the external resultant force?

8.4. A rigid body rotates about a fixed axis. How much does its kinetic energy vary if the angular velocity doubles?

8.5. Two material points of masses m_1 and m_2 are linked by a rigid bar of length L and negligible mass. Find the moment of inertia about a perpendicular axis through the center.

8.6. The density $\rho(r)$ of a cylinder of length L and radius R varies linearly with the distance r from the axis from the value ρ_1 on the axis to the value $\rho_2 = 3\rho_1$ on the lateral surface. Find the moment of inertia about the axis.

8.7. Figure 8.48 represents a thin annular sheet of radii R_1 and R_2. Find the moment of inertia about the a axis.

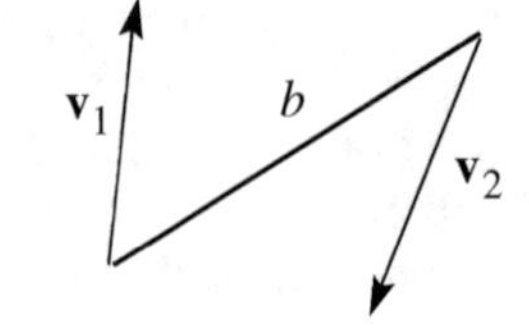

Fig. 8.47 Problem 8.1

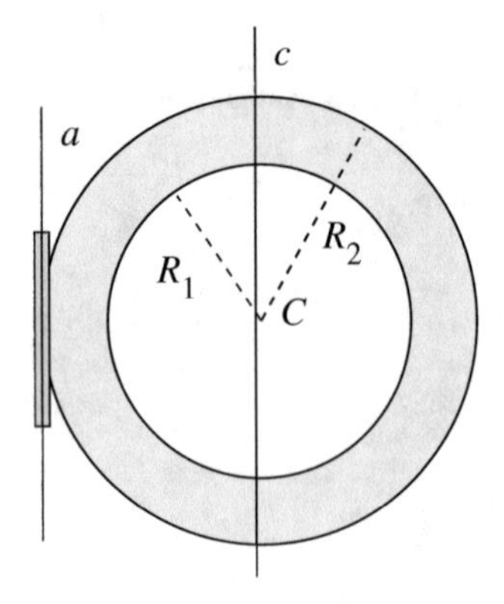

Fig. 8.48 Problem 8.7

8.8. A rigid cylinder rolls on an inclined plane without slipping. Its density is not necessarily uniform. Can the kinetic energy relative to the center of mass be larger than that of the center of mass?

8.9. Two material points of masses m_1 and m_2 are fixed to the extremes of a rigid bar of length L and negligible mass. We want to bring the bar into rotation with angular velocity ω about an axis perpendicular to the bar through one of its points. How should we choose this point so as to have the minimum kinetic energy for the given angular velocity?

8.10. Under which conditions do the angular velocity $\boldsymbol{\omega}$ and the angular momentum $\mathbf{L}$ of a rigid body have the same direction?

8.11. In which cases is equation $\mathbf{M}_O = I d\boldsymbol{\omega}/dt$ valid for a rigid body?

8.12. In which cases is the kinetic rotation energy of a rigid body given by $U_K = I\omega^2/2$?

8.13. A homogeneous sphere of radius R and mass m rotates about an axis through its center C with angular velocity $\boldsymbol{\omega}$. Find the angular momentum about C. Does the angular momentum depend on the pole?

8.14. A homogeneous sphere of radius R and mass m rotates without slipping on a horizontal plane. Its axis advances with velocity $\mathbf{v}$. The points X, Y, and Z in Fig. 8.49 are in the vertical plane containing the center of the cylinder shown in the figure. Their heights are $R/2$ lower than C, equal to C and $R/2$ higher than C, respectively. Find the angular momentum of the cylinder about each of these points.

8.15. A bar of length $l = 3$ m of mass $m = 50$ kg is initially vertical at rest with one extreme O laying on the ground. With O maintained at rest, the bar falls to the ground. Find the angular momentum about O and the velocity of the other extreme at the instant in which the bar hits the ground.

8.16. A rigid homogeneous sphere is set free on a plane inclined at 40° with the horizontal. At which values of the friction coefficient will the sphere roll without slipping?

8.17. A yo-yo (Fig. 8.50), which we consider to be a homogenous cylinder, of mass $m = 100$ g, hangs from a wire wrapped around its axis. The axis is horizontal. Assume the radius of the wrapping to be equal to the radius of the cylinder. The yo-yo is released at rest. (a) What is time t that it takes to drop to $h = 50$ cm? (b) Which is the tension T of the wire during the descent?

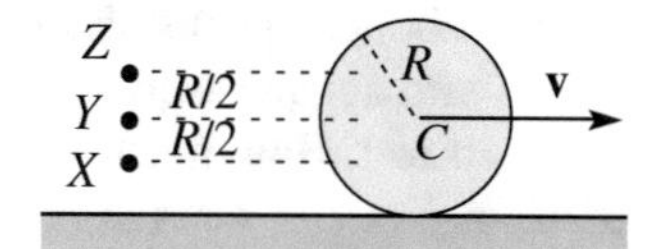

Fig. 8.49 Problem 8.14

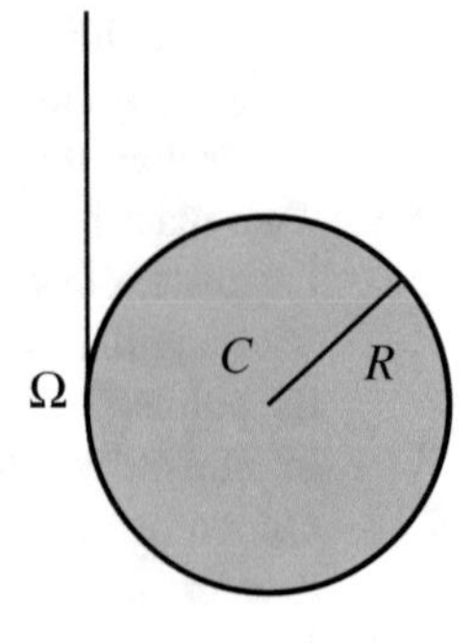

Fig. 8.50 The yo-yo of Problem 8.17

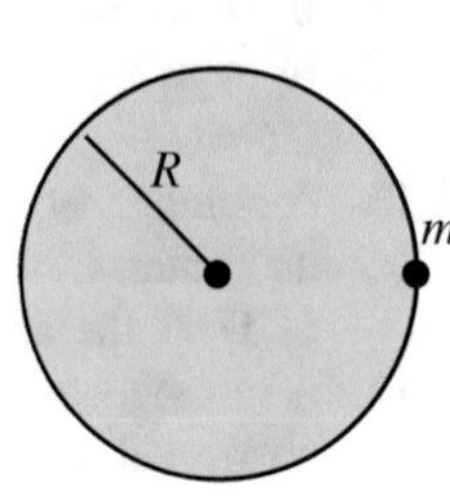

Fig. 8.51 Problem 8.18

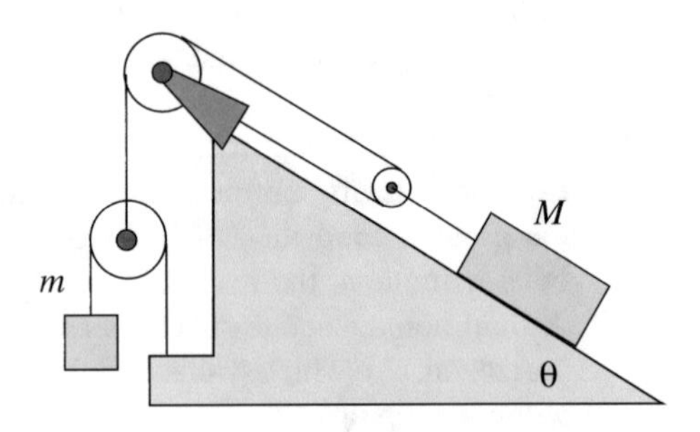

Fig. 8.52 The system of Exercise 8.20

8.18. A homogeneous disk of radius R in a vertical plane can rotate about its geometrical axis (Fig. 8.51). The friction on the axis is not negligible but exerts a torque M_a about the axis, independent of the angular velocity. A particle of mass m sticks to the rim of the cylinder at the level of the axis. The system is released at rest. (a) Which is the minimum value of m for the cylinder to start rotation? (b) Which is the value of m at which it rotates a quarter of a turn and stops?

8.19. A homogeneous disk of radius R and mass m rotates about its geometric axis with angular velocity ω. The frictions of the axis slow it down until it comes to rest. How much work have they done?

8.20. A block of mass M on an inclined plane at the angle θ is held in equilibrium by a set of pulleys, as shown in Fig. 8.52. Using the virtual works principle, find the value of the mass m of the counterweight needed to insure the equilibrium.

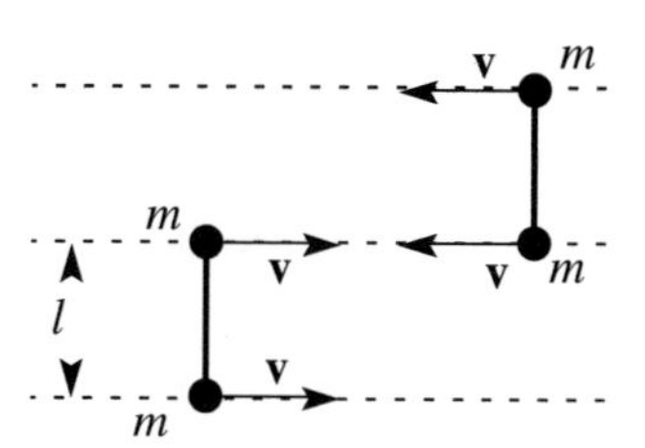

Fig. 8.53 The system of Problem 8.21

8.21. The system in Fig. 8.53 is made of two identical dumbbells. Each of them consists of two small spheres, each of mass $m = 0.3$ kg, separated by a bar of negligible mass of length $l = 1$ m. The dumbbells move on a horizontal plane with negligible friction with equal and opposite velocities $\upsilon = 1$ m/s. Two spheres, as shown in the figure, collide elastically. (a) Describe the motion after the collision. Find the angular velocities (magnitude and direction). (b) How long does the rotation last? (c) Then what happens?

Correction to: Gravitation

Correction to: Chapter 4 in: A. Bettini, *A Course in Classical Physics 1—Mechanics*, Undergraduate Lecture Notes in Physics, https://doi.org/10.1007/978-3-319-29257-1_4

In the original version of the book, the following belated corrections have been incorporated in chapter "Gravitation":

Page 144, line 13:

Change "He found correctly that the distance of the moon is about 60 times the radius of the earth. However, due to the insufficient resolution in the measurement of the angles, he evaluated that the sun is 20 times farther than the moon, rather than about 400 as it is"
Into "Due to the insufficient resolution in the measurement of the angles, he evaluated that the distance of the moon is about 20 times the radius of the earth, instead of 60 as it is, and that the sun is 20 times farther than the moon, rather than about 400 as it is"

Page 144, last line:

Change "from East to West"
Into "from West to East relative to the fixed stars"

Page 146, line 5:

Change "Mars and Venus"
Into "Mercury and Venus"

The updated version of this chapter can be found at
https://doi.org/10.1007/978-3-319-29257-1_4

A. Bettini, *A Course in Classical Physics 1—Mechanics*,
Undergraduate Lecture Notes in Physics, DOI 10.1007/978-3-319-29257-1_9

Page 146, line 3 from bottom:

Erase: "both the equant and"

Page 149, Section 4.3, line 2:

Erase "Both systems make use of the equant"
Change "To be precise, the centre of the Copernicus system is not the sun, but the equant of the earth (what we now know to be the empty focus of her elliptical orbit)."
Into "To be precise, the centre of the Copernicus system is not the sun, but the centre of the elliptical orbit of the earth."

Solutions

1.1. (a) No. (b) Yes, if $\Delta\mathbf{V}$ has the same direction and verse as $\mathbf{V}$.
1.2. $\Delta\mathbf{V} = -2\,\mathbf{V}$, $\Delta V = 0$ and $|\Delta\mathbf{V}| = 2\,V$.
1.3. (a) $\Delta\mathbf{v} = (4,0,3)$, (b) $|\Delta\mathbf{v}| = 5$, (c) $\Delta v = 3.9$.
1.4. (a) $\langle v\rangle = v$, (b) $\langle\mathbf{v}\rangle = 0$.
1.5. (a) $\mathbf{v}(t) = (2\mathbf{i} + 6t\mathbf{j} + \mathbf{k})$ m/s, $\mathbf{a}(t) = 6\,\mathbf{j}$ m/s^2; (b) $v\ (t = 2\ \text{s}) = 12.2$ m/s.
1.6. $R = v^2/a$.
1.7. (a) $\mathbf{v}(t) = -\mathbf{i}A\omega\sin\omega t + \mathbf{j}A\omega\cos\omega t$; $\mathbf{a}(t) = -\mathbf{i}A\omega^2\sin\omega t - \mathbf{j}A\omega^2\cos\omega t$; $v(t) = A\omega$, $a(t) = A\omega^2$; (b) $\mathbf{r}\cdot\mathbf{v} = 0$, velocity is perpendicular to the position vector (c) $\mathbf{r}\cdot\mathbf{a} = -\omega^2 A = -ra$, acceleration is parallel and opposite to the position vector; (d) $x(t)^2 + y(t)^2 = A^2 =$ constant. The trajectory is a circle with center in the origin and radius A. The motion is uniform in anticlockwise direction; (e) direction changes to clockwise.
1.8. (a) The first step in solving this type of problems is drawing the vectors they contain, as in Fig. 1. $\mathbf{v}_1$ is the cyclist velocity, $\mathbf{v}_2$ is the wind velocity relative to ground, $\mathbf{v}_2 - \mathbf{v}_1$ is the wind velocity as felt by the cyclist. Vectors and angle drawn with continuous lines are known. With the sine law we get $|\mathbf{v}_2 - \mathbf{v}_1| = 15.1$ km/h and $\beta = 139.5°$. Consequently, the wind blows from 40.5° from North to East. (b) The new apparent direction of the wind (the velocity of the cyclist is $-\mathbf{v}_1$) is $|\mathbf{v}_2 + \mathbf{v}_1| = 6.7$ km/h and its apparent direction is 35.6° from South to East.
1.9. She will cross at 3 miles towards stern in 18′.
1.10. (a) The rotation axis in the plane xz is at 27° to the x-axis (b) 20 rad. (c) the magnitude of $\boldsymbol{\omega}$ grows proportionally to the square of time, its direction is constant.
1.11. (a) $\alpha = 78.5°$, (b) $t = 11.5$ s (the smaller solution must be chosen); (c) $s = 1.15$ km.
1.12. The motion is the sum of a translation at the velocity $\mathbf{v}$ and a rotation about the wheel axis. Hence, $v_A = (v, v, 0)$; $v_B = (2v, 0, 0)$; $v_C = (v, -v, 0)$.

A. Bettini, *A Course in Classical Physics 1—Mechanics*,
Undergraduate Lecture Notes in Physics, DOI 10.1007/978-3-319-29257-1

Fig. 1 Velocities and relative velocity of problem 1.8

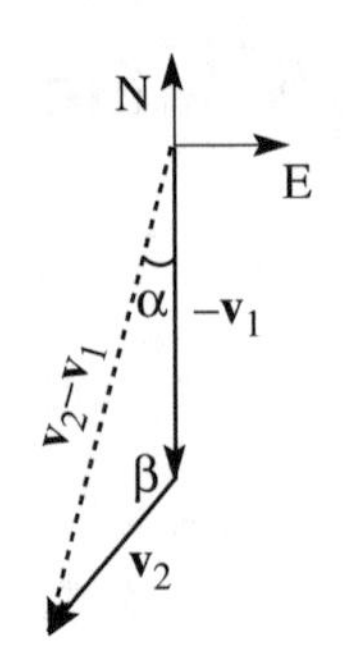

2.3. $T = m_1 m_2/(m_1 + m_2) g \sin\theta$; $a_2 = m_2/(m_1 + m_2) g \sin\theta$
2.4. 3.94 N
2.5. The equation of motion of the body of mass M is $-T + Mg = Ma$ and for the body of mass m is $-T + mg = ma$. Hence $a = g(M - m)/(Mmm)$.
2.6. The kinetic energy of the hammer is $(1/2)mv^2$ when it hits and 0 at the end. The change of kinetic energy is equal to the work done on the nail, which in turn is equal to the mean force times the displacement s. The mean force is then $\langle F \rangle \doteq mv^2/(2s) = 5$ N.
2.7. F, for the action reaction law.
2.8. $-\mathbf{F}$ on each hand, independently of the accelerations of the spheres being different, as a consequence of the action-reaction law
2.9. Statement 1 is, in general, false. Statement 2 is true for the rings on the guides b and c, for energy conservation. For the same reason the statement is false for the guide a, because that ring cannot reach B.
2.10. $\langle F \rangle s = mgh$, $\langle F \rangle = 2 \times 10^4$ N
2.11. The initial kinetic energy transforms into elastic energy of the pole and then in potential gravitational energy of the athlete. $mv^2/2 = mgh$, $h = 5$ m. (NB In practice, the athlete raises even more doing work with his arms.)
2.12. The two ropes have equal tensions. They break at the same time.
2.13. The lighter sphere rises four times more (energy conservation). $h = p_0^2/(2m^2 g)$
2.14. (a) $F(t = 0) = 0.09$ N; (b) $F_{max} = 10$ N.
2.15. If the rotation plane is horizontal, the wire is on a cone at an angle, say θ, with the horizon, in order to balance the weight mg with the vertical component of the tension, $T \sin\theta$. Hence the radius of the circle is $l \cos\theta$. We have two equations $T \sin\theta = mg$, $T\cos\theta = mv_{\max}^2/(l\cos\theta)$. Eliminating θ, we have $v_{\max} = \sqrt{Tl/m\left[1 - (mg/T)^2\right]}$

If the circle is vertical, its radius is l. The tension varies along the circle, reaching its maximum in the lowest point. Draw the situation. In this point $T - mg = mv_{\max}^2/l$, hence $v_{\max} = \sqrt{(l/m)(T - mg)}$.

3.1. In the SI units, $\omega_0 = 10\ \mathrm{s}^{-1}$, $\nu = 1.6\ \mathrm{s}^{-1}$, $T = 0.63$ s.

3.3. Expand Eq. (3.46) in series of the small quantity $\gamma/\omega_0 = 0.02$ as $\omega_1 = \omega_0\sqrt{1-(\gamma/\omega_0)^2/4} = \omega_0\left[1-(\gamma/\omega_0)^2/8\right]$. ω_1 is smaller than ω_0 of five parts in one hundred thousand.

3.4. (b) Each spring expands $x/2$, the restoring force reduces to ½, the proper angular frequency is $\sqrt{2}$ smaller. (c) The proper angular frequency is $\sqrt{2}$ larger than in (a).

3.5. (a) $k = 1$ kN/m; (b) with x in centimeters and t in seconds $x(t) = 5\cos 10t$; (c) in the same units $x(t) = 5\cos 10t + 10\cos 10t$.

3.6. Take the average on a period of Eq. (3.70) and compare the members.

3.8 In a vector diagram like in Figs. 3.7 and 3.8 the two forces are represented by rotating vectors at the same angular velocity. The angle between them, which is the difference between their phases, ϕ, is constant. The phase difference between forces is the same as between displacements. From the geometry we have $A^2 = A_1^2 + A_2^2 + 2A_1A_2\cos\phi$ and ϕ =133°.

3.9. Initial velocity is $\upsilon = 28$ m/s and the kinetic energy $U_k = 390$ kJ. This is the work of the force in 90 m. The magnitude of the force is 4.3 kN (43 % of the weight of the car). With 15% slope, in 100 m the car descends $h = 15$ m and the potential energy decreases by $mgh = 150$ kJ. To stop the car the work of the braking force should be 430 kJ. After 100 m the kinetic energy is reduced to 110 kJ and the velocity is 15 m/s (53 km/h).

3.10. The vertical forces are equal and opposite. The horizontal forces are the tension of the wire T, which is the centripetal force of magnitude $m\upsilon^2/l$ directed towards O and the friction of magnitude $\mu_d mg$ directed opposite to velocity. The magnitudes of both are equal to 4 N. The angle between them is 90°. Hence the magnitude of the resultant is 5.7 N and its direction is at 135° with velocity.

3.11. At the limit velocity υ_{lim} the drag force is equal to the weight mg. If the term proportional to the velocity dominates, $R = C_1 a\upsilon$, $\upsilon_{\mathrm{lim}} = 1.3\times10^8 a^2$ m/s. For $a = 1$ mm $\upsilon_{\mathrm{lim}} = 130$ m/s, for $a = 0.1$ mm $\upsilon_{\mathrm{lim}} = 1.3$ m/s. The term proportional to velocity is dominant only in the second case. If the term proportional to the square velocity of the drag dominates, $\upsilon_{\mathrm{lim}} = 217\sqrt{a}$ m/s. Hence, for $a = 1$ mm, $\upsilon_{\mathrm{lim}} = 6.9$ m/s. Neglecting the term proportional to velocity is justified.

3.12. (a) $T = m\upsilon^2/R - mg$. The centripetal force is the sum of the weight and the tension, which in the considered point have the same direction, vertical downwards. If the velocity is smaller than the critical one the motion is not circular. (b) $T = m\upsilon^2/R + mg$.

3.13. (a) $h = 2R/3$. (b) same on the moon, it does not depend on g.

4.1. 0.5 s
4.2. $g(h_2 - h_1) = \Delta\phi$. Hence 1000/9.8 or about 100 m.
4.3. Neither velocity nor acceleration are constant.
4.4. Mass does not vary, weight diminishes.
4.5. Answers are found putting the centripetal force equal to the gravitational attraction.
4.6. $T_p = (\alpha/2)^{3/2} T_E = 10400$ s (almost 3 h).
4.7. If r is the radius of the orbit of the satellite, its velocity is $\upsilon = \sqrt{G_N M_E / r}$. Remember that $g = G_N M_E / R_E^2$. The period of the satellite is then $T = 2\pi r^{3/2} / (g^{1/2} R_E)$ and $r - R_E = 1500$ km.
4.8. The radius of the spheres is $r = 0.60$ m, the distance between their centers is $d = 1.23$ m. The gravitational force is $F = 4.4 \times 10^{-3}$ N and the shrinking of the spring is 90 μm.
4.9. $M_E = g R_E / G_N = 6 \times 10^{34}$ kg, $\langle\rho\rangle = 5.5 \times 10^3 \text{kg/m}^3$ (notice that this value is much higher than the density of the crust, which is about 2000 kg/m^3 showing that the central part of the earth must be much denser than the average; it is made of iron).
4.10. $M_S = (4\pi^2 / G_N)(r_E^3 / T_E^2) \approx 2 \times 10^{30}$ kg, $\rho_S \approx 1.3 \times 10^3$ kg/m^2, a little denser than water.
4.11. $M_{tot}/M_S = (\upsilon_S/\upsilon_E)^2 (r_S/r_E) = 1.2 \times 10^{11}$.
4.12. The orbital velocity of Io is $\upsilon_I = 18$ km/s and consequently $M_G/M_S = (\upsilon_I/\upsilon_E)^2 (r_I/r_E) = 10^{-3}$.
4.14. $\phi_S(E) = 10^9$ m^{2}s^{-2}; $\phi_E(E)/\phi_S(E) = 0.07$; $\phi_G(E)/\phi_S(E) = 60$. We see that the contributions to the potential of the distant masses are important, differently than for the force. Indeed, the potential decreases as $1/r$, the force as $1/r^2$.

5.1. (a). Vertically. (b) At the angle arc tang (a/g) to the vertical, forward.
5.2. (a) During the braking, the acceleration of the train is $a_t = -3$ ms^{-2}. In the reference frame of the train, the forces acting on the case are the inertial force $-m\, a_t$ and the friction force $-\mu_d mg$. Its acceleration relative to the train is $a_r = 1$ ms^{-2} and the absolute one $a_a = -2$ ms^{-2}. (b) During the time t_b of the braking, the case moves relative to the train with acceleration a_r starting from rest. Its speed is 10 m/s, both relative to the train and the ground (train has stopped). (c) The case travels a first path $s_1 = 50$ m during braking (accelerated relative motion) and a second one s_2 when the train has stopped. In the second path, the acceleration is $a' = -2$ ms^{-2}, taking 5 to stop. The time to stop is $s_2 = 25$ m.
5.3. The acceleration of the lift is 2.8 ms^{-2} upwards. Nothing can be said on velocity.

5.6. The angular velocity is ω = 3.45 rad/s, the centrifugal force at the rim is m 1.8 N, where m is the mass of the insect. The force of static friction is m 0.98 N. It does not make it.

5.7. Not really, because the lateral shift of the point where the ground is hit is about 5 mm.

6.1. Suppose that the direction from the lamp to the mirror is the same as the velocity $\mathbf{v}_{O'}$. (the analysis of the apposite case is quite similar). The time in S taken by the pulse for its round-trip is always $\Delta t_0 = 2\ l/c$. The observer in S sees the clock of S' moving in the direction of the length; this is $l' = l\sqrt{1-\beta_{O'}^2}$. In addition he sees that, while the pulse is travelling, the mirror recedes with speed $\upsilon_{O'}$. Call $\Delta t'_a$ the time to reach the mirror, we have $c\Delta t'_a = l' + \upsilon_{O'}\Delta t'_a$, hence $\Delta t'_a = (l/c)\sqrt{1-\beta_{O'}^2}/(1-\beta_{O'})$. When the pulse comes back, the observer in S' sees the detector approaching. If we call $\Delta t'_a$ the return time, we have $c\Delta t'_a = l' - \upsilon_{O'}\Delta t'_a$, hence $\Delta t'_a = (l/c)\sqrt{1-\beta_{O'}^2}/(1+\beta_{O'})$. The period of the clock is the sum of the two.

6.2. (a) 13.5 μs. (b) 1.9 μs. (c) 560 m.

6.3. The limit for $t \to \infty$ is c. $F = mc^2/x_0$ constant in time.

6.4. υ_f = 0.62 c, M = 2.1 m.

6.5. $\gamma = 10^{10}$, about 5 min.

6.6. p = 1.42 meV/c.

6.7. p = 0.295 meV/c.

6.8. E = 0.852 meV.

6.9. γ = 2.75, β = 0.93.

6.10. $\gamma = 10^5, (c-\upsilon)/c = 1/2\gamma^2 \approx 5 \times 10^{-11}$.

7.1. It is zero.

7.2. It moves with acceleration **g**.

7.3. 1 m/s. Yes, the collision was completely inelastic.

7.4. $i = 5 \times 10^4$ Ns, $F = 10^5$ N.

7.5. $m_2/m_1 = 3$, $\upsilon_{CM}/\upsilon_i = 1/4$.

7.6. (a) $U_K^* = (3/4)U_{Ki}(1)$, (b) $U_{Kf} = (1/4)U_{Ki}(1)$.

7.7. 0.74 R_E.

7.8. (a)$T = 7.3 \times 10^5 R^{3/2}/\sqrt{M}$. (b) 10.

7.9. $m_1 + m_2 = (R^3/T^2)M_S$.

7.10. The velocity after the collision is **V** = (0, 2, 2) equal to the center of mass velocity. (b) 50 J, 30 J, 20 J.

7.11. (a) $\theta = -29°$, $\upsilon_{2f} = 0.37\upsilon_{1i}$. (b) Not.

7.12. (a) (0,0,14) Nm, (b) If α is the angle between vectors **r** and **F**, $\tau = rF\sin\alpha = bF$, hence b = 2.8 m, (c) $\tau = rF_n$, hence F_n = 1.4 N.

7.13. 5 m, 3.2 m, 2.05 m and 1.31 m. The initial energy is $U = mg\ 5$, the following ones are $0.8U$, $(0.8)^2U$, $(0.8)^4U$, $(0.8)^6U$.
7.14. The maximum energy transfer is 15 J, corresponding to Δx = 32 cm.
7.15. (14/6, 11/6, 11/6).
7.16. $\Delta \mathbf{L}_O = (0, -800, 0)$ kg m^2 s^{-1}.
7.17. $\mathbf{M}_O = -mg\upsilon_0(\cos\alpha)t\mathbf{u}_z$, $\mathbf{L}_O = -\frac{1}{2}mg\upsilon_0(\cos\alpha)t^2\mathbf{u}_z$.

8.1. No.
8.2. The external resultant force is zero. The total external moment about one of the support points is zero. F_1 = 590 N, F_2 = 390 N.
8.4. Quadruple.
8.5. $I = \mu L^2$, where μ is the reduced mass.
8.6. $I = (39/70)mR^2$.
8.7. Use the parallel axes theorem. $I_a = m(R_1^2 + 5R_2^2)/4$.
8.8. A positive answer would require $I/R^2 > M$, where M is the mass and R is the radius of the cylinder. Clearly, this is impossible for any distribution of the masses.
8.9. At the distance from m_1 of $r_1 = m_2/(m_1 + m_2)L$, which is the center of mass.
8.11. When both $\mathbf{M}_O$ and $\boldsymbol{\omega}$ have the direction of a principal axis of inertia and when O is fixed or is the center of mass and the three principal axes about it are equal.
8.13. $\mathbf{L}_C = (2/5)mR^2\boldsymbol{\omega}$, which does not depend on the pole, if it is still.
8.14. $L_X = L_C + Rm\upsilon/2 = Rm\upsilon$, $L_y = Rm\upsilon/2$, $L = 0$.
8.15. $L = mg\sqrt{gl/3} = 4.7 \times 10^2$ kg m^{2}s^{-1}, $\upsilon = \sqrt{3gl} = 9.5$ ms^{-1}.$\mu_s = 0.24$
8.17. There are two unknown, the tension of the wire and the acceleration of the center of mass. Use the equations (7.49) and (7.59) and solve them. There are two alternatives for the second equation, namely taking the pole in the center of mass or in the point Ω where the wire detaches from the yo-yo. In the latter case, take into account that the velocity of the pole is parallel to the total linear momentum.
$t = \sqrt{3h/g} = 0.39$ s,$T = mg/3 = 0.33$ N.
8.18. (a) $m > M_a/(gR)$, (b) $m = \pi M_a/(2gR)$
8.19. $W = mR^2\omega^2/4$
8.20. $m = (M/4)\sin\theta$
8.21. (a) Both dumbbells rotate with counter-clockwise angular velocity, and their centers are at rest (angular and linear momentum conservation). The magnitude of the angular velocity $\omega = 2\upsilon/l$ = 2 rad/s. (b) They rotate half a turn, then collide again. It takes $t = \pi/\omega$ = 1.57 s. (c) The second collision, which is symmetric to the first, blocks the rotations and the two dumbbells separate with translations of speed opposite to the initial ones.

Index

A. Bettini, *A Course in Classical Physics 1—Mechanics*,
Undergraduate Lecture Notes in Physics, DOI 10.1007/978-3-319-29257-1

Printed in the United States
by Baker & Taylor Publisher Services